Hierarchical Nanostructures for Energy Devices

RSC Nanoscience & Nanotechnology

Titles in the Series:

1: Nanotubes and Nanowires
2: Fullerenes: Principles and Applications
3: Nanocharacterisation
4: Atom Resolved Surface Reactions: Nanocatalysis
5: Biomimetic Nanoceramics in Clinical Use: From Materials to Applications
6: Nanofluidics: Nanoscience and Nanotechnology
7: Bionanodesign: Following Nature's Touch
8: Nano-Society: Pushing the Boundaries of Technology
9: Polymer-based Nanostructures: Medical Applications
10: Metallic and Molecular Interactions in Nanometer Layers, Pores and Particles: New Findings at the Yoctolitre Level
11: Nanocasting: A Versatile Strategy for Creating Nanostructured Porous Materials
12: Titanate and Titania Nanotubes: Synthesis, Properties and Applications
13: Raman Spectroscopy, Fullerenes and Nanotechnology
14: Nanotechnologies in Food
15: Unravelling Single Cell Genomics: Micro and Nanotools
16: Polymer Nanocomposites by Emulsion and Suspension
17: Phage Nanobiotechnology
18: Nanotubes and Nanowires: 2nd Edition
19: Nanostructured Catalysts: Transition Metal Oxides
20: Fullerenes: Principles and Applications, 2nd Edition
21: Biological Interactions with Surface Charge Biomaterials
22: Nanoporous Gold: From an Ancient Technology to a High-Tech Material
23: Nanoparticles in Anti-Microbial Materials: Use and Characterisation
24: Manipulation of Nanoscale Materials: An Introduction to Nanoarchitectonics

25: Towards Efficient Designing of Safe Nanomaterials: Innovative Merge of Computational Approaches and Experimental Techniques
26: Polymer–Graphene Nanocomposites
27: Carbon Nanotube-Polymer Composites
28: Nanoscience for the Conservation of Works of Art
29: Polymer Nanofibers: Building Blocks for Nanotechnology
30: Artificial Cilia
31: Nanodiamond
32: Nanofabrication and its Application in Renewable Energy
33: Semiconductor Quantum Dots: Organometallic and Inorganic Synthesis
34: Soft Nanoparticles for Biomedical Applications
35: Hierarchical Nanostructures for Energy Devices

How to obtain future titles on publication:
A standing order plan is available for this series. A standing order will bring delivery of each new volume immediately on publication.

For further information please contact:
Book Sales Department, Royal Society of Chemistry, Thomas Graham House, Science Park, Milton Road, Cambridge, CB4 0WF, UK
Telephone: +44 (0)1223 420066, Fax: +44 (0)1223 420247
Email: booksales@rsc.org
Visit our website at www.rsc.org/books

Hierarchical Nanostructures for Energy Devices

Edited by

Seung Hwan Ko
Seoul National University, Seoul, Korea
Email: maxko@snu.ac.kr

Costas P Grigoropoulos
University of California, Berkeley, CA, USA
Email: cgrigoro@me.berkeley.edu

RSC Nanoscience & Nanotechnology No. 35

Print ISBN: 978-1-84973-628-2
PDF eISBN: 978-1-84973-750-0
ISSN: 1757-7136

A catalogue record for this book is available from the British Library

Published by The Royal Society of Chemistry,
Thomas Graham House, Science Park, Milton Road,
Cambridge CB4 0WF, UK

Registered Charity Number 207890

For further information see our web site at www.rsc.org

Printed and bound by CPI Group (UK) Ltd, Croydon, CR0 4YY

Preface

Energy has been the major global issue in our society. Since the Fukushima nuclear disaster in 2011, future renewable energy development has been viewed through the safety prism. Non-nuclear-based, safe and sustainable energy sources have therefore attracted tremendous attention.

Research studies on energy devices have traditionally focused on the development of new materials for components such as anodes, cathodes, dyes, electrolytes and catalysts. However, in the last decade, new material development has been sluggish as it is admittedly very hard to overcome constraints posed by the materials' intrinsic structure. Therefore, researchers have been seeking new ground-breaking approaches by smart design/structuring of known materials through three-dimensional (3D) multi-scale hierarchical nano-architectures comprised of nanoscale building blocks. Recently, research in 3D branched hierarchical nanowire structures has been booming among researchers in various energy device fields, including energy conversion, storage and consumption. 3D branched hierarchical nanowire structures that possess a high surface area and offer direct transport pathways for charge carriers are especially attractive for energy applications. More specifically, 3D branched hierarchical nanowires improve light absorption due to the increased optical path as well as additional light trapping through reduced reflection and multiple scattering in comparison to 1D nanowire arrays, which are beneficial for solar energy harvesting applications. The high surface area can also increase surface activity and electrolyte infiltration in energy storage devices. The direct charge carrier transport pathway in both the trunks and branches boosts the charge collection efficiency. These fascinating properties of branched hierarchical nanowire structures have indeed many ideal characteristics for energy devices.

RSC Nanoscience & Nanotechnology No. 35
Hierarchical Nanostructures for Energy Devices
Edited by Seung Hwan Ko and Costas P Grigoropoulos

Published by the Royal Society of Chemistry, www.rsc.org

This book will focus on the recent developments in hierarchical nanostructuring, especially for highly efficient energy device applications. Hierarchical nanostructures usually entail a combination of multi-scale, multi-dimensional nanostructures such as nanowires, nanoparticles, nanosheets and nanopores. Because of the ability to tailor the architecture, synergistically combine functionalities and thereby specifically tune the transport properties, hierarchical nanostructures are expected to overcome the limitations of single scale nanostructures for achieving enhanced performance. Surface characteristics are of primary concern in most energy devices where maximizing efficiency can be achieved by either new material development or functional structuring. In this respect, hierarchical functional nanostructuring is particularly effective for achieving a surface area increase and favourable electrical properties. The energy devices covered in this book are: (1) energy generation devices (solar cells [DSSCs, OPVs]), fuel cells, piezoelectric, thermoelectric, water splitting *etc.*, (2) energy storage devices (secondary batteries, super capacitors *etc.*), (3) energy efficient electronics (displays, sensors, *etc.*). The hierarchical nanostructuring routes include construction of highly porous metal–organic frameworks, nanoparticle assembly with defined pore size, and synthesis of multiple generation highly branched nanowire trees.

Hierarchical nanostructure research has a bright future in solving the current limitations of energy devices. The ultimate goal is to push energy devices towards practical applications, which requires the development of devices with high efficiency, low cost and long lifespan. We hope this book will provide an account of the state-of-the-art research trends and a perspective on hierarchical nanostructures for energy device applications.

This book would not be possible without the commitment, effort and enthusiasm of all the contributing authors whom we sincerely thank. Our gratitude is extended to the Royal Society of Chemistry (RSC) for giving us a chance to embark on this great adventure and its high standard of support in preparing the book. In particular, we would like to acknowledge the administrative help from Mrs Alice Toby-Brant and Dr Merlin Fox. Finally, Dr Ko wants to thank his family (his wife Hyun Jung Kim and his new born son Suh June Ko) for their warm support and understanding.

Seung Hwan Ko
Seoul National University, Seoul, Korea

Costas P. Grigoropoulos
UC Berkeley, California, USA

Contents

RSC Nanoscience & Nanotechnology No. 35
Hierarchical Nanostructures for Energy Devices
Edited by Seung Hwan Ko and Costas P Grigoropoulos

Published by the Royal Society of Chemistry, www.rsc.org

CHAPTER 1

Introduction: Hierarchical Nanostructures for Energy Devices

SEUNG HWAN KO

Applied Nano and Thermal Science (ANTS) Lab, Department of Mechanical Engineering, Seoul National University, 1 Gwanak-ro, Gwanak-gu, Seoul 151-742, Korea
Email: maxko@snu.ac.kr

1.1 Introduction

Energy has been the hottest social issue for a long time. Energy issues have been related to the problems associated with current major energy sources such as fossil and mineral energy sources: (1) their inevitable exhaustion in the near future,[1] (2) environmental problems such as global warming due to a commensurate increase in CO_2 (a prominent greenhouse gas) emissions,[2] (3) an energy shortage due to a recent dramatic increase in global energy consumption[2] (between 2004 and 2030, the annual global consumption of energy is estimated to rise by more than 50%) and thus a price increase. Renewable energy sources, such as hydroelectric, solar, wind, hydrothermal, biomass and nuclear power, are expected to solve the problems associated with fossil fuels. However, energy issues are becoming more serious global problems in the aftermath of the Fukushima catastrophe.

Despite the projected persistent increases in oil and gas prices, less than 10% of the global energy production in 2030 is predicted to come from renewable energy sources. In order to moderate global reliance on

RSC Nanoscience & Nanotechnology No. 35
Hierarchical Nanostructures for Energy Devices
Edited by Seung Hwan Ko and Costas P Grigoropoulos

Published by the Royal Society of Chemistry, www.rsc.org

exhaustible natural resources and their environmentally hazardous combustion, more scientific efforts should be directed toward reducing the cost of energy production from renewable sources.[2]

Developing sustainable renewable energy sources has been a major research topic in an effort to solve the environmental problems caused by fossil fuels. Significant progress has been made in increasing the efficiency of various renewable energy technologies including solar cells, fuel cells, nuclear energy, wind power and so on.[3] Since the nuclear power plant disasters at Japan and Ukraine, the safety issue has become the most important factor.

1.2 Energy Cycle

Energy devices do not mean only energy generation devices but also include energy storage and energy consumption devices. To fully understand efficient energy usage and to increase the efficiency, the term *Energy Cycle* should be understood. *Energy Cycle* is the complete life of energy from birth to death: energy generation, energy storage and energy consumption (Figure 1.1). Efficiency is a major concern in energy devices and the total efficiency of energy devices is limited by the one with lowest efficiency (just like a chemical reaction rate is dominated by the slowest process). Even though one may develop an extremely efficient energy generation device, if the generated energy is stored in a poor efficiency energy storage device or used for a poor efficiency energy consumption device, the efficiency will be low from the total energy cycle viewpoint. Therefore, to approach the energy

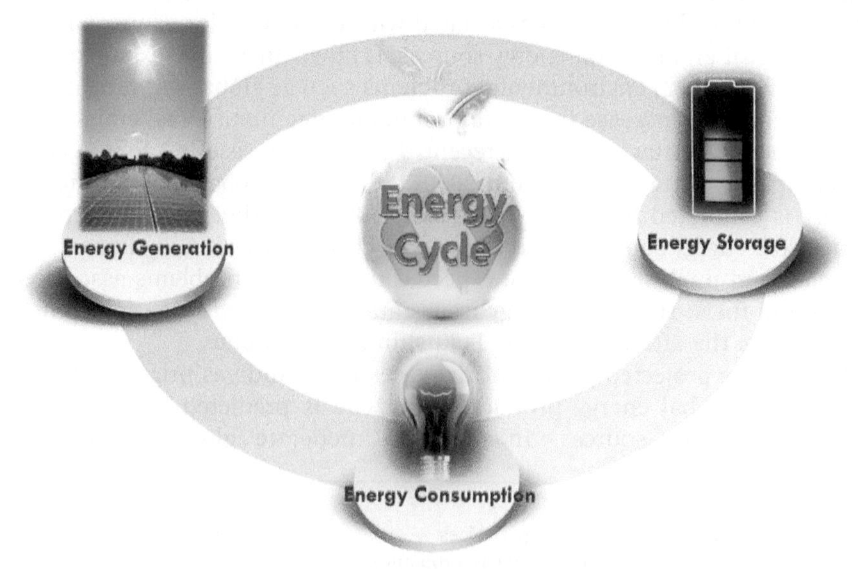

Figure 1.1 The *Energy Cycle* is the complete life cycle of energy from birth to death: energy generation, energy storage and energy consumption.

problem more practically and effectively, the concept of an Energy Cycle should be introduced and the total efficiency of all energy devices involved should be counted systematically.

The most important factor is not just a simple number, such as the efficiency of a single energy device; the balance between many energy devices is very important. This may sound as though researchers in the energy field should know about all different types of energy devices (generation, storage and consumption) to increase an energy device's efficiency in the energy cycle. However, a closer look at the various energy devices may reveal that most of them have similar structures and requirements to make more efficient devices. The structures usually have an active layer sandwiched between two electrodes. The electrodes may be a transparent or non-transparent conductor depending on the application (optoelectronic devices need at least one transparent electrode, such as a solar cell and LED display). Furthermore, most of the energy devices are surface devices (using an interface) and therefore, the efficiency can be increased using a larger surface area. That is where nanomaterials can be useful. However, a larger surface area does not always yield a highly increased efficiency. Additional smart structuring, which can lead to better carrier transport, can boost up the efficiency along with an increased surface area.

1.3 Hierarchical Nanostructures for Efficient Energy Devices

The study of energy device materials is a field full of opportunities for practical and socially significant applications.[2] Many potential renewable energy technologies in the form of solid-state devices and condensed matter phenomena involving the conversion of energy from one form to another exist, and some proceed with efficiency near unity. Within the last couple of decades, there has been an increase in interest in materials with nanometre-scale dimensions. Semiconductor nanowires, a subset of these materials, have received exceptional attention for their unique properties and complex structures. Many nanowire-based materials are promising candidates for energy conversion devices.

However, efficiency increases in the energy devices have been sluggish recently and there has been a need for new groundbreaking approaches, such as the design and fabrication of three-dimensional multifunctional architectures from appropriate nanoscale building blocks, including the strategic use of void space and deliberate disorder as design components to permit a re-examination of devices that produce or store energy.[4] Recently, the importance of nanostructured materials in energy harvesting, conversion and storage technologies has been highlighted in several review articles.[5–10] In particular, 3D branched nanowire structures with high surface areas and direct transport pathways for charge carriers are especially attractive for energy applications.[11,12] For example, 3D branched nanowires improve light absorption due to the increased optical path as well as additional light trapping through reduced reflection and multi-scattering in comparison to 1D nanowire arrays, which are beneficial for solar energy harvesting applications.[5] The high surface area can also increase surface activity and electrolyte infiltration in supercapacitors and batteries, and the direct charge carrier transport pathway in both the trunks and branches boosts the charge collection efficiency.[5] These fascinating properties of 3D branched nanowire structures have therefore stimulated widespread interest in fabricating them. The bottom-up approaches, including vapour phase and solution-based routes, allow fabrication of a wide variety of 3D branched nanowires with diverse functions.[5]

The appropriate electronic, ionic, and electrochemical requirements for such devices may now be assembled into nanoarchitectures on the benchtop through the synthesis of low density, ultraporous nanoarchitectures that meld a high surface area for heterogeneous reactions with a continuous, porous network for rapid molecular flux, for example, the three-dimensional design for batteries in Figure 1.2. Such nanoarchitectures amplify the nature of electrified interfaces and challenge the standard ways in which electrochemically active materials are both understood and used for energy storage. An architectural viewpoint provides a powerful metaphor to guide chemists

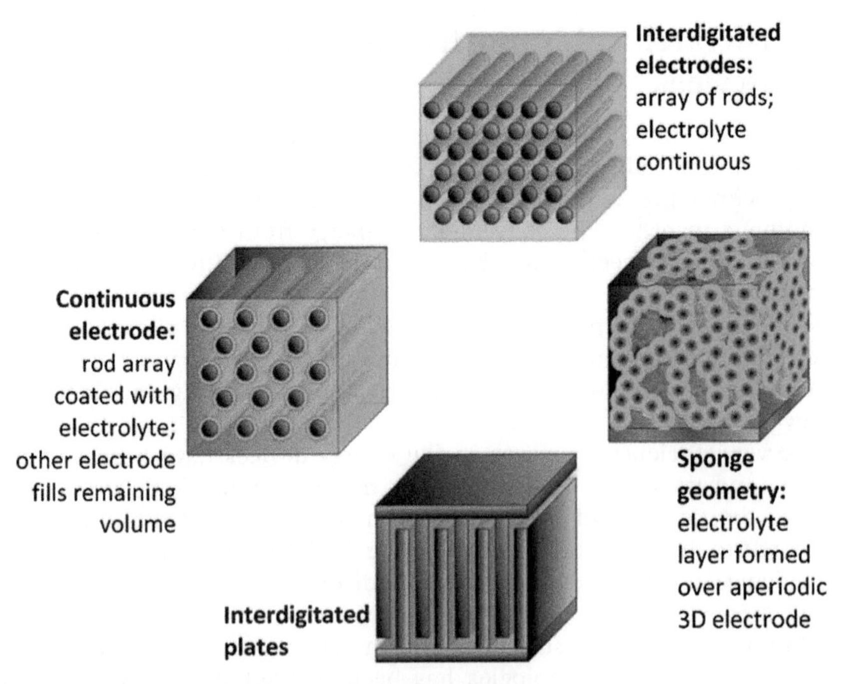

Figure 1.2 Three-dimensional designs for batteries.
(Adapted from ref. 8; reprinted with permission. Copyright 2004, American Chemical Society.) Reproduced by permission of The Royal Society of Chemistry.[4]

and materials scientists in the design of energy-storing nanoarchitectures that depart from the hegemony of periodicity and order with the promise and demonstration of an even higher performance.[3]

This book will focus on recent developments in hierarchical nanostructuring, especially for highly efficient energy device applications. Surface is a primary concern in most energy devices. Maximizing efficiency in energy devices can be achieved by either new material development or functional structuring. Hierarchical functional nanostructuring has rapidly gained interest to achieve increases in surface areas and favourable electrical properties. The energy devices covered in this book are: (1) energy generation devices (solar cells [DSSC, OPV], fuel cells, piezoelectric, thermoelectric, water splitting and so on), (2) energy storage devices (secondary battery, super capacitor, hydrogen storage), and (3) energy efficient electronics (display, sensors, *etc*). The hierarchical nanostructuring includes highly porous metal–organic frameworks, nanoparticle assembly with defined pore size, and multiple generation highly branched nanowire trees. This book is composed of four major parts as follows:

Part 1 (Chapters 1–3): Fundamentals—a general introduction to hierarchical nanostructures, characteristics, synthesis methods, and brief applications.

Part 2 (Chapters 4–8): Hierarchical nanostructures for high efficiency energy harvesting devices. Among the energy devices, this chapter will focus on high efficiency energy generation devices such as PV, fuel cells, thermoelectric devices and piezoelectric devices.

Part 3 (Chapters 9): Hierarchical nanostructures for high efficiency energy storage devices. Among the energy devices, this chapter will focus on high efficiency energy storage devices such as supercapacitors and secondary batteries. Mostly, anode and cathode structures will be discussed.

Part 4 (Chapter 10–13): Hierarchical nanostructures for high efficiency energy consumption devices. Among the energy devices, this chapter will focus on high efficiency energy consumption devices such as field emission devices, sensors and other applications.

These topics are currently among the major issues in society. Energy related issues have increased since the recent energy crisis. However, the widespread use of next generation green energy devices is still limited by efficiency and cost. Most of the energy related books focus only on the development of new materials. This book will cover the fundamentals to state-of-the-art functional hierarchical nanostructuring aspects.

References

1. Q. Zhang and G. Cao, *Nano Today*, 2011, **6**, 91–109.
2. A. I. Hochbau and P. Yang, *Chem. Rev.*, 2010, **110**, 527–546.

3. Q. Zhang, C. S. Dandeneau, X. Zhou and G. Cao, *Adv. Mater.*, 2009, **21**, 4087.
4. D. R. Rolison, J. W. Long, J. C. Lytle, A. E. Fischer, C. P. Rhodes, T. M. McEvoy, M. E. Bourg and A. M. Lubers, *Chem. Soc. Rev.*, 2009, **38**, 226–252.
5. C. Cheng and H. J. Fan, *Nano Today*, 2012, 7, 327–343.
6. C. Surojit, C. Li-Chyong and C. Kuei-Hsien, *NPG Asia Mater.*, 2011, **3**, 74–81.
7. R. Kapadia, Z. Fan, K. Takei and A. Javey, *Nano Energy*, 2012, **1**, 132–144.
8. B. Tian, T. J. Kempa and C. M. Lieber, *Chem. Soc. Rev.*, 2009, **38**, 16–24.
9. L. Li, T. Zhai, Y. Bando and D. Golberg, *Nano Energy*, 2012, **1**, 91–106.
10. R. Yu, Q. Lin, S.-F. Leung and Z. Fan, *Nano Energy*, 2012, **1**, 57–72.
11. M. J. Bierman and S. Jin, *Energy Environ. Sci.*, 2009, **2**, 1050–1059.
12. X. H. Liu, Y. J. Lin, S. Zhou, S. Sheehan and D. W. Wang, Energies (*Basel, Switz.)*, 2010, **3**, 285–300.

CHAPTER 2

Fundamentals of Hierarchical Nanostructures

JINHWAN LEE*[a] AND SEUNG HWAN KO[b]

[a] Department of Mechanical Engineering, Korea Advanced Institute of Science and Technology (KAIST), 291 Daehak-ro, Yuseong-gu, Daejeon 305-701, Korea; [b] Applied Nano and Thermal Science (ANTS) Lab Department of Mechanical Engineering, Seoul National University, 1 Gwanak-ro, Gwanak-gu, Seoul 151-742, Korea
*Email: mir_ljh@kaist.ac.kr

2.1 Introduction

Due to a large surface-to-volume ratio and quantum confinement effects, nanoscale materials show distinctive optical, mechanical, chemical, thermal and electronic properties compared with their bulk counterparts. A large fraction of their atoms are located at the surface. For example, a material that is 5 cm^3 possesses almost 0% ($\sim 10^{-5}$%) surface atoms, but when the cube is divided 24 times into 1 nm-sized cubes, the percentage of surface atoms increases to 80% so that the same mass of nanomaterial will have enough surface area to cover an entire football field. The surface atom percentage explains why the nanomaterials' properties are size dependent. The atoms at the surface have weak bonding (because atoms or molecules on a surface possess fewer nearest neighbors) compared with bulk atoms, which means that the atoms at the surface have a tendency to react easily to external perturbation or energy. Because bulk materials have a very small fraction of surface atoms (almost 0%), they show bulk-atom-dominated material properties, which we know well. However, when the materials

RSC Nanoscience & Nanotechnology No. 35
Hierarchical Nanostructures for Energy Devices
Edited by Seung Hwan Ko and Costas P Grigoropoulos

Published by the Royal Society of Chemistry, www.rsc.org

become nanometre sized, the percentage of surface atoms cannot be ignored anymore and the nanomaterial system starts to show characteristics of both surface atoms and bulk atoms. Depending on the size, the ratio between the surface atoms and bulk atoms and the resultant characteristics of the nanomaterials also show size-dependent properties. As the nanomaterial becomes smaller, the system shows more surface-atom-dominated characteristics with huge surface energy.

In nanomaterial systems, we need to decide when the surface atom becomes dominant and size-dependent phenomena start to show. However, this is not a simple problem because the length reference size scale to decide whether a system is small enough to be called a nanosystem depends on which characteristics you are interested in (*e.g.* optical, electrical, mechanical, thermal, chemical *etc.*) and material type.

In this chapter, we will explore the unique characteristics of nanomaterials and the simple physics behind them. These characteristics include thermal, electrical, phonon transport, mechanical, optical and magnetic properties.

2.2 Unique Characteristics of Nanomaterials

In light of the down-sizing trend in microelectronics, nanomaterials have received tremendous interest from various fields due to their unique characteristics.[1–8] One may find various examples of miniaturization including magnetic and optical storage components with critical dimensions as small as tens of nanometres,[9] size-dependent excitation or emission,[10–13] quantized (or ballistic) conductance,[14,15] Coulomb blockade (or single-electron tunneling, SET),[16,17] and metal–insulator transition.[18] It is generally accepted that quantum confinement of electrons by the potential wells of nanometre-sized structures may provide one of the most powerful (and versatile) means to control the electrical, optical, magnetic, and thermoelectric properties of solid-state functional materials.[19] Additionally, some remarkable specific properties are related to other origins: for example, (i) large fraction of surface atoms, (ii) large surface energy, (iii) spatial confinement, and (iv) reduced imperfections.[20] The following are a few examples suggested by G. Cao:[20]

(1) Nanomaterials may have a significantly lower melting point or phase transition temperature and appreciably reduced lattice constants, due to a huge fraction of surface atoms in the total amount of atoms.
(2) Mechanical properties of nanomaterials may reach the theoretical strength, which is one or two orders of magnitude higher than that of single crystals in the bulk form. The enhancement in mechanical strength is simply due to the reduced probability of defects.
(3) Optical properties of nanomaterials can be significantly different from bulk crystals. For example, the optical absorption peak of a semiconductor nanoparticle shifts to a short wavelength due to an

increased band gap. The color of metallic nanoparticles may change with their size due to surface plasmon resonance.

(4) Electrical conductivity decreases with a reduced dimension due to increased surface scattering. However, electrical conductivity of nanomaterials could also be enhanced appreciably due to the better ordering in microstructure.

(5) The magnetic properties of nanostructured materials are distinctly different from those of bulk materials. The ferromagnetism of bulk materials disappears and transfers to superparamagnetism on the nanometre scale due to the huge surface energy.

(6) Self-purification is an intrinsic thermodynamic property of nanostructures and nanomaterials. Any heat treatment increases the diffusion of impurities, intrinsic structural defects and dislocations, and one can easily push them to the surface nearby. Increased perfection would have an appreciable impact on the chemical and physical properties.

Many such properties are size dependent. In other words, the properties of nanostructured materials can be tuned considerably simply by adjusting the size, shape or extent of agglomeration.

2.2.1 Thermodynamic Properties and Thermal Stability

Due to their large surface area, all nanomaterials possess a huge surface energy and thus are thermodynamically unstable or metastable. Nanomaterials are found to have lower melting temperatures compared with their bulk counterparts when the systems' sizes decrease below a certain critical size. This melting point drop was found a long time ago by various researchers. Buffat and Borel[21] presented results that are related to an essentially thermodynamic size effect, *i.e.* the reduction of the melting point of a small gold aggregate as a function of decreasing particle size as shown in Figure 2.1. The melting point drop is generally explained by the increasing surface energy relationship with decreasing system size. The decrease in the phase transition temperature can be attributed to a change in the ratio of surface energy to volume energy as a function of particle size. It is not always clear to determine or define the melting temperature of nanomaterials. For example, the vapor pressure of a small particle is significantly higher than that of its bulk counterpart, and the surface properties of nanomaterials are very different from those of the bulk materials. Evaporation from the surface would result in an effective reduction of nanomaterial size and thus would affect the melting temperature. For some materials, increased surface reactivity due to a large surface to volume ratio may promote the oxidation of the surface layer and thus change the chemical composition on the nanomaterial's surface through the surface chemical reaction, leading to a change of melting temperature. Buffet and Borel[21] proposed clever experimental criterion to determine the size-dependent melting of

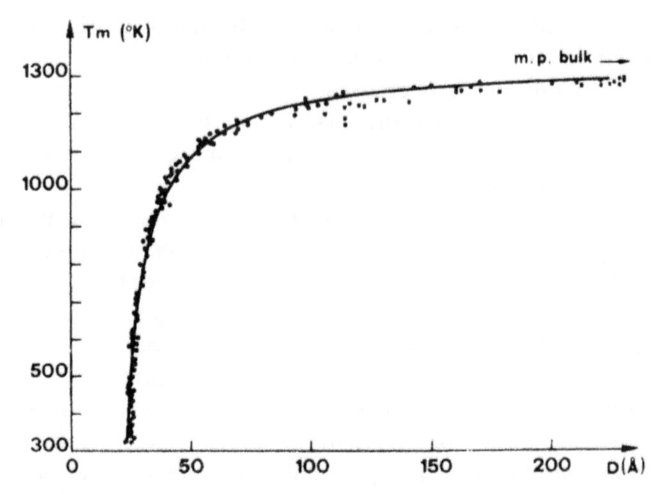

Figure 2.1 Experimental (dots) and theoretical (solid line) values of the melting point temperature of gold particles. Note that the melting point of bulk gold is 1337 K and decreases rapidly for nanoparticles with diameters below 5 nm.
Reproduced with permission from ref. 21.

nanomaterials: (i) the disappearance of the state of order in the solid, (ii) the sharp variation of some physical properties, such as evaporation rate, and (iii) the sudden change in particle shape. Bulk gold has a melting point of 1337 K and it decreases rapidly for nanoparticles with sizes below 5 nm as shown in Figure 2.1. Such size dependence has also been found in other materials such as copper,[22] tin,[23] indium,[24] lead and bismuth,[25] barium titanate ($BaTiO_3$),[26] lead titanate ($PbTiO_3$)[27] in the forms of particles and films.

Various theories have been proposed to explain the size-dependent melting temperature drop of nanomaterials. A classical thermodynamics theory, such as the Gibbs model to account surface area, can be exploited to explain this phenomenon to in the nanosystems as follows (eqn (2.1)):[21,28]

$$T_b - T_m = \left[\frac{2T_b}{\Delta H \rho_s \gamma_s}\right]\left[\gamma_s - \gamma_l\left(\frac{\rho_s}{\rho_l}\right)\right]^{2/3} \tag{2.1}$$

where T_b and T_m are the melting points of a bulk material and a nanoparticle respectively, γ_s is the radius of the particle, ΔH is the molar latent heat of fusion, and γ and p are surface energy and density respectively. This simple relationship (eqn (2.1)) is based on many assumptions to explain the size-dependent melting temperature drop in nanoparticles. Besides nanoparticles, the size-dependent melting temperature drop has been observed in nanowires including gold nanorods[29] and Ge nanowires.[30,31] After melting, nanowires may spontaneously break up into smaller and shorter segments and finally form small micro/nano beads due to Rayleigh instability[32] to reduce the high surface energy of nanowires or nanorods, when their diameters are sufficiently small or the bonding between constituent atoms is too weak.[19]

2.2.2 Electrical Properties (Electron Transport)

A reduction in the dimension to below a certain critical size (*i.e.* electron de Broglie wavelength) will result in a change of electronic structure and thus the electrical properties of nanomaterials. As the critical dimension of an individual device becomes smaller, the electron transport properties of its components become an important issue to study.[19] Studies from a number of groups indicate that some metal nanowires might undergo a transition to become semiconducting as their diameters are reduced below certain values. For instance, two-probe measurements by Dresselhaus and co-workers[33] on arrays of single-crystalline Bi nanowires indicated that these nanowires underwent a metal-to-semiconductor transition at a diameter of ~52 nm. Two-probe measurements performed by Choi and co-workers[34] on individual single-crystalline Bi nanowires of ~40 nm in diameter showed that these nanowires were semiconductors or insulators because their resistances increased with decreasing temperature. As a result of quantum confinement, it was proposed that the external conduction sub-band and valence sub-band of this system moved in opposite directions to open up a band gap in this particular confinement along the long axis of a wire and by surface imperfection.[19] Gold is another metal whose electron-transport properties have been extensively studied in the form of short nanowires as thin as a single, linear chain of atoms.[35] Because these wires are extremely short in length (usually a few atoms across, sometimes also referred to as point contacts), their conductance has been shown to be in the ballistic regime with the transverse momentum of electrons becoming discrete.[19] The transport phenomena (*e.g.* conductance quantization in units of $2e^2\ h^{-1}$) observed in this kind of 1D system are found to be independent of material.[36] As for semiconductors, recent measurements on a set of nanoscale electronic devices indicated that GaN nanowires as thin as 17.6 nm could still function properly as a semiconductor.[37,38] In a related study, transport measurements by Heath and co-workers[39] suggested that Si nanowires with a thickness of 15 nm had become insulating.

The effects of size on electrical conductivity of nanostructures and nanomaterials are complex. The mechanism can be generally grouped into four categories: surface scattering including grain boundary scattering, quantized conduction including ballistic conduction, Coulomb charging and tunneling, widening and discrete band gap, and charge of microstructure. In addition, increased perfection (reduced impurities, structural defects and dislocations) would affect the electrical conductivity of nanostructured materials.[20]

Scaling down in microelectronics is very important to realize high performance electronics. The ongoing miniaturization technology in electronics is approaching the point where the fundamental issues are expected to limit the performance enhancement. Therefore, the aforementioned unique electrical properties observed in nanomaterials are becoming very important in micro/nano electronics fabrication development. There are several appealing features for this "bottom-up" approach to nanoelectronics.[19] First, the size of the nanowire building blocks can be readily tuned to sub-100 nm and smaller, which should lead to a high density of devices on a chip. Second, the material systems for the nanowires are essentially unlimited, which should give researchers great flexibility to select the right materials for the desired device functionality. It is obvious that great progress has been made along the direction of using nanowire building blocks for various device applications. Nevertheless, one has to admit that ultimately achieving the goal of "bottom-up" manufacturing in the future will still require substantial work, including, for example, the development of 3D hierarchical assembly processes, as well as the improvement of material synthesis.

2.2.3 Phonon Transport (Thermo-electric Properties)

While electron transport in nanomaterials has been extensively studied, investigation of phonon transport in nanostructures was relatively less studied until very recently even though electron transport and phonon transport are closely connected. As the dimension of a 1D nanostructure is reduced to the range of phonon mean free paths (MFPs), the thermal conductivity will be reduced due to scattering by boundaries.[19] Theoretical studies suggested that as the diameter of a silicon nanowire became smaller than 20 nm, the phonon dispersion relation might be modified (as a result of the phonon confinement) such that the phonon group velocities would be significantly less than the bulk value.[40,41] Molecular dynamics (MD) simulations also showed that the thermal conductivities of Si nanowires could be two orders of magnitude smaller than that of bulk silicon in the temperature range from 200 K to 500 K.[42] The reduced thermal conductivity is desirable in applications such as thermoelectric cooling and power generation, but is not preferable for other applications such as electronics and photonics. Dresselhaus and co-workers[43,44] have theoretically predicted that the thermoelectric figure of merit could be substantially enhanced for thin nanowires by carefully tailoring their diameters, compositions, and carrier concentrations. This prediction still needs to be validated experimentally by measuring the thermal conductivities, Seebeck coefficients, carrier mobilities, and electrical conductivities of different nanowire systems. Good thermoelectrical systems include nanowires made of Bi, BiSb alloy and Bi_2Te_3.[45,46] Recently, Si/SiGe superlattice nanowires,[47] Si nanowires[48] and rough Si nanowires[49] have expanded the list of potential high performance thermoelectric materials with high *ZT* values. As shown in Figure 2.2, Si nanowires added another interesting twist to this direction of work, noting the possible interfacial phonon scattering within such highly complex 1D nanostructures.

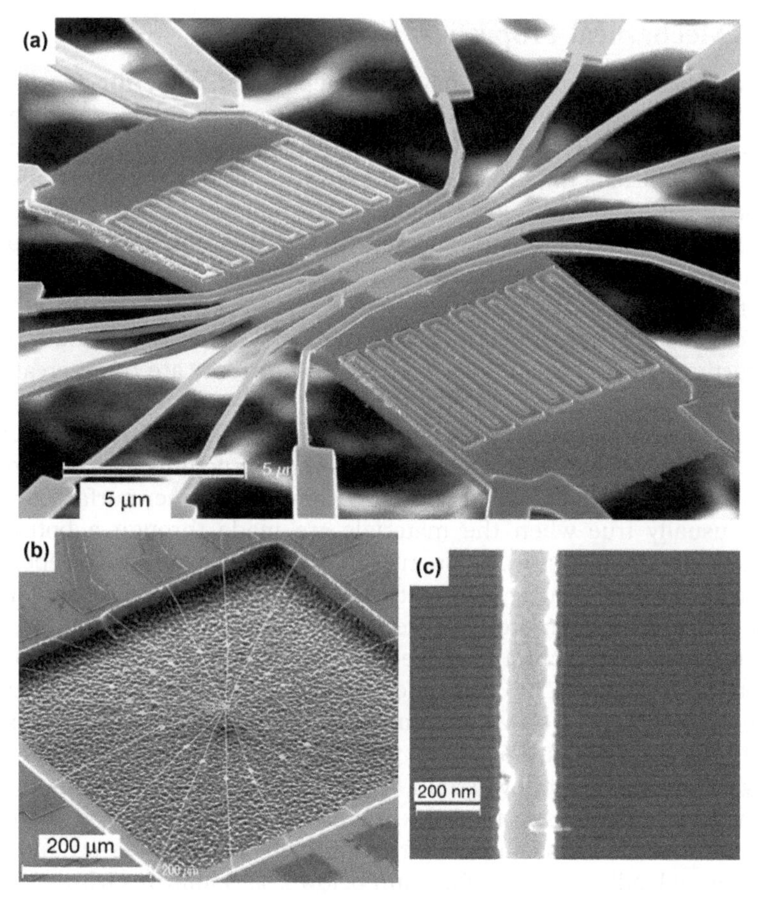

Figure 2.2 (a) This false-color image of a suspended platform shows all electrical connections. The central green area is the Si nanowire array, which is not resolved at this magnification. The four-lead yellow electrodes are used for thermometry to quantify the temperature difference across the nanowire array. The thermal gradient is established with either of the two Joule heaters (the right-hand heater is colored red). The yellow and blue electrodes are combined to carry out four-point electrical conductivity measurements on the nanowires. The grey region underlying the nanowires and the electrodes is the 150-nm-thick SiO_2 insulator that is sandwiched between the top Si(100) single-crystal film from which the nanowires are fabricated, and the underlying Si wafer. The underlying Si wafer has been etched back to suspend the measurement platform, placing the background of this image out of focus. (b) Low-resolution micrograph of the suspended platform. The electrical connections radiate outwards and support the device. (c) High-resolution image of an array of 20-nm-wide Si nanowires with a Pt electrode.
Reproduced with permission from ref. 48. Nature Publishing Group.

2.2.4 Mechanical Properties

Nanomaterials are known to have size-dependent, superior mechanical properties over their bulk counterparts. The strength enhancement of the nanomaterials is due to their being near single crystalline with reduced internal and surface imperfections (impurities, structural defects and dislocations). The smaller the cross-section of the nanosystems, the less likely it is to find internal imperfections such as dislocation, micro-twins, impurity precipitates *etc.*[50] Thermodynamically, imperfections in crystals are highly energetic and should be eliminated from the perfect crystal structure. Such imperfections are easier to eliminate in smaller systems. Additionally, some imperfections in bulk materials, such as dislocations, are often created to accommodate stresses generated in the synthesis and processing of bulk materials due to temperature gradient and other inhomogeneities. Such stresses are unlikely to exist in small structures, particularly in nanomaterials.[20] Additionally, smaller systems have fewer surface defects. This is usually true when the materials are made through a bottom-up approach due to less growth fluctuation. To predict the size-dependent mechanical strength on a nanometre scale, the conventional Hall–Petch model is not valid anymore.[51] Currently, various combined experimental–computational approaches have been studied with *in situ* transmission electron microscopy (TEM), atomic force microscopy (AFM) characterization and molecular dynamic (MD) simulations. Figure 2.3 shows the experimental stress–strain response of penta-twinned silver nanowires with varying diameters.[52] At low strain, the nanowires are found to deform elastically with the elastic modulus increasing with decreasing diameter. The elastic modulus was computed from the slope of the stress–strain curves for the initial loading regime of a strain below 2%. Nanowires with a smaller diameter were found to exhibit moduli of up to 1.5 times the bulk value.[52] The initial yield strain, corresponding to the first plateau in the stress–strain curve, shows a weak dependence on nanowire diameter from the nucleation theory.[53] However, the deformation behavior of the nanowires after the initial yielding is significantly affected by the nanowire diameter. Thinner nanowires show a more pronounced strain hardening effect.[52]

The penta-twinned nanowires' superelastic behavior with high yield strength is attributed to the role of twin boundary confinement.[54,55] The unique strain hardening and multiple plastic zone formation in small diameter nanowires ($D<100$ nm) is explained by dislocation nucleation from a local stress concentrator, which leads to the formation of a linear chain of stacking fault decahedrons (SFDs). By confining dislocation activity to SFD chain propagation, the internal twin boundaries cause local hardening of thin nanowires, resulting in defect insensitive structures with significantly enhanced flow stress, ductility and strength. As nanowire diameter becomes larger, the number of local plastic zones decreases due to the earlier onset of necking.[52] The superior mechanical properties of

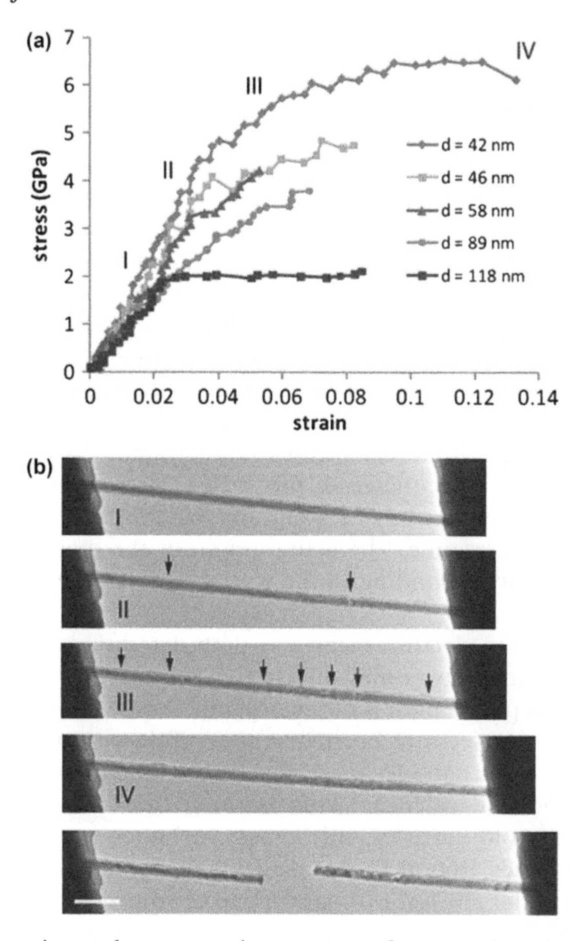

Figure 2.3 Experimental stress–strain response of penta-twinned Ag NWs. (a) Experimental stress–strain curves of Ag NWs of varying diameter. (b) Sequential bright field TEM images recorded during the tensile testing of a 42 nm diameter Ag NW. The arrows indicate localized regions of plastic deformation nucleated during the tensile test. The roman numerals indicate the point on the stress–strain curve in (a) at which each image was recorded. The bottom image was recorded after the NW fractured. The length of the scale bar is 250 nm.
Reproduced with permission from ref. 52.

nanomaterials opened up a new research area that requires high mechanical compliance such as flexible or stretchable electronics.[56–75]

2.2.5 Optical Properties

The brilliant and tunable optical properties of metal nanoparticles have been intensively researched for the past hundred years. The size-dependent optical properties of nanomaterials can be generally explained by the increased energy level spacing as the systems become more confined, and the surface plasmon resonance.[19]

As for quantum dots, size-confinement also plays an important role in determining the energy levels of a nanomaterial once its size has been reduced below a critical value which is the Bohr radius.[20] Korgel and co-workers[76–78] found that the absorption edge of Si nanowires was significantly blue-shifted compared with the indirect band gap (−1.1 eV) of bulk silicon. They also observed sharp, discrete features in the absorption spectra and relatively strong "band-edge" photoluminescence. These optical features most likely originated from quantum confinement effects, although surface states might also make additional contributions.[79] There is an important transition as surface effects begin to perturb the periodicity of the "infinite" lattice that constitutes the crystal.[80] As this transition regime is reached, the material properties alter dramatically. These changes are known collectively as quantum size effects and typically occur in the 1–10 nm size regime. The actual origin of the quantum size effects depends on the type of bonding in the crystal suggested by Lulvaney:[80]

(a) For metals, the electron mean free path determines the thermal and electrical conductivity and affects the color of the metal. For most metals, this path-length is of the order of 50–500 Å. As one crosses this threshold, the electrons begin to scatter off the crystal surface, and the resistivity of the particles increases. For very small metal particles, the conduction and valence bands begin to break down into discrete levels. For gold particles, this causes a change in color from red to orange at sizes around 1.5 nm. Below 1 nm, gold metal is best treated as a molecular cluster. Ultimately, one obtains gold atoms, which are colorless.

(b) For an ionic crystal such as NaCl, the Madelung constant determines the overall Coulomb attraction between the cations and anions that constitute the lattice. It depends, in a complex fashion, on inter-ion separation and crystal structure, but overall, the Madelung constant directly determines crystal cohesion, the Young's modulus of the crystal, and the solubility of the crystal, as well as other fundamental parameters. As ionic crystals become smaller, the total electrostatic energy, which determines the crystal cohesion, decreases, the crystal solubility increases, and the equilibrium bond length increases.

(c) For semiconductors, which are usually more covalent than salt crystals such as NaCl, the exciton radius, which typically lies between 20 Å and 100 Å, determines the size regime in which the band gap (analogous to the HOMO–LUMO gap in molecular structures) begins to widen, and molecular properties begin to emerge from the band structure. This results in drastic changes in crystal color and may also lead to spectacular changes in luminescence.

(d) These effects mean that nanocrystals of almost any material will have unique physical and optical properties that are determined by the nanocrystal size and shape. In this article, we concentrate on gold as just one example of a material that can be synthesized with a range of sizes and optical properties.

One of the key concepts in the field of nanomaterial optical properties is the localized surface plasmon (LSP) resonance, which can be understood as a collective resonant oscillation of all of the conduction electrons of the nanoparticle in response to an incident optical field.[81] A nanoparticle supports a range of LSPs, but the optical response is usually dominated by dipolar modes. The oscillation frequencies are highly dependent on the shape of the nanoparticle, but they also crucially depend on the absolute size and dielectric environment.[82] The oscillation can be localized on a single nanoparticle, or it can involve many coupled nanoparticles. Because of the strong and tunable particle–light interaction, several interesting effects and applications are possible.[82] Examples include surface-enhanced Raman scattering (SERS),[83,84] optical waveguides[85,86] and biochemical sensors.[87–90] Figure 2.4 shows the size-dependent optical property of single nanoparticles. Figure 2.4(a) shows scattering spectra of single isolated silver particles with diameters of $D = 50$, 100, 150, and 200 nm, together with Lorentzian curve fits and calculations.[81] The intensity axis in Figure 2.4(a) is proportional to scattering efficiency. This allows for convenient comparisons of particles of different sizes and scattering cross sections. Figure 2.4(b) shows that there is a linear shift of the resonance as a function of particle size, as expected from theory. From the slope of the line, it is found that the peak position shift is approximately 27 nm per 10 nm change in diameter for the geometry used. The experimental and theoretical slopes agree very well (26.4 and 27.5 nm, respectively).[81]

Recent advances in particle synthesis and nanofabrication technology have made it possible to produce well-defined metal nanostructures, enabling systematic studies of their optical characteristics.[81] A number of groups have studied the optical properties of gold and silver nanoparticle arrays.[91–95] The results show that interparticle coupling effects give rise to pronounced shifts of the LSP resonance frequency compared to isolated particles. Competing factors that give rise to either blue or red shifts with decreasing interparticle distances have been identified.[91,96,97] The electromagnetic coupling between particles involves both very short-distance interactions, due to evanescent fields, and long-range interactions, mainly due to propagating dipolar fields.[91] A thorough understanding of these local field and coupling effects is crucial for nanoparticle wave-guiding applications, where one wants to achieve efficient energy transport between particles while minimizing far-field scattering losses.[91]

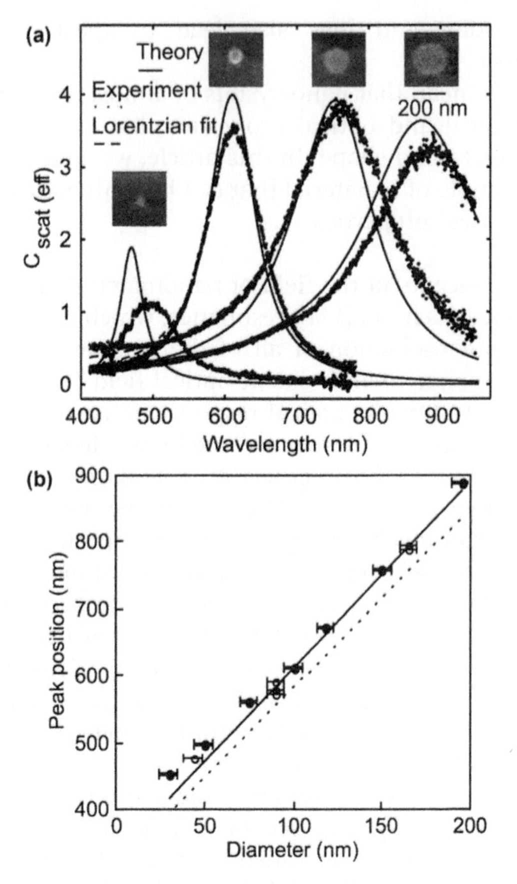

Figure 2.4 (a) Unpolarized dark-field scattering spectra and corresponding SEM images of single isolated particles of different diameters ($D = 50$, 100, 150 and 200 nm). The dashed lines show Lorentzian fits of the experimental data according to eqn 8, and solid lines show scattering spectra calculated on the basis of the MLWA polarizability. The experimental and theoretical spectra have been normalized to the geometrical cross section of each particle so that the intensity axis is proportional to the relative scattering efficiency. (b) Experimental LSP position *vs.* particle diameter as determined by SEM analysis. The resolution of the SEM is approximately 10 nm, which is indicated by error bars in the plot. The solid and dashed lines show MLWA results for oblate spheroids with heights of 20 and 25 nm, respectively. The substrate is taken into account through an effective refractive index of 1.25.

Near-field coupling becomes dominant for separations on the order of a particle radius and leads to a strong enhancement of the local electric field in the gap between the particles. This field confinement is believed to be the main contributor to single-molecule SERS, which has been observed by several groups over the past few years.[98–101] It might also lead to strong optical forces directed toward the gap between particles,[102] which might enable optical nanopositioning of molecules for advanced biosensing applications.[81]

2.2.6 Magnetic Properties (Superparamagnetism)

Ferromagnetic particles become unstable when the particle size is reduced below a certain size, because the surface energy provides sufficient energy for domains to spontaneously switch polarization directions.[20] As a result, ferromagnetic particles become paramagnetic particles. However, nanometer-sized ferromagnetic particles tuned to paramagnetic particles behave in a different way to conventional paramagnetic particles and are referred to as superparamagnetic particles. Superparamagnetism is a type of magnetism that occurs in small ferromagnetic or ferrimagnetic nanoparticles; this implies sizes around a few nanometres to a couple of tenths of nanometres, depending on the material. Additionally, these nanoparticles are single-domain particles. In a simple approximation, the total magnetic moment of the nanoparticle can be regarded as one giant magnetic moment, composed of all the individual magnetic moments of the atoms that form the nanoparticle.[103]

Magnetism is highly volume and temperature dependent because this property arises from the collective interaction of atomic magnetic dipoles.[104] When the size of a ferro- or ferrimagnet decreases to a certain critical value r_c, the particles change from a state with multiple magnetic domains to one with a single domain. As shown in Figure 2.5, if the size continues to decrease to a value r_0, the thermal energy becomes comparable with that required for spin to flip directions, leading to the randomization of the magnetic dipoles in a short period of time.[104] Such small particles do not have permanent magnetic moments in the absence of an external field but can respond to an external magnetic field. They are referred to as

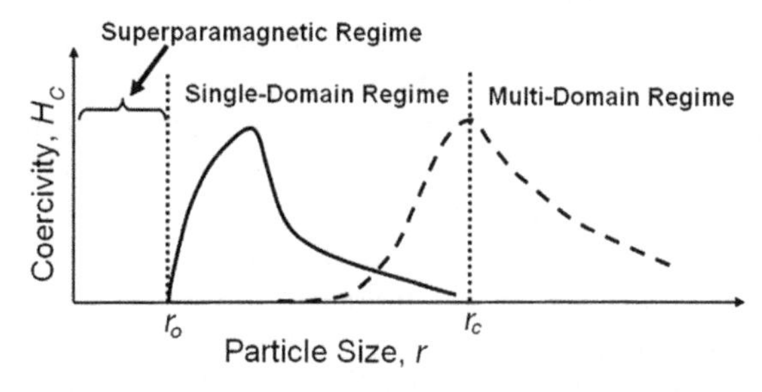

Figure 2.5 Schematic illustrating the dependence of magnetic coercivity on particle size. In the single-domain regime, the coercivity can follow either the solid curve for non-interacting particles or the dashed line for particles that have coupling between them. The coercivity falls to zero for superparamagnetic colloidal particles.
Reproduced with permission from ref. 104. Copyright 2013, Wiley-VCH.

superparamagnetic colloids. The critical radius r_c for a spherical particle with the stability defined by the flipping probability of the magnetic moment of <10% over one second can be estimated using the following equation:

$$r_c = (6k_B T/K_u)^{1/3} \tag{2.2}$$

where k_B is the Boltzmann constant, T is temperature, and K_u is the crystalline magnetoanisotropy. Depending on k_u, the critical radii of nanoparticles can be 3–4 nm for very hard magnets and over *ca.* 20 nm for soft magnets.[104]

Superparamagnetic colloids have great potential for many other applications related to biomedical research.[104] They have found widespread use in many traditional areas including magnetic data storage, ferrofluid technology, magnetorheological polishing, and energy storage. To this end, superparamagnetic colloids have been exploited for labeling and separation of DNA, proteins, bacteria, and various biological species, as well as being applied to magnetic resonance imaging (MRI), guided drug delivery, and the hyperthermia treatment of cancer. Even certain types of molecular interactions can also be probed *in vivo* using specially designed magnetic probes.[105] In recent years, a wealth of magnetic materials have been prepared as superparamagnetic colloids using a variety of chemical methods. In addition to the aqueous system, there is rapid progress in the synthesis of superparamagnetic nanoparticles using nonhydrolytic solvents.[106–108] Different precursors, solvents, and capping agents have all been systematically examined for producing monodisperse superparamagnetic particles.

2.3 Summary

The properties of nanomaterials are substantially different from those of their bulk counterparts. In this chapter, the unique characteristics of nanomaterials including thermal, electrical, phonon transport, mechanical, optical and magnetic properties were discussed.

The unique characteristics arise from many different aspects, for example, the huge surface area of the nanomaterial is responsible for the reduction in thermal stability and the supermagnetism; the increased surface scattering is responsible for the reduced electrical conductivity; size confinement results in a change of both electronic and optical properties; and the reduction in size decreases the defects and increases the perfection and mechanical properties.

Hierarchical nanostructures still hold the unique characteristics of the nanomaterials and they can even be amplified by smart design of the functional nanostructures. Especially, the hierarchical nanostructures are designed to possess larger surface areas with enhanced electrical transport characteristics for fast carrier transport. That is the reason this book focuses more on the energy devices than any other applications such as bioengineering. All of the energy devices explained in this book will basically be based on the unique characteristics explained in this chapter.

References

1. G. A. Ozin, *Adv. Mater.*, 1992, **4**, 612.
2. P. Alivisatos, P. F. Barbara, A. W. Castleman, J. Chang, D. A. Dixon, M. L. Line, G. L. McLendon, J. S. Miller, M. A. Ratner, P. J. Rossky, S. I. Stupp and M. I. Thompson, *Adv. Mater.*, 1998, **1**, 1297.
3. R. Dagani, *Chem. Eng. News*, 2000, October, 27.
4. W. Schulz, *Chem. Eng. News*, 2000, May, 41.
5. A. Thiaville and J. Miltat, *Science*, 1999, **284**, 1939.
6. *Handbook of Nanostructured Materials and Nano-technology*, ed. H. S. Nalwa, Academic Press, New York, 2000.
7. *Nano-structured Materials: Clusters, Composites, and Thin Films*, ed. V. M. Shalaev and M. Moskovits, American Chemical Society, Washington, DC, 1997.
8. *Nanomaterials: Synthesis, Properties, and Applications*, ed. A. S. Edelstein and R. C. Cammarata, Institute of Physics, Philadelphia, PA, 1996.
9. C. Ross, *Annu. Rev. Mater. Sci.*, 2001, **31**, 203.
10. C. B. Murray, C. R. Kagan and M. G. Bawendi, *Annu. Rev. Mater. Sci.*, 2000, **3**, 545.
11. A. P. Alivisatos, *Science*, 1996, **271**, 933.
12. L. Brus, *J. Phys. Chem.*, 1994, **98**, 575.
13. M. G. Bawendi, M. L. Steigerwald and L. E. Brus, *Annu. Rev. Phys. Chem.*, 1990, **41**, 477.
14. J. M. Krans, J. M. van Rutenbeek, V. V. Fisun, I. K. Yanson and L. J. de Jongh, *Nature*, 1995, **375**, 767.
15. B. L. Al'tshuler and P. A. Lee, *Phys. Today*, 1988, December, 36.
16. K. K. Likharev and T. Claeson, *Sci. Am.*, 1992, June, 80.
17. K. K. K. Likharev, *IBM J. Res. Dev.*, 1988, **32**, 144.
18. G. Markovich, C. P. Collier, S. E. Henrichs, F. Remacle, R. D. Levine and J. R. Heath, *Acc. Chem. Res.*, 1999, **32**, 415.
19. Y. Xia, P. Yang, Y. Sun, Y. Wu, B. Mayers, B. Gates, Y. Yin, F. Kim and H. Yan, *Adv. Mater.*, 2003, **15**, 353–389.
20. G. Cao, *Nanostructures and Nanomaterials: Synthesis, Properties and Applications*, Imperial College Press, MA, 2004.
21. Ph. Buffat and J.-P. Borel, *Phys. Rev. A*, 1976, **13**(6), 2287.
22. N. T. Gladkich, R. Niedermayer and K. Spiegel, *Phys. Status Solidi*, 1966, **15**, 181.
23. M. Blackman and A. E. Curzon, Structure and Properties of Thin Films, Wiley, New York, 1959.
24. B. T. Boiko, A. T. Pugachev and Y. M. Mratsykhin, *Sov. Phys. Sol. State.*, 1969, **10**, 2832.
25. M. Takagi, *J. Phys. Soc. Jpn.*, 1954, **9**, 359.
26. R. E. Newnham, K. R. Udayakumar and S. Trolier-McKinstry, in *Chemical Processing of Advance Materials*, ed. L. L. Hench and J. K. West, John Wiley and Sons, New York, 1992, p. 379.

27. K. Ishikawa, K. Yoshikawa and N. Okada, *Phys. Rev.*, 1988, **B37**, 5852.
28. K. J. Hanszen, *Z. Phys.*, 1960, **157**, 523.
29. S. Link, C. Burda, M. B. Mohamed, B. Nikoobakht and M. A. El-Sayed, *Phys. Rev.*, 2000, **B61**, 6086.
30. Y. Wu and P. Yang, *Appl. Phys. Lett.*, 2000, 77, 43.
31. Y. Wu and P. Yang, *Adv. Mater.*, 2001, **13**, 520.
32. D. Quere, J.-M. D. Meglio and F. Brochard-Wyart, *Science*, 1990, **249**, 1256.
33. Z. Zhang, X. Sun, M. S. Dresselhaus and J. Y. Ying, *Phys. Rev. B*, 2000, **61**, 4850.
34. S. H. Choi, K. L. Wang, M. S. Leung, G. W. Stupian, N. Presser, B. A. Morgan, R. E. Robertson, M. Abraham, S. W. Chung, J. R. Heath, S. L. Cho and J. B. Ketterson, *J. Vac. Sci. Technol., A*, 2000, **18**, 1326.
35. A. I. Yanson, G. R. Bollinger, H. E. van den Brom, N. Agrait and J. M. van Ruitenbeek, *Nature*, 1998, **395**, 783.
36. H. van Houten and C. Beenakker, *Phys. Today*, 1996, July, 22.
37. Y. Wang, X. Duan, Y. Cui and C. M. Lieber, *Nano Lett.*, 2002, **2**, 101.
38. Y. Cui and C. M. Lieber, *Science*, 2001, **291**, 851.
39. S. W. Chung, J. Y. Yu and J. R. Heath, *Appl. Phys. Lett.*, 2000, **76**, 2068.
40. K. Schwab, E. A. Henriksen, J. M. Worlock and M. L. Roukes, *Nature*, 2000, **44**, 974.
41. A. Buldum, S. Ciraci and C. Y. Fong, *J. Phys. Condens. Mater.*, 2000, **12**, 3349.
42. S. G. Volz and G. Chen, *Appl. Phys. Lett.*, 1999, **75**, 2056.
43. J. Heremans and C. M. Thrush, *Phys. Rev. B*, 1999, **59**, 12579.
44. G. Dresselhaus, M. S. Dresselhaus, Z. Zhang and X. Sun, International Conference on Thermoelectrics, Nagoya, Japan, 1998, IEEE, Piscataway, NJ 1998, p. 43.
45. R. Venkatasubramanian, E. Silvola, T. Colpitts and B. O'Quinn, *Nature*, 2001, **413**, 597.
46. C. B. Vining, *Nature*, 2001, **413**, 577.
47. Y. Wu, R. Fang and P. Yang, *Nano Lett.*, 2002, **2**, 83.
48. A. I. Boukai, Y. Bunimovich, J. Tahir-Kheli, J.-K. Yu, W. A. Goddard III and J. R. Heath, *Nature*, 2008, **451**, 168.
49. A. I. Hochbaum, R. Chen, R. D. Delgado, W. Liang, E. C. Garnett, M. Najarian, A. Majumdar and P. Yang, *Nature*, 2008, **451**, 163.
50. V. G. Lyuttsau, M. Fishman and I. L. Svelov, *Sov. Phys. Crystallogr.*, 1966, **10**, 707.
51. R. W. Siegel, *Mater. Sci. Eng.*, 1993, **A168**, 189.
52. T. Fillester, S. Ryu, K. Kang, J. Yin, R. A. Bernal, K. Sohn, S. Li, J. Huang, W. Cai and H. D. Espinosa, *Small*, 2012, **8**, 2986.
53. T. Zhu, J. Li, A. Samanta, A. Leach and K. Gall, *Phys. Rev. Lett.*, 2008, **100**, 1.

54. B. Wu, A. Heiderberg, J. J. Boland, J. E. Sader, X. M. Sun and Y. D. Li, *Nano Lett.*, 2006, **6**, 468.
55. M. Lucas, A. M. Leach, M. T. McDowell, S. E. Hunyadi, K. Gall, C. J. Murphy and E. Riedo, *Phys. Rev. B*, 2008, 77, 245420.
56. P. Lee, J. H. Lee, H. M. Lee, J. Yeo, S. Hong, K. H. Nam, S. S. Lee and S. H. Ko, *Adv. Mater.*, 2012, **24**, 3326.
57. J. A. Rogers, T. Someya and Y. Huang, *Science*, 2010, **26**, 1603.
58. R. F. Service, *Science*, 2003, **301**, 909.
59. J. A. Rogers, Z. Bao, K. Baldwin, A. Dodabalapur, B. Crone, V. R. Raju, V. Kuck, H. Katz, K. Amundson, J. Ewing and P. Drzaic, *Proc. Natl. Acad. Sci. U. S. A.*, 2001, **98**, 4835.
60. S. H. Ko, H. Pan, C. P. Grigoropoulos, C. K. Luscombe, J. M. J. Fréchet and D. Poulikakos, *Nanotechnology*, 2007, **18**, 345202.
61. Y. Ahn, E. B. Duoss, M. J. Motala, X. Guo, S.-I. Park, Y. Xiong, J. Yoon, R. G. Nuzzo, J. A. Rogers and J. A. Lewis, *Science*, 2009, **323**, 1590.
62. Y. Son, J. Yeo, H. Moon, T. W. Lim, K. H. Nam, C. P. Grigoropoulos, S. Yoo, D.-Y. Yang and S. H. Ko, *Adv. Mater.*, 2011, **23**, 3176.
63. H. Pan, S. H. Ko, N. Misra and C. P. Grigoropoulos, *Appl. Phys. Lett.*, 2009, **94**, 071117.
64. T. Sekitani, Y. Nouguchi, K. Hata, T. Fukushima, T. Aida and T. Someya, *Science*, 2008, **321**, 1468.
65. S. H. Ko, H. Pan, D. Lee, C. P. Grigoropoulos and H. K. Park, *Jpn. J. Appl. Phys.*, 2010, **49**, 05EA12.
66. K. Takei, T. Takahashi, J. C. Ho, H. Ko, A. G. Gillies, P. W. Leu, R. S. Fearing and A. Javey, *Nat. Mater.*, 2010, **9**, 821.
67. X. Wang, H. Hu, Y. Shen, X. Zhou and Z. Zheng, *Adv. Mater.*, 2011, **23**, 3090.
68. D. S. Hecht, L. Hu and G. Irvin, *Adv. Mater.*, 2011, **23**, 1482.
69. A. A. Argun, A. Cirpan and J. R. Reynolds, *Adv. Mater.*, 2003, **15**, 1338.
70. D. Zhang, K. Ryu, X. Liu, E. Polikarpov, J. Ly, M. E. Tompson and C. Zhou, *Nano Lett.*, 2006, **6**, 1880.
71. Z. Yu, X. Niu, Z. Liu and Q. Pei, *Adv. Mater.*, 2011, **23**, 3989.
72. G. Eda, G. Fanchini and M. Chhowalla, *Nat. Nanotechnol.*, 2008, **3**, 270.
73. K. S. Kim, Y. Zhao, H. Jang, S. Y. Lee, J. M. Kim, K. S. Kim, J.-H. Ahn, P. Kim, J.-Y. Choi and B. H. Hong, *Nature*, 2009, **457**, 706.
74. D. C. Hyun, M. Park, C. Park, B. Kim, Y. Xia, J. H. Hur, J. M. Kim, J. J. Park and U. Jeong, *Adv. Mater.*, 2011, **23**, 2946.
75. M. G. Kang, M. S. Kim, J. S. Kim and L. J. Guo, *Adv. Mater.*, 2008, **20**, 4408.
76. X. Lu, T. Hanrath, K. P. Johnston and B. A. Korgel, *Nano Lett.*, 2003, **3**, 93.
77. T. T. Hanrath and B. A. Korgel, *J. Am. Chem. Soc.*, 2001, **124**, 1424.

78. J. D. Holmes, K. P. Johnston, R. C. Doty and B. A. Korgel, *Science*, 2000, **287**, 1471.
79. M. V. Wolkin, J. Jorne, P. M. Fauchet, G. Allan and C. Delerue, *Phys. Rev. Lett.*, 1999, **82**, 197.
80. P. Lulvaney, *MRS Bull.*, 2001, **26**, 1009.
81. L. Gunnarsson, T. Rindzevicisu, J. Prikulis, B. Kasemo, M. Kall, S. Zou and G. C. Schatz, *J. Phys. Chem. B*, 2005, **109**, 1079.
82. K. L. Kelly, E. Coronado, L. Zhao and G. C. Schatz, *J. Phys. Chem. B*, 2003, **107**, 668.
83. M. Moskovits, *Rev. Mod. Phys.*, 1985, **57**(3), 783.
84. G. C. Schatz and R. P. Van Duyne, *Handbook of Vibrational Spectroscopy*, ed. J. M. Chalmers and P. R. Griffiths, Wiley, New York, 2001, pp. 759–774.
85. S. A. Maier, P. G. Kik, H. A. Atwater, S. Meltser, E. Harel, B. C. Koel and A. R. G. Requicha, *Nat. Mater.*, 2003, **2**, 229.
86. M. Quinten, A. Leitner, J. R. Krenn and F. R. Aussenegg, *Opt. Lett.*, 1998, **23**(17), 1331.
87. J. J. Storhoff, R. Elghanian, R. C. Mucic, C. A. Mirkin and R. L. Leitsinger, *J. Am. Chem. Soc.*, 1998, **120**, 1959.
88. A. D. McFarland and R. P. Van Duyne, *Nano Lett.*, 2003, **3**, 1057.
89. L. Olofsson, T. Rindzevicius, I. Pfeiffer, M. Kall and F. Hook, *Langmuir*, 2003, **19**, 10414.
90. A. Haes and R. P. V. Duyne, *J. Am. Chem. Soc.*, 2002, **124**, 10596.
91. B. Lamprecht, G. Schider, R. T. Lechner, J. R. Krenn and F. R. Aussenegg, *Phys. Rev. Lett.*, 2000, **84**, 4721.
92. C. L. Haynes, A. D. McFarland, L. L. Zhao, R. P. Van Duyne, G. C. Schatz, L. Gunnarsson, J. Prikulis, B. Kasemo and M. Kall, *J. Phys. Chem. B*, 2003, **107**, 7337.
93. W. Gotschy, K. Vonmetz, A. Leitner and F. R. Aussenegg, *Appl. Phys. B*, 1996, **63**, 381.
94. M. Kahl and E. Voges, *Phys. Rev. B*, 2000, **61**(20), 14078.
95. T. R. Jensen, M. D. Malinsky, C. L. Haynes and R. P. Van Duyne, *J. Phys. Chem. B*, 2000, **104**, 10545.
96. M. Meier, A. Wokaun and P. F. Liao, *J. Opt. Soc. Am. B*, 1985, **2**, 931.
97. L. L. Zhao, K. L. Kelly and G. C. Schatz, *J. Phys. Chem. B*, 2003, **107**, 7343.
98. K. Kneipp, Y. Wang, H. Kneipp, L. T. Perelman, I. Itzkan, R. R. Dasari and M. S. Feld, *Phys. Rev. Lett.*, 1997, **78**, 1667.
99. S. Nie and S. R. Emory, *Science*, 1997, **275**, 1102.
100. H. Xu, J. Bjerneld, M. Kall and L. Borjesson, *Phys. Rev. Lett.*, 1999, **83**, 4357.
101. J. Jiang, K. Bosnick, M. Milliard and L. Brus, *J. Phys. Chem. B*, 2003, **107**, 9964.
102. H. Xu and M. Kall, *Phys. Rev. Lett.*, 2002, **89**, 246802.

103. Q. A. Pankhurst, J. Connolly, S. K. Jones and J. Dobson, *J. Phys. D: Appl. Phys.*, 2003, **36**, R167–R181.
104. U. Jeong, X. W. Teng, Y. Wang, H. Yang and Y. N. Xia, *Adv. Mater.*, 2007, **19**, 33–60.
105. D. E. Ingber, *Annu. Rev. Physiol.*, 1997, **59**, 575.
106. C. B. Murray, S. Sun, H. Doyle and T. Betley, *MRS Bull.*, 2001, **26**, 985.
107. T. Hyeon, *Chem. Commun.*, 2003, 927.
108. S. Sun, *Adv. Mater.*, 2006, **18**, 393.

CHAPTER 3

Nanotechnology's Wonder Material: Synthesis of Carbon Nanotubes

JUNG BIN IN*[a] AND ALEKSANDR NOY*[b]

[a] Mechanical Engineering, University of California Berkeley, Berkeley, California 94720, U.S.A.; [b] Physics and Life Sciences Directorate, Lawrence Livermore National Laboratory, Livermore, California 94550, U.S.A.
*Email: jbin@berkeley.edu; noy1@llnl.gov

3.1 Introduction

3.1.1 Carbon Nanotubes: General Overview

By now the carbon nanotube (CNT) has firmly established its place as a nanoscience icon and the foundation of many scientific and (increasingly nowadays) technological breakthroughs.[1] Despite having a very simple chemical composition—it is just another allotropic form of carbon—nanotubes have an astonishing variety of unique properties. A carbon nanotube is simply a nanometre-sized rolled-up graphene sheet that forms a perfect seamless cylinder (Figure 3.1) capped at the ends by fullerene caps. The structure of a simple one-shell carbon nanotube is fully defined by its roll-up vector (n,m), called chirality or helicity, which defines the position of the matched carbon rings during the roll-up of the graphene sheet.[2] Significantly, this roll-up vector fully defines the nanotube morphology,

RSC Nanoscience & Nanotechnology No. 35
Hierarchical Nanostructures for Energy Devices
Edited by Seung Hwan Ko and Costas P Grigoropoulos

Published by the Royal Society of Chemistry, www.rsc.org

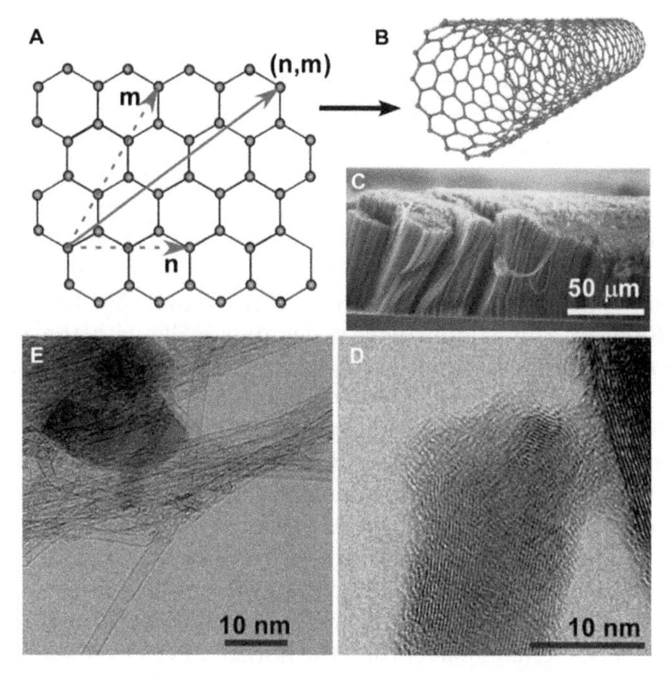

Figure 3.1 Structure and morphology of carbon nanotubes. (A) Schematic representation of a graphene sheet and a carbon nanotube roll-up vector. The roll-up vector is perpendicular to the axis of carbon nanotube. (B) A 3D model of a single-wall carbon nanotube. (C) A SEM image of a vertically-aligned array of multi-wall carbon nanotubes grown on a silicon substrate. (D, E) TEM images of single-wall (E) and multi-wall (D) carbon nanotubes. (SEM and TEM images: Lawrence Livermore National Laboratory).

diameter, and most importantly, its electronic properties. For example, an (n,m) carbon nanotube has an inner diameter, d_{in}, of:[2]

$$d_{in} = \frac{a}{\pi}\sqrt{n^2 + m^2 + nm} - 2r_c \tag{3.1}$$

where a is the lattice parameter of graphene (2.5 Å) and r_c is the van der Waals radius of the carbon atom (1.7 Å).

A carbon nanotube can have one (as in the case of a single-walled carbon nanotube, Figure 3.1E), or several concentric graphitic shells (as in the case of multi-walled nanotubes, Figure 3.1D). Remarkably, despite its nanometre diameter, the length of a single nanotube can reach or exceed a centimetre, giving it an aspect ratio of greater than 10^7. Typically, nanotubes have lengths in the range of several microns, which still represents an aspect ratio of 10^3 or greater. Typically, carbon nanotubes are synthesized in three forms: loose powders, densely-packed vertical "forest" arrays (Figure 3.1C), or individual nanotubes grown on prepatterned surfaces.

3.1.2 Carbon Nanotubes in Science and Technology

Almost every physical property of the carbon nanotube invokes a superlative. They are the strongest fiber with Young's moduli exceeding 1TPa, yet they can buckle and spring back without incurring damage, just like a drinking straw. The inner pores of the nanotubes represent the smoothest pore known in nature, and the combination of this smoothness with the hydrophobic nature of the pore surface supports water transport rates that are 1000 times higher than the transport rates through conventional pores of a similar size.[3] The electrical conductivity of the nanotubes exceeds that of copper. Moreover, this conductivity varies drastically with the nanotube's structure: depending on chirality, the nanotube can be a metal or a semiconductor. In a randomly distributed chirality sample, 1/3 of the nanotubes are metallic, and 2/3 are semiconducting.

This remarkable set of properties drives the intense interest of the science community in the different applications of nanotubes, including nanoelectronic transistors, highly permeable membranes for liquid and gas separations, structural composite reinforcements, and supercapacitors for energy storage applications.[4] Yet, despite tremendous efforts and the progress made in many of those fields, the mass-scale industrial applications of carbon nanotubes remain just beyond the horizon. Even a cursory look at the problems that researchers encountered in trying to realize these applications traces their routes back to the synthesis. Reliable electronic applications require large-scale production of nanotubes of uniform size and chirality; current synthesis techniques still produce a mixture of different roll-up vector tubes. Carbon nanotube membranes need large-scale uniform nanotube arrays that are free from defects over a significant length and area; current synthesis techniques still struggle to achieve this goal. Structural reinforcement applications require the defined placement and alignment of long carbon nanotubes; most of the common synthetic approaches for long nanotubes produce low-density and often tangled samples. Despite ongoing efforts, carbon nanotube arrays cannot reach heights above several centimetres. These examples underscore the imperative to develop new synthesis techniques and to understand the physics and chemistry of the existing growth processes.

3.2 Carbon Nanotube Synthesis

3.2.1 CNT Growth Technologies

Much of the technological progress in the carbon nanotube field was spurred by the ability to make nanotubes of required quantity, and to a lesser extent of required quality, structure, and purity. Since the pioneering works demonstrated synthetic feasibility,[5] researchers have come up with a suite of different fabrication methods. Notably, we can trace most of this progress to the desire to achieve controllability of the atomic structure and nanotube properties. As we noted in the previous section, the structural configuration (*i.e.* chirality, diameter, and the number of walls) can cause a substantial change in the physical properties of the produced nanotubes; thus, the development of CNT synthesis techniques has always pursued *selective* growth. The accompanying synthetic quality factors, such as defects and non-graphitic by-product generation, can also degrade the performance of the material in different applications, therefore control of these parameters is also an important task. Production yield and scalability go to the forefront of the CNT synthesis challenges for the commercial use of CNTs.

At its most general, nanotube growth requires a source of carbon atoms as well as significant energy input to dissociate the carbon precursors and form ordered graphitic cylindrical walls. This description is indeed surprisingly general: researchers were able to grow nanotubes from gas,[6,7] vapor,[8–10] or solid precursors[11] and they used widely varying methods to deliver the activation energy that is necessary to overcome the energetic barriers for CNT growth.

The early stage of CNT synthesis saw arc discharge and laser ablation as the methods of choice for producing high quality single- or multi-walled carbon nanotubes.[12] The short duration and high energy density of electric arc and laser pulses supply the burst of energy that vaporizes a solid carbon source (*i.e.* graphite or metal–carbon composite) and enables the generation of CNTs.[13,14] However, the same high energy density used in the process hampers efforts to control the growth; for instance, making long carbon nanotubes using these techniques is highly problematic. Moreover, the nanotube product typically coats the chamber walls, making subsequent harvesting a challenge and complicating the subsequent efforts to use nanotubes in any bottom-up fabrication process. Despite the potential to use ablation processes for large-scale production, the production yield remains low ($\sim$30%) due to formation of significant by-products such as amorphous carbon and other carbon allotropes.

The breakthrough in carbon nanotube synthesis came in 1998 when Smalley's group demonstrated a synthesis of carbon nanotubes using chemical vapor deposition (CVD).[15] 15 years later CVD synthesis dominates the CNT growth field. CVD has attractive features that overcome the disadvantages inherent to laser ablation and arc discharge methods. Most importantly, the CVD process allows researchers much better control over the energy delivery to the system, and thus much finer control of the rates and chemistry of carbon precursor pyrolysis. The reaction environment (temperature, pressure, gas flow rate, *etc.*) of CVD is highly controllable and spatially uniform (in properly designed reactors); hence, CVD methods enable easy scale-up of the growth substrate size. Conventional epitaxial film growth by metal–organic CVD (MOCVD), which has many similarities to the CNT growth, has proven the technical benefits of CVD in large-scale fabrication of high quality crystalline materials and well-established knowledge on traditional CVD processes can translate to CNT synthesis. CVD methods can also benefit from process compatibility with other conventional microfabrication processes for electronics applications. The high growth temperatures of CNTs (600–900 °C) remain an obstacle to full process integration, but researchers keep lowering the growth temperature for nanotube CVD growth.[16] It is safe to say that CVD remains the most plausible way to incorporate CNTs in industrial manufacturing processes and bottom-up fabrication strategies. The next section will introduce several kinds of CVD methods for CNT growth and discuss their technical features.

3.2.1.1 Chemical Vapor Deposition for CNT Growth

As the name indicates, CVD relies on vapor or a gaseous carbon feedstock. Common CVD precursors include a number of hydrocarbons or alcohols that differ in their chemical reactivity and compatibility with gas delivery systems. CVD apparatus usually contains equipment to control and monitor gas flow and process pressure. As the sample size gets bigger, the heating uniformity takes center stage and the machines get increasingly sophisticated and acquire additional features such as multiple heating zones.

CVD machines also fall roughly into three principal designs: (1) hot-wall reactor, such as a tube furnace, where hot walls are the primary means of heating the process gases, (2) cold-wall reactor where only the sample gets heated, and the process gas cracks upon contact with the sample, and (3) hot-filament reactor where the gas cracks on a hot filament situated close to the sample surface. As spatial uniformity of the mass transport process becomes paramount to achieve uniform growth, the shower-head type cold-wall CVD reactors are by far the most practical type of reactor to obtain uniform growth on samples larger than 10 cm. However, due to cost considerations as well as historical reasons, the hot-wall reactors (tube furnaces) dominate the carbon nanotube growth literature, and most of the results that we discuss were obtained in a hot-wall reactor. We also note parenthetically that it is one of the major reasons that the results from the research literature are often difficult to translate to industrial scale growth processes.

The process used to deliver the cracking energy also distinguishes several forms of CVD (Figure 3.2); mostly, they are categorized into thermal CVD, which utilizes Joule heating, or plasma enhanced CVD (PECVD). Both processes can be implemented in cold-wall and hot-wall reactors.

In order to enhance the energy efficiency of thermal CVD, the hot-wall reactor furnace is insulated by ceramic walls, and the chemical reactions for growth occur inside a leak-free quartz reactor (Figure 3.2A). The thermal energy in this case is delivered to the growth substrate by thermal conduction from the heating element, or by radiation of infrared (IR) light from the hot walls. Each thermal contribution depends on the mechanical design of the furnace system and spectral characteristics of the quartz glasses in IR ranges at growth temperatures.

In a cold-wall reactor, the heated area is confined to the immediate vicinity of the substrate. In addition, reduced thermal capacitance of the cold-wall

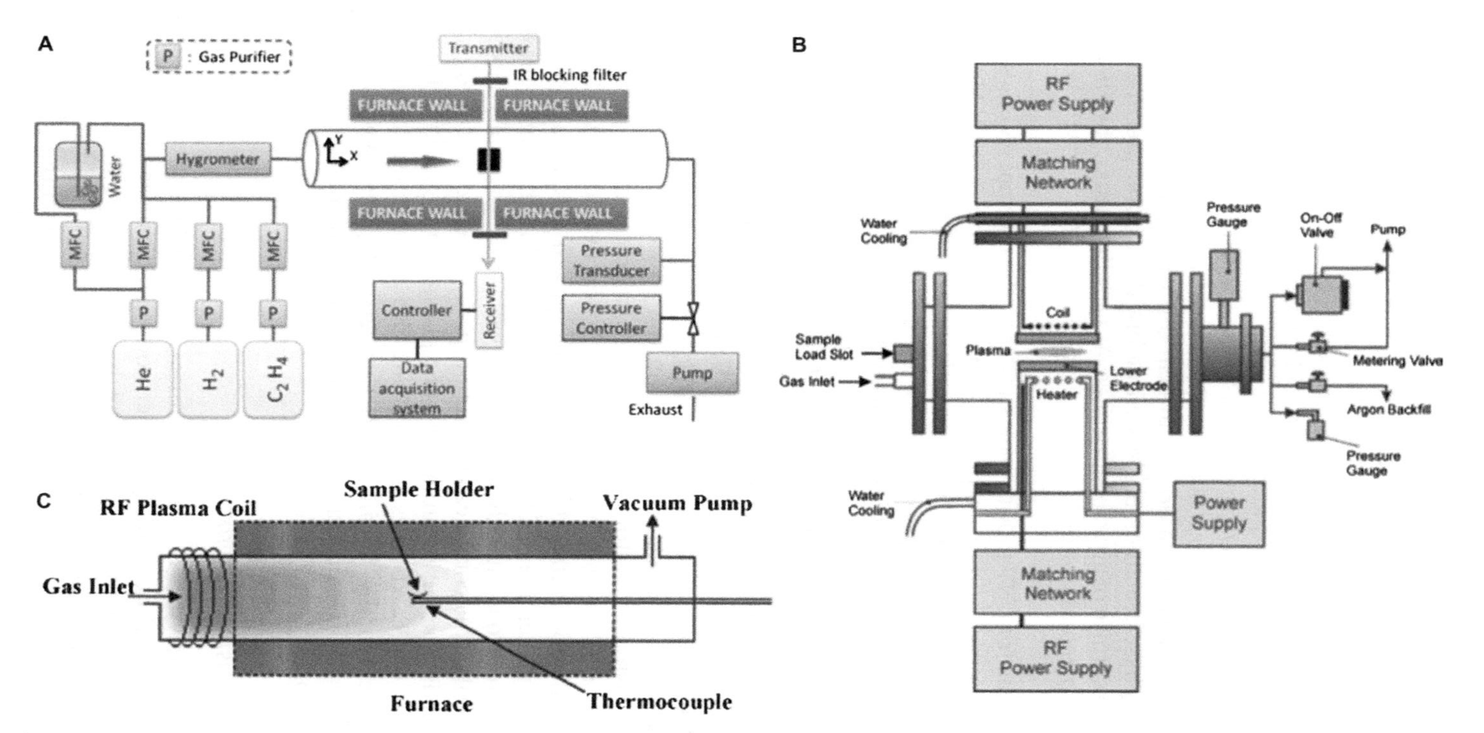

Figure 3.2 Common types of CVD systems. (A) Thermal CVD (hot-wall) equipped with an *in situ* growth monitoring system. Reprinted with permission from ref. 47. Copyright 2011, AIP Publishing LLC. (B) PECVD setup. Reprinted with permission from ref. 21. Copyright 2002, American Institute of Physics. (C) Remote PECVD. Reprinted with permission from ref. 23. Copyright 2004 American Chemical Society.

type system enables a rapid ramp of the substrate temperature, which can mitigate unwanted thermally activated processes (*i.e.* Ostwald ripening of catalyst particles). Hart *et al.*[17,18] pioneered this type of reactor for CNT growth by using a doped silicon susceptor as a Joule heating resistor. Several other designs also take advantage of inductive heating or focused laser irradiation, implementing laser-assisted CVD (LACVD).[19,20]

Besides heating uniformity, another technical difference between hot- or cold-wall reactors in the CVD growth of CNTs is the degree and duration of gas-phase reactions (GPR). In a hot-wall CVD reactor, the temperature of the gas feedstock becomes increased due to heat diffusion from the wall. Consequently, gas-phase reactions begin well before the carbon precursors actually reach the surface of growth substrates. As we will discuss, the effects of this feedstock preconditioning can be significant in the growth kinetics of CNTs.

PECVD systems for CNT growth look similar to the thermal CVD systems, except they contain an additional plasma generator (Figure 3.2B).[21] Methods that can generate plasma for CNT growth include RF capacitance and microwaves. Due to the aggressive reactivity of gas precursors in plasma, most PECVD systems use a cold-wall reactor and low process pressure. The most important feature of PECVD is that reactive carbon precursors generated by collision with energetic electrons in plasma enable CNT growth at remarkably lower temperatures. For instance, Hofmann *et al.*[22] demonstrated that multi-walled CNTs can grow at 120 °C by PECVD, compared to general growth temperatures (600–900 °C) by thermal CVD.

While PECVD lowers the thermal budget of CNT growth processes, the plasma can also cause structural defects on the growing nanotubes. Alternatively, the plasma generator can be installed at an upstream position to avoid such damage. This type of CVD is called *remote* PECVD (Figure 3.2C).[23] Zhang *et al.*[24] demonstrated the reproducible growth of vertically aligned SWCNT arrays by a remote PECVD. Radicals generated in the *remote* plasma zone are likely to evolve quickly by colliding with other gas molecules and the chamber wall. This complicated reaction path can obscure the role of a specific gas feedstock, but it can mitigate the plasma damage without losing the advantages of PECVD. Researchers can also achieve similar types of growth enhancement by flowing gases through a high-temperature flow cell.[17] In this case, thermal energy replaces plasma to pre-condition the carbon precursors.

3.2.1.2 Catalyst Nanoparticles

In principle, a catalyst accelerates chemical reaction kinetics by lowering energetic barriers, neither modifying thermodynamics nor being consumed during reactions. Carbon nanotube growth in a CVD-type process almost always uses a catalyst and thus represents an example of catalytic CVD (CCVD). Particle-based heterogeneous catalysis is an essential part for the low temperature growth of CNTs, and it has been consistently demonstrated that catalysts enhance growth yields remarkably. The best way to understand the role of the catalyst is to consider the energetics of carbon–carbon or carbon–hydrogen bonding. When hydrocarbon molecules are used as a carbon feedstock, the overall chemical reaction of nanotube growth can be described by the following expression:

$$C_xH_y \rightarrow xC_{NT} + \frac{y}{2}H_2(g) \tag{3.2}$$

where the subscript NT stands for a nanotube. When methane (CH_4) gas is used as a carbon source, four C–H bonds need to break. However, the energy of each bond is too high (439 kJ mol^{-1}) to be overcome by the pure thermal energy of CVD. At general CVD temperatures, the feedstock dissociation and carbon polymerization in a gaseous state are very limited; researchers found only ppm levels of polyaromatic hydrocarbon (PAH) products from such pyrolysis.[25] With a catalyst, however, adsorbed precursor molecules undergo multiple dissociation steps. Not only did the energetic modification of catalysis help the dissociation of the adsorbed precursor molecules, but also the electron-abundant surface of a metal catalyst facilitates carbon polymerization by holding the carbon radicals for an extended residence time.

Catalyst nanoparticles also play a crucial role in determining the morphologies of the resulting nanotubes. The curvature of a nanoparticle surface provides favorable nucleation sites to overcome the strain energy necessary for hemispherical nanotube cap formation. Indeed, CVD techniques for graphene growth use a flat metal foil of copper or nickel as catalytic substrates.[26,27] The confined domain size of a catalyst nanoparticle has significant correlation with the diameter of the nanotube grown from that catalyst. Yamada *et al.*[28] conducted a systematic study to explore correlation of particle size (or catalyst film thickness) with diameter, and number of walls. The data clearly revealed a straightforward relation: larger catalysts produce nanotubes with larger diameters and more walls. Lin *et al.*[29] also observed by *in situ* TEM that the diameter ratio of a nanotube to its catalyst particle falls within a finite range (0.5–1).

When it comes to selection of a catalyst composition, each element shows unique characteristics depending on the detail of the chemical reaction, operation temperatures and pressures, support materials, *etc.* In general, carbide-forming transition metals such as Fe, Co and Ni make the most efficient catalytic materials for CNT growth. Binary alloys of these metals also catalyze nanotube growth. A co-catalyst is not an active catalyst by itself, but it can contribute to enhancing growth activity when combined with active (Fe, Co, Ni) catalysts. For instance, molybdenum (Mo) is widely used as a co-catalyst for SWNT growth. While the definitive role of an auxiliary Mo catalyst remains elusive, several publications proposed that Mo helps the main catalysts by promoting carbon delivery to active catalyst particles,[30,31] preventing silicate formation on a silicon oxide support,[32] or improving dispersion of catalyst particles.[10,33]

Recent studies demonstrated that CNT growth by CVD is feasible even from non-metallic catalysts such as semiconductors[34] and oxide nanoparticles.[35,36] Considering the compatibility requirement of CMOS (complementary metal–oxide–semiconductor) fabrication, these semiconducting or ceramic catalysts could be technologically important. However, the production yields of these catalysts have not yet matched those of conventional metallic catalysts. Precious metals such as gold, silver, palladium, and platinum have been widely used in various catalysis processes using hydrocarbon gases. However, these precious metals are either less active for growth or only co-catalytic to other active catalysts. For instance, gold is a well-known catalytic material for the growth of inorganic nanowires, but without an additional activation treatment,[37] its application to CNT growth was not successful. Ding *et al.*'s study gives a fundamental clue to this elemental selection of catalysts. Based on their MD simulation results,[38] they proposed that the inefficiency of these metals is attributed to the weak bond strength between a nanotube and its catalyst (Figure 3.3). Indeed, their calculations on various metal elements are consistent with the empirical preference: Fe, Co, Ni better than Pt, Pd.

Researchers have expended a large amount of effort on designing catalysts for the selective growth of CNTs and assembling them on the substrate surface. First, prepared catalyst nanoparticles can be directly deposited on a substrate from solution-state dispersions. This wet process has an advantage in that the areal density of particles can be easily controlled by adjusting the concentration of the solution or coating speed. Besides, in this straightforward approach, researchers can control the particle size during the particle preparation step.

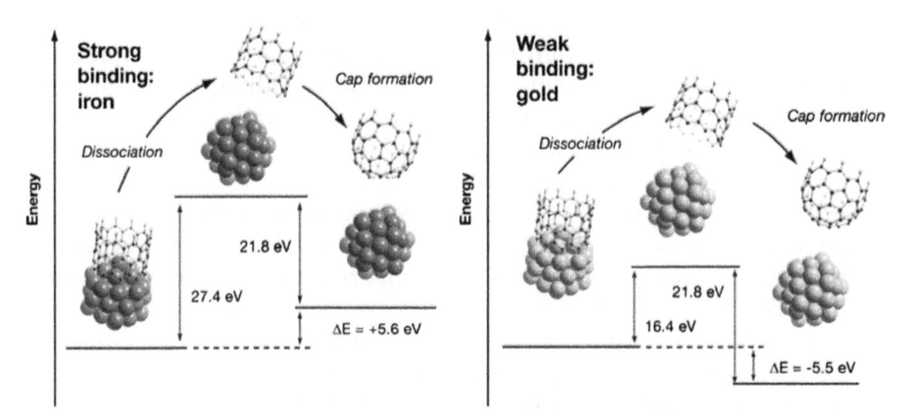

Figure 3.3 Energy diagrams describing detachment of a SWNT from a catalyst nanoparticle and the subsequent nanotube opening closure. The growth stability of the iron catalyst is attributed to the stronger adhesion of the CNT–iron (left) system. Relatively weak binding between CNT–gold (right) energetically favors separation of a nanotube from its catalyst. Reprinted with permission from ref. 38. Copyright 2008 American Chemical Society.

An alternative method for catalyst delivery to the surface relies on vapor phase transport in a so-called floating catalyst method. Catalyst precursors, such as ferrocene ($Fe(C_5H_5)_2$) and iron chloride ($FeCl_3$), get vaporized and then mixed with process gases. When this mixture of gases reaches the hot zone of the CVD furnace, catalyst particles precipitate on the substrate *in situ* and immediately initiate CNT growth. This method combines catalyst preparation and CNT growth processes, simplifying the overall synthesis processes. It also enables the continuous production of CNTs, whereas conventional CVD is a batch process that needs replacement of the substrate. The drawback is that the floating catalyst method can cause contamination of the CVD chamber since the catalyst gets deposited everywhere in the reactor. For the same reason, catalyst patterning on a desired area of a substrate is difficult to achieve.

Lastly, catalyst nanoparticles can be formed *in situ* during heat treatment of a thin catalyst film (Figure 3.4). Thin film deposition methods, such as sputtering and electron beam evaporation, can create an extremely thin (less than 10 nm) catalyst film on a substrate. When annealed at a high temperature, the metal film minimizes its surface energy by breaking into discrete islands and dewetting on the underlying oxide support. Researchers have adopted this method widely because of its simplicity in preparation and the ability to change the resulting catalyst size by controlling the initial film thickness. The spontaneous particle formation, however, produces a range of catalyst size distributions and consequently, nanotubes with a range of different diameters. The Ostwald ripening process changes the particle size distribution even more;[39] therefore, the dynamics of catalyst evolution also represents one of the critical parameters for CNT growth control.

An area density of nanotubes is proportional to that of the catalyst particles. Interestingly, an increased catalyst density leads to an abrupt change in the collective morphology of the nanotubes. As the inter-nanotube distance decreases, mechanical interactions start to work and the crowded-out nanotubes start to support each other, resulting in self-aligned nanotube arrays. The growth direction lies perpendicularly to the substrate surface, so the nanotubes of this particular structure are called vertically aligned carbon nanotubes (VACNTs). Because of their near-straight alignment, VACNTs are invaluable templates for device fabrication as well as for fundamental research on CNT growth mechanisms.

3.2.1.3 Process Gases

Process gases for CVD growth of CNTs contain a carbon precursor, a carrier gas, and sometimes a reducing gas. Light hydrocarbon gases are the most common carbon sources owing to their resistance to non-catalytic carbon polymerization. Several studies showed that alcohols in a vapor phase could also be used as a carbon feedstock. Hydrogen (H_2) gas is often present in the process feed gases to reduce the oxidized catalyst as well as for adjusting the chemical equilibrium of the gas-phase reaction. Argon (Ar) and helium (He)

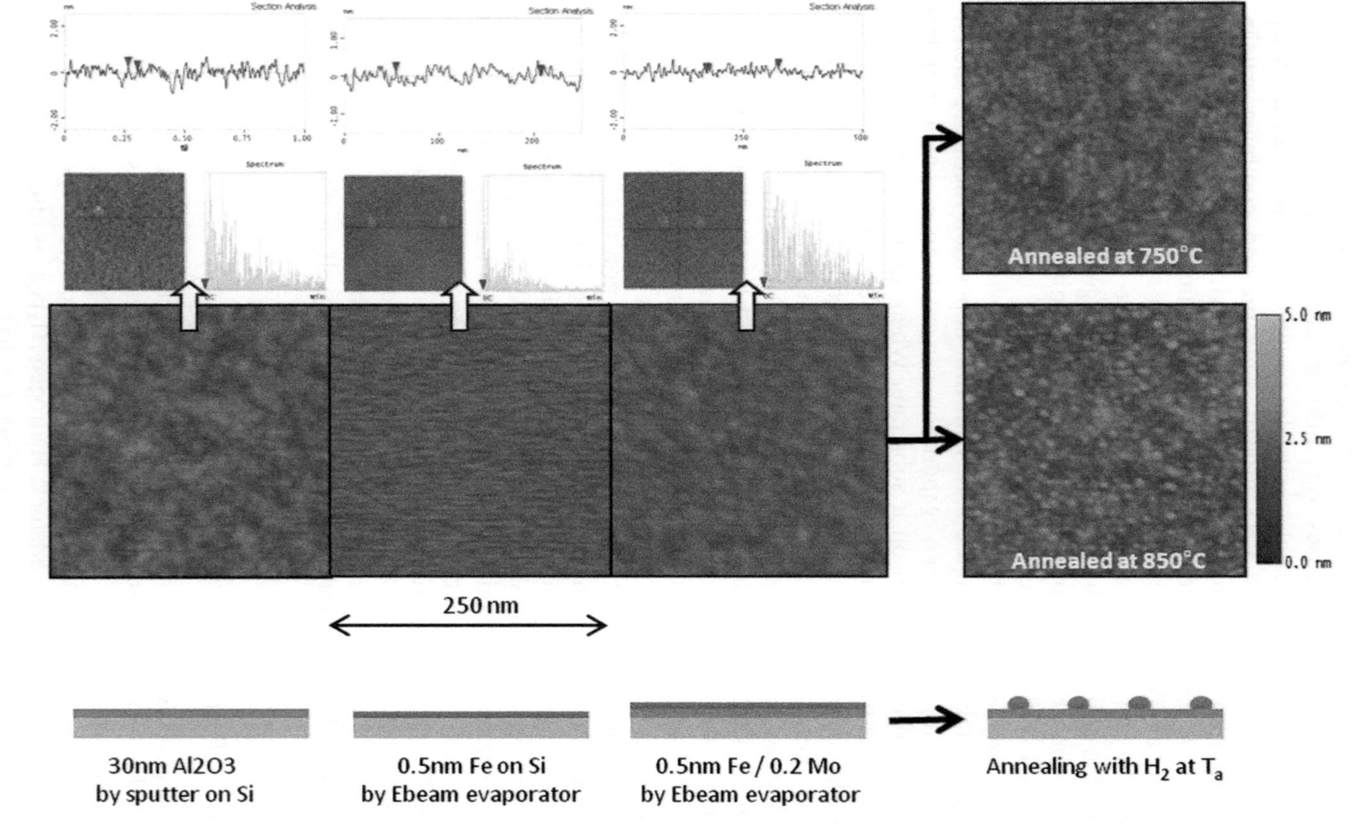

Figure 3.4 Atomic force microscopy images of as-deposited and annealed catalyst films. Thermal annealing breaks the thin film into dense catalyst nanoparticles.

are the most common diluting/carrier gases. These inert gases are not supposed to chemically participate in the decomposition of the carbon precursor molecules. Interestingly, a recent study by Harutyunyan *et al.*[40] demonstrated that a particular inert gas can induce restructuring of catalyst particles that enable a chiral-selective growth of nanotubes.

Whereas the surface chemistry of catalysts has been a major subject for growth mechanism studies, recent reports unveiled the significant effects of gas-phase reactions (GPRs) on growth kinetics and nanotube structure. Several GPR products, such as benzene,[41] acetylene,[42–44] and vinyl acetylene,[45] were found to be more efficient than the original methane or ethylene carbon feedstock. These species showed higher reactivity to accelerate the growth speed with improved growth yields. Meshot *et al.*[17] proved the striking effects of GPRs in a more obvious way by decoupling gas heating from substrate heating of their cold-wall reactor.

This new aspect of the CVD growth of CNTs raises an important question: what is the real (or most effective) carbon precursor? The answer should be valuable not only for growth mechanism studies but also for the efficient use of carbon gases and energy resources.[25] In a landmark study, Eres *et al.*[6] examined various precursor gases to compare their growth efficiencies (Table 3.1). They used a molecular jet growth setup at a relatively low temperature (650 °C) to minimize possible GPRs before the precursor molecule adsorbs on catalysts. They found that acetylene (C_2H_2) was the most efficient carbon precursor among the selected gases. It is interesting that ethylene (C_2H_4) and methane (CH_4) gases could not yield CNTs in their growth condition, although these gases have been widely used for nanotube growth but at rather high temperatures. Indeed, acetylene has been commonly used for low temperature growth of CNTs.

The most common CVD gases are supplied from gas cylinders and inevitably contain traces of gas impurities. In general, ultra-high purity (UHP) grade gases contain ppm-level impurities such as water vapor, oxygen and other process-related by-products. Gas impurity effects on the CVD growth of CNTs had not drawn much attention before a recent report from In *et al.*[46,47] They found that even minute levels of gas impurities could have striking effects on CNT growth kinetics; interestingly, the impurities significantly enhanced growth yields. Especially, by using inline gas purifiers (deoxo units, Figure 3.5A), they revealed that the growth enhancement was due to oxygen-containing impurity molecules such as water (H_2O), oxygen (O_2), and oxides of carbon (CO and CO_2) as shown in Figure 3.5B.

From another point of view, this enhancing effect of gas impurities can actually be utilized to promote growth. For instance, water is a well-known growth promoter. A controlled amount (approximately 10–200 ppm) of a water vapor mixed with other feedstock gases can dramatically improve catalyst lifetime, growth speed, and graphitic qualities of nanotubes; this is the essence of the so-called "supergrowth" process.[48,49] Many studies show that there exists an optimal level of water as a growth promoter, and excessive water is deleterious to growth most likely by oxidizing metal

Table 3.1 Growth efficiency of various gases in a molecular jet growth system. Reprinted (adapted) with permission from ref. 6. Copyright 2005 American Chemical Society.

Carbon source	Formula	Carrier gas	Temp,[a] °C	Deposit	MWCNT[b]	SWCNT[c]	Yield	VA-SWCNT[d]
Methane[e]	CH_4	10% H_2, 88% He	540–740	N				
Ethylene[e]	C_2H_4	10% H_2, 88% He	650	N				
Acetylene[e]	C_2H_2	10% H_2, 88% He	540–740	Y	N	Y	10	strong
Propylene[e]	CH_3CHCH_2	10% H_2, 88% He	540–740	Y	Y	Y	0.1	no
Benzene[f]	C_6H_6	10% H_2, 90% He	650	N				
Xylene[f]	$C_6H_4(CH_3)_2$	10% H_2, 90% He	650	N				
Acetone[f]	CH_3COCH_3	10% H_2, 90% He	540–740	Y	N	Y	1	weak
Diethyl ether[f]	$C_2H_5OC_2H_5$	10% H_2, 90% He	650	Y	N	Y	0.5	no
Methanol[f]	CH_3OH	10% H_2, 90% He	650	Y	N	Y	0.5	no
Ethanol[f]	CH_3CH_2OH	10% H_2, 90% He	540–740	Y	N	Y	0.5	no
Propanol[f]	$CH_3CH_2CH_2OH$	10% H_2, 90% He	650	Y	N	Y	0.5	no
Carbon monoxide[e]	CO	10% H_2, 88% He	540–740	N				

[a]Temperature refers to the temperature of the substrate.
[b]MWCNT = multiwall carbon nanotube.
[c]SWCNT = single-wall carbon nanotube.
[d]VA-SWCNT = vertically aligned SWCNT.
[e]These experiments were performed in a range of incidence rates from 1.5×10^{18} to 5.3×10^{18} molecules/(cm^2 s).
[f]Entrained in carrier gas using a bubbler. The beam composition is determined by the room-temperature vapor pressure.

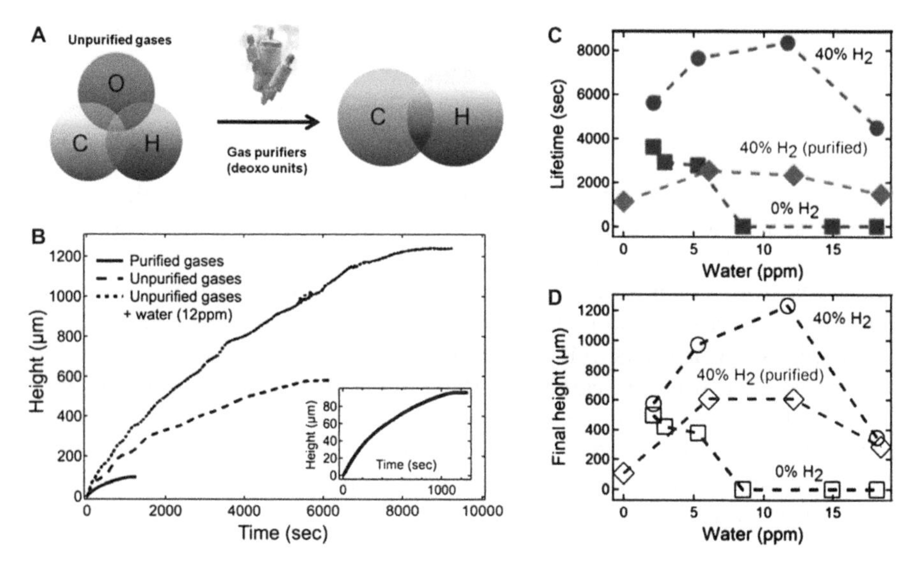

Figure 3.5 Gas purification effect on CNT growth kinetics. (A) Schematic describing the action of *deoxo* gas purifiers. (B) Kinetic curves of CNT growth with purified or unpurified gases (inset: enlarged plot of the purified gas data). (C) Plot of catalyst lifetime as a function of water concentration under different gas purification conditions. (D) Final height data corresponding to the data in (C). (B, C, D) Reprinted with permission from ref. 47. Copyright 2011, AIP Publishing LLC.

catalysts.[47,50,51] Other oxygen-containing molecules including alcohol carbon precursors exhibit similar effects.

Unfortunately, In *et al.* reported that the performance of a water promoter differs when the water vapor is combined with other gas impurities.[47] The lifetime and growth height data in Figure 3.5C and D show that the impurities still influence nanotube growth in spite of a quantitatively dominant water introduction. This result implies that the multiple oxygen-containing species make the growth chemistry complicated by promoting or deactivating the catalyst depending on their molecular forms and concentrations.[43,52] Moreover, their roles in the gas-phase reactions have not yet been explored rigorously. Therefore, gas purification should be a *required* step to avoid this complication and improve growth controllability and reproducibility.

3.2.2 Growth Mechanism

In this section, we describe the CNT growth mechanism in detail. In order to understand this sophisticated physical and chemical process, we first outline the overall growth process and then go over the details of each growth step. For brevity, we skip the catalyst pretreatment process and focus on the transport and chemical conversion events of carbon from its gaseous state to the atomic form in a nanotube. Therefore, we start by assuming that the catalyst nanoparticle is in its "ready for growth" state.

3.2.2.1 CVD Growth Process

Regardless of the exact types of process gases and catalysts, nanotube growth by CVD shares several common mechanistic steps (Figure 3.6). A controlled amount of carbon feedstock enters the process vessel and undergoes gas-phase reactions depending on the thermal environment of the reactor. Then the produced active carbon species are delivered to the gas–catalyst interface by gas-phase diffusion as well as by convective flow of the gases. The arriving molecules adsorb to the catalyst surface and subsequently dissociate into carbon adsorbates. The hydrogen products desorb back into the gas stream. Now the carbon adatoms diffuse through the catalyst to the open end of the nanotube, or nucleate into a graphitic cage to initiate growth if there is no pre-existing nanotube.

3.2.2.2 Gas-phase Reactions

Modern CVD literature contains ample proof that gas-phase reactions (GPRs) have significant effects on deposition kinetics and film morphology.[53] GPRs (or pyrolysis) of light hydrocarbons are not negligible at CNT growth temperatures,[54,55] and their profound effects on the growth of CNTs are critical for understanding the growth mechanism. As mentioned before, recent literature has revealed the existence of efficient pyrolysis products for CNT growth. However, pyrolysis of hydrocarbon species involves many elementary steps comprising initiation, radical propagation, production of intermediate species, and reaction termination with final products. Unfortunately, the complicated self-pyrolysis chemistry is also coupled with fluid dynamics of the gas flow and mass transport of the products to the

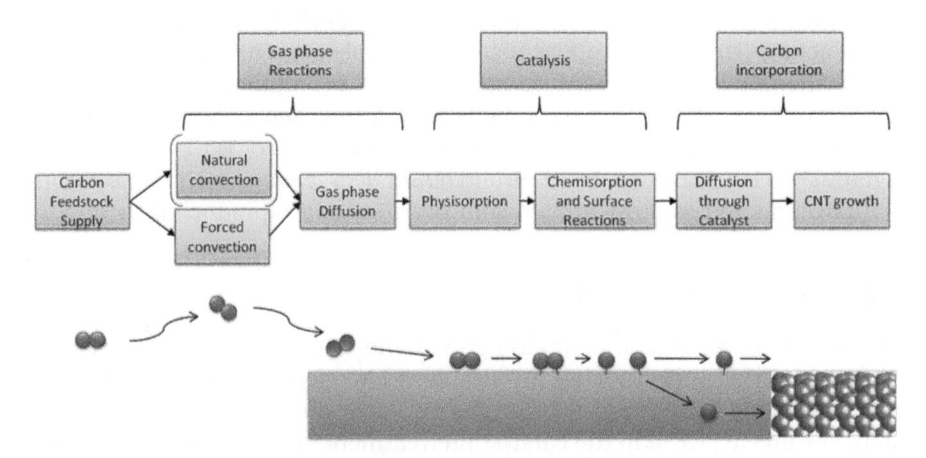

Figure 3.6 Main stages of the CNT growth process.

catalysts. Computer simulations may help in the understanding of this coupled non-linear numerical problem, yet reasonable solutions are still highly challenging to obtain. Therefore, rather than pursuing a complete analysis, most research focuses on the phenomenological aspects of GPRs in the CVD growth of CNTs.

In general CVD, non-equilibrium processes are dominant in GPRs of hydrocarbon pyrolysis, therefore the reaction time or residence time (τ_r) of gas molecules is a key parameter along with other thermodynamic properties. The residence time can be estimated from the following expression:

$$\tau_r = \frac{V_{rt}}{Q} \tag{3.3}$$

where V_{rt} is the volume of the hot zone in CVD, and Q is the flow rate of gases. Whereas a fast gas flow (or large Q) enhances the mass transport of carbon precursors, it also sweeps gas molecules in a short time from the hot zone where GPRs occur. Thus, the effect of the mass transfer enhancement by a faster gas flow, and the effect of the GPRs, need to be decoupled.

Excessive pyrolysis can also generate non-graphitic carbon contaminants such as aromatic soot precursors and amorphous carbon. These by-products can be deleterious to the nanotube quality and yield.[56] This effect coupled with the positive effect of GPRs leads to the existence of an optimum residence time (Figure 3.7).[57] In the case of a cold-wall reactor with a separated gas pretreatment cell,[17] adjusting the pretreatment temperature can be an effective way to optimize GPRs due to their high activation energy (~3.36 eV for ethylene).[54]

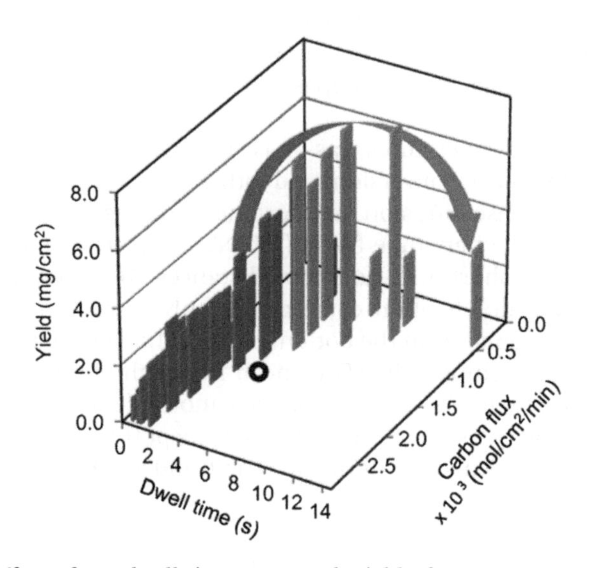

Figure 3.7 Effect of gas dwell time on growth yield of SWNT.
Reprinted with permission from ref. 57. Copyright 2011 American Chemical Society.

3.2.2.3 Mass Transport of Gas Precursors

In the CVD process, the main routes for supplying the carbon precursors to the growth sites on the catalyst are convective and diffusive transport. The convective flow of gases inside a long flow channel is a well-established classical problem of a fully developed internal flow. As the molecules approach the substrate, however, the convective transport decelerates and a rather complex flow pattern starts to develop. The detailed description of a flow pattern near the substrate is not simple since it depends on the geometric shapes of the reactor, substrate, and other objects inside the reactor. Nevertheless, a widely accepted notion has a thin boundary layer on the substrate surface, across which temperature, flow velocity, and gas composition change with an abrupt gradient.

Diffusive mass transport through this boundary layer is an important gas-phase mechanism of carbon delivery. The deposition surface can be modeled as a mass sink, while the free stream acts as a continuous supplier of carbon precursor molecules. The concentration gradient acts as a driving force of this transport process, and the transport rate can be expressed using Fick's law:

$$J_c = D_g \frac{(N_g - N_c)}{\delta_c} \tag{3.4}$$

where J_c is the molecular flux of the carbon precursors, D_g is the mass diffusivity of the precursor gas, δ_c is the thickness of the (concentration) boundary layer, and N_g and N_c are molecular concentrations of the precursors in the free stream and very near the catalysts, respectively. Here, we note that in this lumped relation, the boundary layer thickness (δ_c) is an implicit parameter that can encompass all aspects of mass transport near the surface.

At common CVD temperatures, gas-phase diffusion is considered very fast compared to chemical conversion and other mass transport processes. However, the gas-phase diffusion may become slower as the nanotubes grow longer; especially, the nanotube forest (VACNTs) can act as a diffusion barrier due to the very short inter-nanotube distances (10–100 nm).[58] We note that this diffusion deceleration corresponds only to root growth of CNTs where the catalyst remains on the bottom of VACNTs. The length scale of the interstitial space is comparable to the mean free path of the gas molecules, resulting in more frequent collisions of a gas molecule with CNT walls than with other gas molecules. Then the diffusive transport undergoes a different transfer mechanism: Knudsen diffusion. The Knudsen diffusivity (D_K) takes the following form:

$$D_K = 97 r_e \frac{\epsilon_v}{\tau} \left(\frac{T}{M}\right)^{1/2} \tag{3.5}$$

where r_e: pore radius, ϵ_v: porosity, τ: tortuosity, T: temperature, and M: molecular mass. The above two diffusion mechanisms can be combined into an effective diffusion coefficient (D_e):

$$\frac{1}{D_e} = \frac{1}{D_g} + \frac{1}{D_K} \tag{3.6}$$

Xiang *et al.* evaluated possible growth deceleration by a VACNT diffusion barrier.[58] Based on a quantitative analysis with the reported diffusion and surface reaction parameters, they suggested that the growth deceleration could be caused by the aforementioned diffusional barrier. However, the estimated surface reaction parameters can be significantly modified depending on possible prior gas-phase reactions and the identity of the corresponding active carbon precursors. Therefore, a more thorough analysis seems to be necessary for the rate-limiting effect of gas-phase diffusion to be assured.

3.2.2.4 Adsorption and Surface Reactions

After undergoing a series of conversions in the gas phase, an active carbon precursor species impinges on the surface and forms a transient adsorbate. As typical physisorption energy is small compared to the thermal energy at a typical growth temperature, most of the species that reside on the surface must be chemisorbed. In the case of CNT growth, chemisorption involves dehydrogenation of the hydrocarbon adsorbate and scission of an intercarbon bond to produce an atomic carbon. Carbon-to-catalyst bond formation then compensates for the high cost of the reconfiguration of the molecular structure of the adsorbate.

This adsorption-related argument can explain qualitatively the differential reactivity of various carbon precursor species. For instance, methane (CH_4) is a saturated hydrocarbon that has a very stable molecular structure (4 σ-bonds), thus the only way it can interact with the surface is through dissociative adsorption. In contrast, the π-bonds of unsaturated hydrocarbons (ethylene and acetylene) enable adsorption with no need for dissociation. Indeed, it has been consistently reported that methane needs a higher temperature than ethylene or acetylene to grow CNTs by thermal CVD.

The surface coverage of adsorbates is an important parameter that can account for the reaction rate and the adsorption rate occurring on the catalyst surface. The Langmuir isotherm provides a fundamental model to describe equilibrium surface coverage on a catalyst.[59] According to gas kinetics, the molecular impingement is expressed by:

$$F = \frac{N_A P}{\sqrt{2\pi MRT}} \tag{3.7}$$

where N_A: Avogadro's number, P: precursor partial pressure, M: molecular mass, R: ideal gas constant, T: temperature. Then the surface coverage (θ) by

Langmuir isotherm describes the surface coverage in terms of pressure of the gaseous precursor.

$$\theta = \frac{\sigma}{\sigma_0} = \frac{bP}{1 + bP} \tag{3.8}$$

where σ is the number of occupied sites and σ_0 is the maximum number of available sites. b is a constant that is related to residence time of the adsorbate on the surface.

According to Langmuir–Hinshelwood kinetics, the surface reaction rate is proportional to the surface coverage (θ). Then, the reaction order comes between 0 and 1 with respect to the precursor pressure. Apparently, the surface reaction becomes linear with respect to P when the pressure is low ($P \ll 1$) or the residence time of the adsorbed molecule is very short ($b \ll 1$). In contrast, when the pressure is very high, the molecules cover a large area, eventually saturating the catalyst surface with the adsorbates. If this is the case, the reaction order approaches zero, thereby the surface reaction becomes constant regardless of a further increase in the precursor pressure.

3.2.2.5 *Diffusion through Catalyst and Carbon Incorporation*

Carbon adatoms can find their way to the growing carbon nanotube end *via* two pathways: bulk diffusion and surface diffusion. Generally, surface diffusion has a lower activation barrier than bulk diffusion, so surface diffusion should be prevalent.[60] However, bulk diffusion is not negligible at high temperatures and especially for the growth of multi-walled carbon nanotubes (MWCNTs) since the multiple inner nanotube cylinders are not easily accessible to surface carbon adatoms.[61]

The diffusivity and solubility of carbon in a catalyst greatly depends on the physical phase of the catalyst during growth, which has been a controversial topic in CNT growth research. The vapor–liquid–solid (VLS) mechanism[62,63] postulates that precursor gas molecules adsorb and form a solid solution at the eutectic point. As carbon accumulates in the catalyst, the carbon content reaches the point of supersaturation. Subsequently, an island of carbon cap nucleates and grows into the tip of a nanotube. Then, continuous feeding of carbon lifts the tip cage and drives elongation of the nanotube.

While it has been widely accepted that the other types of catalyst-assisted growth of nanowires such as silicon nanowires follows the VLS mechanism, the state of the nanotube catalysts during growth is not obvious. The melting temperatures of general nanotube catalysts (Fe: 1538 °C, Co: 1495 °C, Ni: 1455 °C) are much higher than growth temperatures. However, the high surface-to-volume ratio of the catalyst nanoparticles can modify the thermodynamics of the melting process, reducing the melting temperatures.[64] Researchers also argued that eutectic mixtures formed by carbon addition can also contribute to an early change in the phase of the catalyst in several simulation studies.[65,66] Harutyunyan *et al.*[67] claimed that CNT growth is highly preferred above the eutectic point with an iron catalyst, and solidification of the catalyst can potentially terminate growth. Recent development of *in situ* TEM techniques has enabled the direct observation of the catalyst phase during growth, revealing that CNTs grow from catalyst particles remaining in a solid phase at relatively low temperatures (600–650 °C).[22,68,69] Although these studies unambiguously demonstrated that the VLS mechanism might not hold for CNT growth, there remains the possibility that CNTs can grow by the VLS mechanism at higher growth temperatures. Up to this point, VLS growth of CNTs has not been directly observed yet, so the debate about this growth mechanism continues.

3.2.3 Growth Kinetics

Despite the large amount of empirical observations describing different aspects of the growth process, the understanding of the growth mechanism of CNTs is still incomplete. However, studies on the growth kinetics can tell a lot about the underlying mechanism since each growth mechanism produces its own kinetic trend. Researchers typically assume that the mechanism contains a rate-limiting step.[70] Fortunately, the longitudinal length of CNT arrays has some correlation with growth kinetics (at least for the longest CNTs in the array) and it is relatively easy to observe. Moreover, the growth rate observed in the CVD growth of CNTs is generally within the measureable range (up to a few micrometres per second) for *in situ* growth monitoring instruments. The following section introduces kinetic models for the CVD growth of CNTs and discusses their validity based on the experimental evidence.

3.2.3.1 Kinetic Model of CNT Growth

The driving force for the conversion of the carbon from the gas phase into the carbon nanotube is equal to the chemical potential difference ($\Delta\mu = \mu_g - \mu_{NT}$) between the gaseous carbon (μ_g) and the carbon in a nanotube crystal (μ_{NT}). To build a generalized model of the CNT growth we can use an equivalent electric circuit (Figure 3.8). Here the electric current corresponds to the mass flux of carbon. This circuit includes contributions from three different resistances: the gas-phase diffusion resistance of the CNT forest to the catalyst site (R_1); resistance associated with the carbon adsorption to the catalyst surface and any further reactions happening there (R_2); finally, a diffusion resistance of carbon adatoms through the catalyst particle to the CNT growth site (R_3). In practice, carbon can be consumed into non-graphitic carbon such as amorphous carbon deposits on the catalyst. We can incorporate this process into the equivalent circuit as a current leak. However, for a well-tuned CNT growth, the amount of such a leak is negligible compared to the main current (*i.e.* the production of amorphous carbon is much slower than the amount that gets incorporated into the growing nanotube array), thus the following analysis will not consider it.

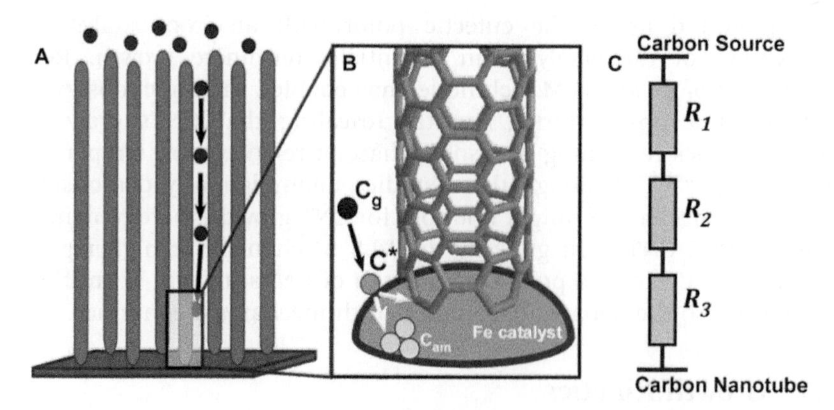

Figure 3.8 A simplified model of carbon nanotube growth. (A) Diffusion through the CNT array followed by (B) reaction at the catalyst and carbon incorporation into a growing nanotube. (C) Equivalent electrical circuit representing the growth model.
Reprinted with permission from ref. 46. Copyright 2011 American Chemical Society.

When the growth reaches a steady state, the growth rate (V_g) can be expressed as:

$$v_g = \frac{\dfrac{\Delta\mu}{k_B T} \cdot n_0 \cdot \omega}{R_1 + R_2 + R_3} \tag{3.9}$$

where $\Delta\mu/k_BT$ is the driving force for the CNT growth, n_0 is the initial concentration of active sites on a catalyst, and ω presents a kinetic coefficient. If the diffusion terms are expressed explicitly,

$$R_1 \sim \frac{h(t)}{D_e} = \frac{\int_0^t v_g(t)dt}{D_e} \tag{3.10}$$

$$R_3 \sim \frac{d_c}{D_c} \tag{3.11}$$

where $h(t)$ is the height of the CNT array, d_c is the effective size of the catalyst particle, and D_e is the effective carbon diffusion coefficient in the gas phase, and D_c is another effective diffusivity that accounts for mass transport both by diffusion in bulk and by surface diffusion. Finally, the general growth kinetics correspond to the following relation:

$$v_g = \frac{\dfrac{\Delta\mu}{k_B T} \cdot n_0 \cdot \omega}{\dfrac{\int_0^t v_g(t)dt}{D_e} + \pounds + \dfrac{d_c}{D_c}} \tag{3.12}$$

where £ is a function that describes the adsorption resistance.

Eqn (3.12) presents several interesting features. First, not surprisingly, it predicts that the growth speed is linearly proportional to the driving force of the CNT growth. Second, it can predict the maximum growth rate that is theoretically achievable. For instance, when the adsorption resistance is much smaller than the diffusion resistance by precursor activation of plasma, and gas-phase diffusion is facilitated with a very short nanotube forest, the maximum growth rate can be obtained as:

$$v_{\max} = \frac{\frac{\Delta\mu}{kT} \cdot n_0 \omega D}{\frac{d_c}{D_c}} \tag{3.13}$$

Third, the general model predicts several kinetic regimes for the growth process depending on the relative dominance of the denominator terms of eqn (3.9). For instance, when $R_1 \gg R_2$, R_3, the equation reduces to the Deal–Grove equation that is often used to describe diffusion-limited CNT array growth.[71] When R_1, $R_3 \ll R_2$, then the growth is primarily limited by the adsorption process.

Note that the general model above assumes in its derivation that the carbon coverage (θ) on the catalyst is negligibly small. If this assumption is not valid, an alternative model should be developed. Tibbetts *et al.*[72] proposed the following relation:

$$v_g = \frac{\Omega}{1 - \left(\frac{r_i}{r_0}\right)^2} \cdot \frac{N_s^0 - N_0}{N_s^0} \cdot \frac{J_c}{1 + \frac{J_c d_c}{D_c N_s^0}} \tag{3.14}$$

where the growth rate increases nonlinearly with respect to carbon flux, resembling the Langmuir adsorption model.

Unfortunately, the models above do not explicitly account for GPRs in their expression despite the critical kinetic effects on CNT growth. Recently, In *et al.* proposed a simplified model to explain GPR kinetics based on their *in situ* growth data.[46] Figure 3.9 shows the calculated high activation energy that is indicative of GPRs. This high activation suggests that ethylene pyrolysis consumes most of the driving force in their growth condition, thereby the GPRs of ethylene are limiting CNT growth. They also showed a straightforward relation between growth rates and partial pressure of ethylene and hydrogen. The following model was suggested to effectively describe the kinetics:

$$\frac{d[C_a]}{dt} = k_1[C_2H_4] - k_2[H_2][C_a] \tag{3.15}$$

$$v_g^0 \sim [C_a] = K \cdot \frac{[C_2H_4]}{[H_2]}\left(1 - e^{-k_2\tau_r[H_2]}\right) \tag{3.16}$$

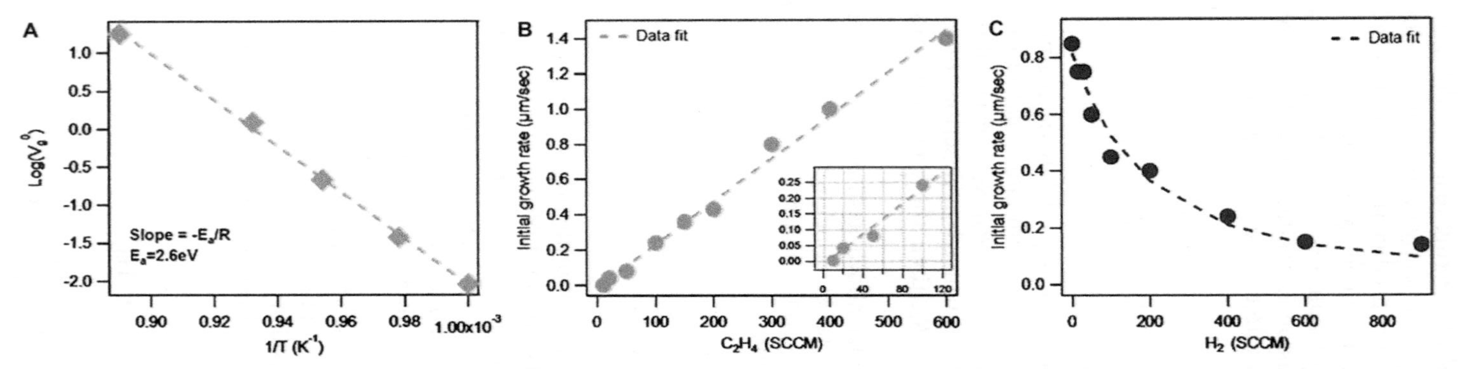

Figure 3.9 Kinetics of CNT growth with purified feed gases. (A) Arrhenius plot of initial growth rates. The activation energy was calculated from the slope of the line. (B) Plot of initial growth rates (40% constant hydrogen) with respect to carbon precursor (ethylene) variation. The inset shows the enlarged plot near the origin. (C) Plot of initial growth rates (10% constant ethylene) with respect to hydrogen variation. In these experiments, the total flow rate was maintained at 1000 SCCM under 1 atm. Reprinted with permission from ref. 46. Copyright 2011 American Chemical Society.

where $K = k_1/k_2$ and τ_r is the residence time of the ethylene gas in the GPR reaction zone, determined mostly by the dynamics of gas flow through the CVD reactor.

3.2.4 Termination of CNT Growth

Experimental observations have consistently shown that CNT growth slows down and eventually terminates. The termination mechanism is ambiguous, and remains yet another active topic in the CNT growth field. Understanding this phenomenon should be highly beneficial in improving growth yields and controllability of the CVD process for CNT growth. We discuss several potential growth termination mechanisms in the next section.

3.2.4.1 Termination Mechanisms

It is clear that the catalyst deactivation mechanism plays a critical role in CNT growth termination. Particularly, this mechanism should be able to explain the abrupt growth cessation that is a very intriguing feature of the growth termination of VACNTs.[18] One of the biggest innovations was the development of growth enhancers such as water, alcohol, and other oxygen-containing molecules. Indeed, numerous experimental reports demonstrated that these enhancers remarkably increase the CVD growth time and thereby enable millimetre-scale (or even cm-scale) growth of carbon nanotubes. Futaba *et al.*[43] examined the effects of various enhancers on the growth kinetics and concluded that their growth-promoting properties are indeed related to the presence of the oxygen-containing molecules in the gas mixture. Therefore, understanding the effect of enhancers will help to elucidate growth termination mechanisms.

The prevalent opinion in the literature is that these mild oxidants remove amorphous carbon from the catalyst to sustain the catalytic activity. Modern catalysis studies reveal that hydrocarbon decomposition leaves carbonaceous by-products that have longer residence time on the metal catalyst and easily get stabilized at high temperatures.[59] Strongly adsorbed carbon residues could poison catalysts by blocking their active sites. A similar approach was accepted by Hata *et al.*;[73] they proposed that catalyst nanoparticles can be deactivated by accumulation of amorphous carbon on the catalyst surface during the CVD growth of CNTs. Therefore, they suggested that the role of water vapor is mainly restricted to removal of such poisonous carbon.

Yet, a similar effect with water could result from a different mechanism. Based on an iron–alumina catalyst system, Amama *et al.*[49] suggested that Ostwald ripening of small iron catalyst particles could be the main cause of termination and the role of water is related to prevention of ripening by surface hydroxylation of the alumina support. This hypothesis can also explain different efficiencies of various catalyst support layers that have different catalyst–support interactions.[74–77] Other studies underscore the importance of the support layer in preventing growth termination. Kim *et al.*[78] demonstrated the migration of catalyst nanoparticles into the underlying alumina, which would explain an irreversible loss of catalytic activity.

If a catalyst maintains its activity during growth, different deactivation mechanisms can become significant. For instance, diffusion-limited growth kinetics can manifest itself in growth deactivation by starving the active catalyst of carbon. Zhong *et al.*[71] proposed that nanoporous nanotube arrays can form the main diffusion barrier (R1 in eqn (3.9)). Alternatively, Patole *et al.*[79] suggested that the catalyst itself can retard carbon diffusion by being gradually oxidized or carbonized. This case corresponds to the increase in R_3 in eqn (3.9). Lastly, it was proposed that steric hindrance can contribute to growth termination. CNTs are mechanically coupled to the neighboring CNTs,[80–82] so a nanotube forest represents an interconnected structure. Due to an imbalance in the growth kinetics of each nanotube, strain energy builds up with an increase of growth length and eventually the growth stops. The often-reported observations of the "curly" nanotube geometry near the bottom of the terminated CNT arrays seem to support this hypothesis.

Despite the abundance of data on CNT growth kinetics,[18,50,83,84] the definitive mechanism of growth termination remains elusive. Possibly, this ambiguity originates from the different growth conditions in each study. Therefore, a more thorough study that encompasses a wide parametric window could reveal this comprehensive termination mechanism.

3.2.4.2 Catalyst Deactivation Kinetics

Before discussing termination kinetics, we can start by defining *catalytic activity* α at time t:

$$\alpha(p_{\rm i}, T, t) = \frac{R(p_{\rm i}, T, t)}{R_0(p_{\rm i}, T)} \tag{3.17}$$

where R_0: an *initial* (or non-deactivating) reaction rate, R: a reaction rate at time t, $p_{\rm i}$: partial pressures of reactants.[85] R_0 can be obtained from the initial growth rate of kinetic data. Then we can calculate the catalytic activity by normalizing the observed growth rate data by the initial growth rate.

An influential paper by Iijima *et al.*[50] proposed the exponential decay model based on growth deactivation by amorphous carbon encapsulation. This model fitted their *ex situ* kinetic curves successfully. However, this model relied on heuristic derivation rather than on analysis of the underlying deactivation mechanism. The mathematical expression takes the following form:

$$v_{\rm g} = v_0 e^{-t/\tau_0} \tag{3.18}$$

$$h(t) = v_0 \tau_0 \left(1 - e^{-t/\tau_0}\right) \tag{3.19}$$

where v_0 is the initial growth rate and τ_0 is a time constant related to growth termination. Then applying the definition of catalytic activity of eqn (3.17):

$$\alpha = \exp\left(-\frac{t}{\tau_0}\right) \tag{3.20}$$

In contrast, Puretzky *et al.*[86] explored a deactivation model based on the growth kinetics of a poisonous carbon platelet. Their model involves several differential equations with a large number of parameters. In brief, the poisonous carbon layers are assumed to be nucleated and grow by carbon adatom addition. The essential governing equation accounting for the spread of the poisonous layer states that:

$$\frac{dN_L(t)}{dt} = k_C N_C(t) \tag{3.21}$$

where N_L is the number of carbon atoms that make up the poisoning carbon platelet, whereas N_C is the number of carbon atoms produced by adsorption of hydrocarbon feedstock gas. Recently, Lee *et al.*[84] suggested a simplified model based on the same governing equations of Puretzky *et al.* Interestingly, the final form turned out to be the exponential decay that is actually consistent with Iijima's model (eqn (3.18), (3.19)).

While the exponential decay model has been widely adopted to describe CNT growth kinetics, researchers should be aware that this model does not consider the nucleation kinetics of the poisonous carbon platelet. Rather, it describes immediate surface poisoning, for instance, by strongly adsorbing gas molecules. Moreover, one of the characteristics of this model is relatively gradual termination, therefore it fails to predict the abrupt termination behavior that was consistently observed in a number of studies.[18,82,83,87–89]

An alternative treatment based on the Kolmogorov–Mehl–Avrami model to describe the growth kinetics of the poisoning carbon platelets can reproduce the abrupt termination behavior.[83,89] This model accounts for sporadic events of carbon platelet nucleation, and the resultant expression gives the following deactivation kinetics:

$$\alpha = \exp\left[-\left(\frac{t}{\tau}\right)^n\right] \tag{3.22}$$

In the case of 2D islands of surface carbon, n become 3 with $\tau^{-3} = \pi J_a v_a^2/3$ where J_a indicates nucleation rate and v_a indicates linear expansion speed of the platelet. Applying this model requires that the critical size (r_c) of a platelet nucleus is sufficiently smaller than the catalyst size. Thus SWNT growth from very small catalysts ($\sim$1 nm) may not follow this model.

Ostwald ripening, which is another reason for growth termination, can be described by the following empirical relation:[80]

$$\frac{d\alpha}{dt} = -k_{rp}(\alpha - \alpha_\infty)^n \tag{3.23}$$

where α_∞ is catalyst activity at a steady state, and k_{rp} is a kinetic constant. This model accounts for sintering kinetics of metal particles supported on an oxide surface. Assuming the ripening is the main growth termination mechanism, Latorre *et al.*[80] applied this equation to the description of the deactivation kinetics. Nonetheless, it is still not clear how the collective growth behavior of VACNTs can be related to dynamic changes in the size of catalyst nanoparticles and their density.

3.2.5 Lessons to the Experimenters from the Growth Models

The previous sections discussed various CNT growth models. While no single model can explain the totality of experimental observations, we can still draw several conclusions. First, it is clear that CNT growth and termination is a complicated process that involves a number of different pathways, and thus a simple universal explanation or mechanism may not exist. Second, the complicated nature of the process places a high requirement on experiments. For example, the kinetic data can mislead the analyst by failing to capture a sufficient number of data points (Figure 3.10). The low-resolution data give convincing evidence for the exponential decay model, yet

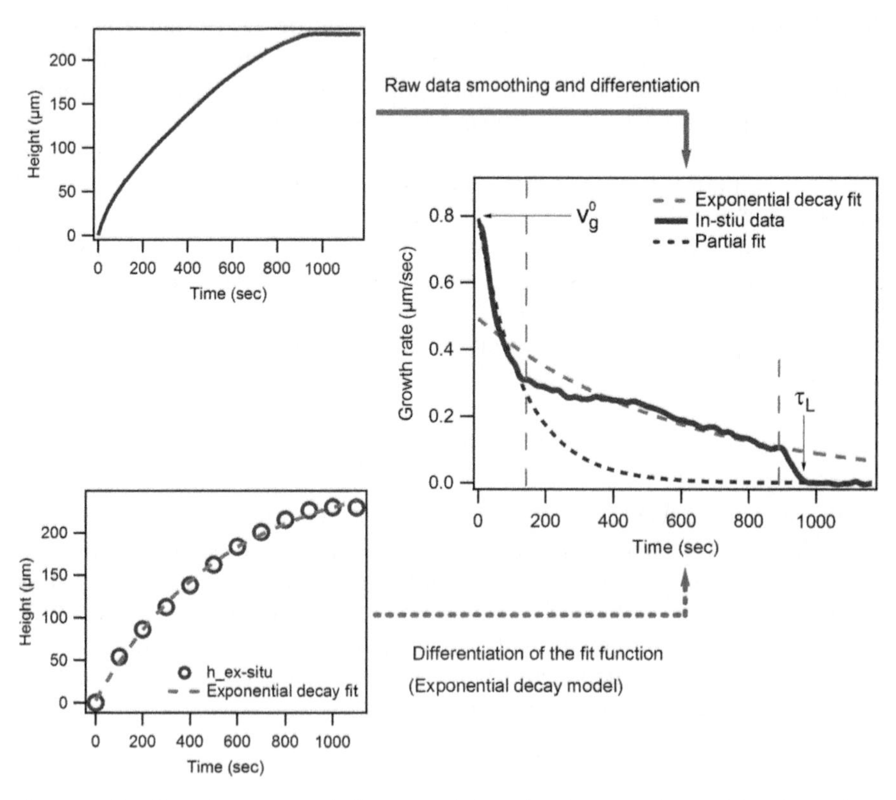

Figure 3.10 Sampling rate and CNT growth kinetics. A comparison of the data taken at high and low sampling rates: circles indicate data points chosen every 100 sec from *in situ* data to simulate *ex situ* data collection. Dashed lines indicate the fit to the exponential decay model and its first derivative for growth rate calculation. The fit to the simulated *ex situ* data is missing several key features of the growth kinetics.

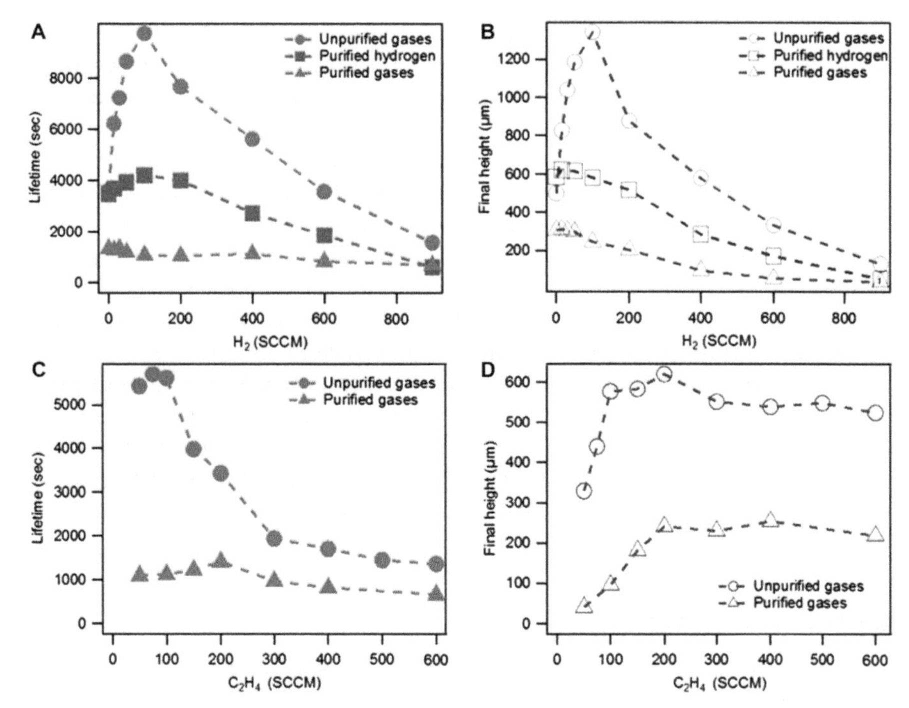

Figure 3.11 Growth termination data obtained with different compositions of unpurified and purified ethylene and hydrogen feedstock. The gas purification results in striking differences in the catalyst lifetime and growth yield.
Reprinted with permission from ref. 46. Copyright 2011 American Chemical Society.

high-resolution data clearly show that this model does not provide an adequate description.

Third, the complicated nature of the process places very stringent requirements on the experimental setup. For instance, purification to remove gas impurities is necessary for reproducible CNT growth. This is because the oxygen-containing species can make the growth chemistry very complicated by promoting or demoting catalysts depending on their molecular forms and concentrations.[43,52] Indeed, different gas compositions produce strikingly different trends of growth deactivation (Figure 3.11), which could in turn produce misleading analysis. Finding the most efficient gas precursor could be a valuable complement to these efforts as it would simplify the GPR kinetics.[6,42,44,45]

3.3 Conclusion

Carbon nanotube synthesis represents a striking example of a simple system that can produce rather complicated behavior. Given the amount of effort that has gone into understanding CNT growth to date, it is remarkable how complicated and intractable this problem remains. It is rather fitting that one of the leading commercial systems for CNT growth produced by Aixtron, AG, is called Black Magic, reflecting on the near-alchemical status of the process. Nevertheless, progress in understanding and controlling CNT growth has been remarkable. Researchers now rely on growth enhancers and in-line gas purifiers to achieve high yields and a high reproducibility of CNT growth. Another remarkable achievement has been to understand and control the evolution of CNT growth catalysts. Today, the CNT growth catalysts are much more reliable and last a lot longer than the catalysts used a decade ago. Another achievement from the past five years of CNT research was the realization that efforts to improve growth yield and quality should be squarely focused on catalyst evolution and treatment. Still, the goal of producing an "endless" carbon nanotube remains elusive, and does not seem to be anywhere near being attainable. But today's CNT growers, armed with sophisticated growth reactors and catalyst treatment techniques, are getting close to the goal of producing CNT arrays required for the commercial applications of CNT materials.

Acknowledgements

Parts of this work was supported by the National Science Foundation NIRT CBET-0709090 (JBI) and the U. S. Department of Energy, Office of Basic Energy Sciences, Division of Materials Sciences and Engineering (AN). Work at the Molecular Foundry was supported by the Office of Science, Office of Basic Energy Sciences, of the U.S. Department of Energy under Contract No. DE-AC02-05CH11231. Part of the work was performed under the auspices of the U. S. Department of Energy by the Lawrence Livermore National Laboratory under Contract DE-AC52-07NA27344.

References

1. M. Terrones, *Annu. Rev. Mater. Res.*, 2003, **33**, 419–501.
2. M. S. Dresselhaus and R. Saito, *Physical Properties of Carbon Nanotubes*, Imperial College Press, London, UK, 1998.
3. J. K. Holt, H. G. Park, Y. M. Wang, M. Stadermann, A. B. Artyukhin, C. P. Grigoropoulos, A. Noy and O. Bakajin, *Science*, 2006, **312**, 1034–1037.
4. M. F. L. De Volder, S. H. Tawfick, R. H. Baughman and A. J. Hart, *Science*, 2013, **339**, 535–539.
5. S. Iijima, *Nature*, 1991, **354**, 56–58.
6. G. Eres, A. A. Kinkhabwala, H. Cui, D. B. Geohegan, A. A. Puretzky and D. H. Lowndes, *J. Phys. Chem. B*, 2005, **109**, 16684–16694.
7. S. Fan, M. G. Chapline, N. R. Franklin, T. W. Tombler, A. M. Cassell and H. Dai, *Science*, 1999, **283**, 512–514.
8. M. Picher, E. Anglaret, R. Arenal and V. Jourdain, *Nano Lett.*, 2009, **9**, 542–547.

9. E. Einarsson, Y. Murakami, M. Kadowaki and S. Maruyama, *Carbon*, 2008, **46**, 923–930.
10. H. Sugime, S. Noda, S. Maruyama and Y. Yamaguchi, *Carbon*, 2009, **47**, 234–241.
11. M. Kusunoki, T. Suzuki, T. Hirayama, N. Shibata and K. Kaneko, *Appl. Phys. Lett.*, 2000, **77**, 531–533.
12. S. Iijima and T. Ichihashi, *Nature*, 1993, **363**, 603–605.
13. T. W. Ebbesen and P. M. Ajayan, *Nature*, 1992, **358**, 220–222.
14. T. Guo, P. Nikolaev, A. Thess, D. T. Colbert and R. E. Smalley, *Chem. Phys. Lett.*, 1995, **243**, 49–54.
15. J. H. Hafner, M. J. Bronikowski, B. R. Azamian, P. Nikolaev, A. G. Rinzler, D. T. Colbert, K. A. Smith and R. E. Smalley, *Chem. Phys. Lett.*, 1998, **296**, 195–202.
16. M. Cantoro, S. Hofmann, S. Pisana, V. Scardaci, A. Parvez, C. Ducati, A. C. Ferrari, A. M. Blackburn, K. Y. Wang and J. Robertson, *Nano Lett.*, 2006, **6**, 1107–1112.
17. E. R. Meshot, D. L. Plata, S. Tawfick, Y. Zhang, E. A. Verploegen and A. J. Hart, *ACS Nano*, 2009, **3**, 2477–2486.
18. E. R. Meshot and A. J. Hart, *Appl. Phys. Lett.*, 2008, **92**, 113107.
19. C. Zhuo, W. Yang, L. Chunxiang, J. Kaili, Z. Lina, L. Qunqing, S. Fan and G. Jiancun, *Appl. Phys. Lett.*, 2007, **90**, 133108.
20. D. Bäuerle, *Laser Processing and Chemistry*, Springer, Berlin–Heidelberg, Germany, 4th edn, 2011.
21. L. Delzeit, I. McAninch, B. A. Cruden, D. Hash, B. Chen, J. Han and M. Meyyappan, *J. Appl. Phys.*, 2002, **91**, 6027–6033.
22. S. Hofmann, C. Ducati, J. Robertson and B. Kleinsorge, *Appl. Phys. Lett.*, 2003, **83**, 135–137.
23. Y. Li, D. Mann, M. Rolandi, W. Kim, A. Ural, S. Hung, A. Javey, J. Cao, D. Wang, E. Yenilmez, Q. Wang, J. F. Gibbons, Y. Nishi and H. Dai, *Nano Lett.*, 2004, **4**, 317–321.
24. G. Zhang, D. Mann, L. Zhang, A. Javey, Y. Li, E. Yenilmez, Q. Wang, J. P. McVittie, Y. Nishi, J. Gibbons and H. Dai, *Proc. Natl. Acad. Sci. U. S. A.*, 2005, **102**, 16141–16145.
25. D. L. Plata, A. J. Hart, C. M. Reddy and P. M. Gschwend, *Environ. Sci. Technol.*, 2009, **43**, 8367–8373.
26. X. Li, W. Cai, L. Colombo and R. S. Ruoff, *Nano Lett.*, 2009, **9**, 4268–4272.
27. S. Bhaviripudi, X. T. Jia, M. S. Dresselhaus and J. Kong, *Nano Lett.*, 2010, **10**, 4128–4133.
28. T. Yamada, T. Namai, K. Hata, D. N. Futaba, K. Mizuno, J. Fan, M. Yudasaka, M. Yumura and S. Iijima, *Nat. Nanotechnol.*, 2006, **1**, 131–136.
29. M. Lin, J. P. Y. Tan, C. Boothroyd, K. P. Loh, E. S. Tok and Y.-L. Foo, *Nano Lett.*, 2006, **6**, 449–452.
30. A. M. Cassell, J. A. Raymakers, J. Kong and H. Dai, *J. Phys. Chem. B*, 1999, **103**, 6484–6492.

31. H. Ago, N. Uehara, N. Yoshihara, M. Tsuji, M. Yumura, N. Tomonaga and T. Setoguchi, *Carbon*, 2006, **44**, 2912–2918.
32. H. Yoshida, T. Shimizu, T. Uchiyama, H. Kohno, Y. Homma and S. Takeda, *Nano Lett.*, 2009, **9**, 3810–3815.
33. M. Hu, Y. Murakami, M. Ogura, S. Maruyama and T. Okubo, *J. Catal.*, 2004, **225**, 230–239.
34. D. Takagi, H. Hibino, S. Suzuki, Y. Kobayashi and Y. Homma, *Nano Lett.*, 2007, 7, 2272–2275.
35. S. Huang, Q. Cai, J. Chen, Y. Qian and L. Zhang, *J. Am. Chem. Soc.*, 2009, **131**, 2094–2095.
36. B. Liu, W. Ren, L. Gao, S. Li, S. Pei, C. Liu, C. Jiang and H. M. Cheng, *J. Am. Chem. Soc.*, 2009, **131**, 2082–2083.
37. D. Takagi, Y. Homma, H. Hibino, S. Suzuki and Y. Kobayashi, *Nano Lett.*, 2006, **6**, 2642–2645.
38. F. Ding, P. Larsson, J. A. Larsson, R. Ahuja, H. M. Duan, A. Rosen and K. Bolton, *Nano Lett.*, 2008, **8**, 463–468.
39. G. D. Nessim, A. J. Hart, J. S. Kim, D. Acquaviva, J. Oh, C. D. Morgan, M. Seita, J. S. Leib and C. V. Thompson, *Nano Lett.*, 2008, **8**, 3587–3593.
40. A. R. Harutyunyan, G. Chen, T. M. Paronyan, E. M. Pigos, O. A. Kuznetsov, K. Hewaparakrama, S. M. Kim, D. Zakharov, E. A. Stach and G. U. Sumanasekera, *Science*, 2009, **326**, 116–120.
41. N. R. Franklin and H. Dai, *Adv. Mater.*, 2000, **12**, 890–894.
42. M. S. Bell, R. G. Lacerda, K. B. K. Teo, N. L. Rupesinghe, G. A. J. Amaratunga, W. I. Milne and M. Chhowalla, *Appl. Phys. Lett.*, 2004, **85**, 1137–1139.
43. D. N. Futaba, J. Goto, S. Yasuda, T. Yamada, M. Yumura and K. Hata, *Adv. Mater.*, 2009, **21**, 4811–4815.
44. G. Zhong, S. Hofmann, F. Yan, H. Telg, J. H. Warner, D. Eder, C. Thomsen, W. I. Milne and J. Robertson, *J. Phys. Chem. C*, 2009, **113**, 17321–17325.
45. D. L. Plata, E. R. Meshot, C. M. Reddy, A. J. Hart and P. M. Gschwend, *ACS Nano*, 2010, **4**, 7185–7192.
46. J. B. In, C. P. Grigoropoulos, A. A. Chernov and A. Noy, *ACS Nano*, 2011, **5**, 9602–9610.
47. J. B. In, C. P. Grigoropoulos, A. A. Chernov and A. Noy, *Appl. Phys. Lett.*, 2011, **98**, 153102.
48. K. Hata, D. N. Futaba, K. Mizuno, T. Namai, M. Yumura and S. Iijima, *Science*, 2004, **306**, 1362–1364.
49. P. B. Amama, C. L. Pint, L. McJilton, S. M. Kim, E. A. Stach, P. T. Murray, R. H. Hauge and B. Maruyama, *Nano Lett.*, 2009, **9**, 44–49.
50. D. N. Futaba, K. Hata, T. Yamada, K. Mizuno, M. Yumura and S. Iijima, *Phys. Rev. Lett.*, 2005, **95**, 056104.
51. S. Yasuda, D. N. Futaba, T. Yamada, J. Satou, A. Shibuya, H. Takai, K. Arakawa, M. Yumura and K. Hata, *ACS Nano*, 2009, **3**, 4164–4170.
52. C. L. Pint, S. T. Pheasant, A. N. G. Parra-Vasquez, C. Horton, Y. Xu and R. H. Hauge, *J. Phys. Chem. C*, 2009, **113**, 4125–4133.

53. *Chemical Vapour Deposition: Precursors, Processes and Applications*, ed. A. C. Jones and M. L. Hitchman, Royal Society of Chemistry, Cambridge, UK, 2008.
54. J. M. Roscoe, A. R. Bossard and M. H. Back, *Can. J. Chem.*, 2000, **78**, 16–25.
55. K. Norinaga and O. Deutschmann, *Ind. Eng. Chem. Res.*, 2007, **46**, 3547–3557.
56. S. Yasuda, T. Hiraoka, D. N. Futaba, T. Yamada, M. Yumura and K. Hata, *Nano Lett.*, 2009, **9**, 769–773.
57. S. Yasuda, D. N. Futaba, T. Yamada, M. Yumura and K. Hata, *Nano Lett.*, 2011, **11**, 3617–3623.
58. R. Xiang, Z. Yang, Q. Zhang, G. Luo, W. Qian, F. Wei, M. Kadowaki, E. Einarsson and S. Maruyama, *J. Phys. Chem. C*, 2008, **112**, 4892–4896.
59. G. A. Somorjai, *Introduction to surface chemistry and catalysis*, Wiley, New York, US, 1994.
60. S. Hofmann, G. Csanyi, A. C. Ferrari, M. C. Payne and J. Robertson, *Phys. Rev. Lett.*, 2005, **95**, 036101.
61. A. Gamalski, E. S. Moore, M. M. J. Treacy, R. Sharma and P. Rez, *Appl. Phys. Lett.*, 2009, **95**, 233109.
62. R. S. Wagner and W. C. Ellis, *Appl. Phys. Lett.*, 1964, **4**, 89–90.
63. R. T. K. Baker, M. A. Barber, R. J. Waite, P. S. Harris and F. S. Feates, *J. Catal.*, 1972, **26**, 51–62.
64. P. Buffat and J. P. Borel, *Phys. Rev. A*, 1976, **13**, 2287–2298.
65. F. Ding, K. Bolton and A. Rosen, *J. Vac. Sci. Technol., A*, 2004, **22**, 1471–1476.
66. Y. Shibuta and T. Suzuki, *Chem. Phys. Lett.*, 2007, **445**, 265–270.
67. A. R. Harutyunyan, T. Tokune and E. Mora, *Appl. Phys. Lett.*, 2005, **87**, 051919.
68. R. Sharma, E. Moore, P. Rez and M. M. J. Treacy, *Nano Lett.*, 2009, **9**, 689–694.
69. H. Yoshida, S. Takeda, T. Uchiyama, H. Kohno and Y. Homma, *Nanotubes, Nanowires, Nanobelts and Nanocoils – Promise, Expectations and Status,*, 2009, 3–8|ix+235.
70. G. C. Bond, *Metal-catalysed reactions of hydrocarbons*, Springer, New York, US, 2005.
71. G. Zhong, T. Iwasaki, J. Robertson and H. Kawarada, *J. Phys. Chem. B*, 2007, **111**, 1907–1910.
72. G. G. Tibbetts, M. G. Devour and E. J. Rodda, *Carbon*, 1987, **25**, 367–375.
73. T. Yamada, A. Maigne, M. Yudasaka, K. Mizuno, D. N. Futaba, M. Yumura, S. Iijima and K. Hata, *Nano Lett.*, 2008, **8**, 4288–4292.
74. C. Mattevi, C. T. Wirth, S. Hofmann, R. Blume, M. Cantoro, C. Ducati, C. Cepek, A. Knop-Gericke, S. Milne, C. Castellarin-Cudia, S. Dolafi, A. Goldoni, R. Schloegl and J. Robertson, *J. Phys. Chem. C*, 2008, **112**, 12207–12213.
75. C. Zhang, F. Yan, C. S. Allen, B. C. Bayer, S. Hofmann, B. J. Hickey, D. Cott, G. Zhong and J. Robertson, *J. Appl. Phys.*, 2010, 024311.

76. Y. Wang, Z. Luo, B. Li, P. S. Ho, Z. Yao, L. Shi, E. N. Bryan and R. J. Nemanich, *J. Appl. Phys.*, 2007, **101**, 124310.
77. P. B. Amama, C. L. Pint, S. M. Kim, L. McJilton, K. G. Eyink, E. A. Stach, R. H. Hauge and B. Maruyama, *ACS Nano*, 2010, **4**, 895–904.
78. S. M. Kim, C. L. Pint, P. B. Amama, D. N. Zakharov, R. H. Hauge, B. Maruyama and E. A. Stach, *J. Phys. Chem. Lett.*, 2010, **1**, 918–922.
79. S. P. Patole, H. Kim, J. Choi, Y. Kim, S. Baik and J. B. Yoo, *Appl. Phys. Lett.*, 2010, **96**, 094101.
80. N. Latorre, E. Romeo, F. Cazana, T. Ubieto, C. Royo, J. J. Villacampa and A. Monzon, *J. Phys. Chem. C*, 2010, **114**, 4773–4782.
81. J. H. Han, R. A. Graff, B. Welch, C. P. Marsh, R. Franks and M. S. Strano, *ACS Nano*, 2008, **2**, 53–60.
82. P. Vinten, P. Marshall, J. Lefebvre and P. Finnie, *Nanotechnology*, 2010, **21**, 035603.
83. M. Stadermann, S. P. Sherlock, J. B. In, F. Fornasiero, H. G. Park, A. B. Artyukhin, Y. M. Wang, J. J. De Yoreo, C. P. Grigoropoulos, O. Bakajin, A. A. Chernov and A. Noy, *Nano Lett.*, 2009, **9**, 738–744.
84. D. H. Lee, S. O. Kim and W. J. Lee, *J. Phys. Chem. C*, 2010, **114**, 3454–3458.
85. J. B. Butt and E. E. Petersen, *Activation, Deactivation, and Poisoning of Catalysts*, Academic Press Inc., London, UK, 1988.
86. A. A. Puretzky, D. B. Geohegan, S. Jesse, I. N. Ivanov and G. Eres, *Appl. Phys. A: Mater.*, 2005, **81**, 223–240.
87. A. A. Puretzky, G. Eres, C. M. Rouleau, I. N. Ivanov and D. B. Geohegan, *Nanotechnology*, 2008, 055605.
88. M. Bedewy, E. R. Meshot, H. Guo, E. A. Verploegen, W. Lu and A. J. Hart, *J. Phys. Chem. C*, 2009, **113**, 20576–20582.
89. X. Wang, Y. Feng, H. E. Unalan, G. Zhong, P. Li, H. Yu, A. I. Akinwande and W. I. Milne, *Carbon*, 2011, **49**, 214–221.

CHAPTER 4

Hierarchical Nanostructures for Solar Cells

JUNYEOB YEO* AND SEUNG HWAN KO

Applied Nano and Thermal Science (ANTS) Lab, Department of Mechanical Engineering, Seoul National University, 1 Gwanak-ro, Gwanak-gu, Seoul 151-742, Korea
*Email: nakaz79@snu.ac.kr

4.1 Introduction

Developing sustainable renewable energy sources has been a major research topic to solve the environmental and limited resource problems of fossil fuels. Significant progress has been made to increase efficiencies in various renewable energy technologies including solar cells, fuel cells, nuclear energy, wind power and so on.[1] Among them, solar energy has been regarded as the most reliable clean energy source due to a large abundance of solar energy.[2] Solar cell devices have already been developed over the five decades.[1] However, the widespread use of solar cells is limited by two major challenges: conversion efficiency and cost.[1] The single crystalline bulk silicon solar cell was the first generation solar cell and it is the most widely used on the market due to its high conversion efficiency. However, because of the considerably high material costs, second generation solar cells (thin film solar cells) have been developed to address production costs.[1] To lower the production costs further, third generation solar cells (nanostructured inorganic–organic hybrid and organic solar cells) have emerged as a promising approach to efficient solar energy conversion.[1] The nanostructured inorganic–inorganic hybrid (dye-sensitized solar cell (DSSC)) and organic

RSC Nanoscience & Nanotechnology No. 35
Hierarchical Nanostructures for Energy Devices
Edited by Seung Hwan Ko and Costas P Grigoropoulos

Published by the Royal Society of Chemistry, www.rsc.org

solar cells (OPV) are produced in a cost-effective way by using a simple solution process and they possess a moderately high efficiency due to their large surface areas. The compatibility and easy applicability of the solution-based process to nanostructured materials have made the nanostructures the most promising materials in realizing high efficiency third generation solar cells. Therefore, to show the applicability of hierarchical nanostructures for solar cell applications, we will focus mainly on third generation DSSC and OPV solar cells.

Since the successful high efficiency DSSC demonstration by Grätzel in 1991[3] with a highly porous TiO_2 nanocrystalline film, the development of new materials (such as dyes, electrolytes, catalysts *etc.*) has been the main research topic in high efficiency solar cell development. Research into the conversion efficiency enhancement that could be obtained from the development of new materials was slow, so researchers turned their attention to smart nanomaterial structuring, which could allow a dramatic photo-conversion efficiency even with the same materials. One of the major research trends in smart nanomaterial structuring was the introduction of nanowires, which were expected to have more favorable electron transport with reduced recombination loss than nanoparticle-based solar cells. The nanowire-based solar cell research started from a simple structure consisting of a vertically-grown nanowire forest, and progressed to more complex structures including 2D and 3D hierarchical nanostructures to achieve high photo-conversion efficiencies. The major objectives of hierarchical nanostructuring in solar cells are: (1) high carrier mobility (mostly electron mobility in photoanodes) along the nanowire structures with less recombination, (2) a large surface area to capture more sunlight and adsorb more dye molecules for DSSCs, and (3) a light scattering layer to capture the sunlight more efficiently by multiple scattering.

In this chapter, various research trends will be introduced including how smart material structuring will lead to a photo-conversion efficiency increase in solar cells, especially by introducing hierarchical nanostructures.

4.2 Hierarchical Nanostructured Solar Cell

Compared with conventional silicon-based solar cells, excitonic solar cells,[4] in which small molecules, polymers, or quantum dots are used as light absorbing materials (for example DSSC and OPV), can benefit greatly from functional nanostructured materials that orthogonalize the geometry (make the junction normal to the substrate).[4] The critical dimension for these photovoltaic devices is the exciton diffusion length, which is typically much shorter—roughly 10 nm or less in polymers and up to a micrometre in high-quality small molecule films—than the minority carrier diffusion length in silicon.[4] This orthogonalized device structure maximizes the volume of the absorber material and/or interfacial area, which contribute to charge generation while providing high-mobility channels through which these charges can be extracted.[4] Excitonic cells are promising due to their being comprised of inexpensive organic materials such as dyes and polymers, and those that employ inorganic components are generally made by scalable solution syntheses.[4] Obtaining the size, spacing and aspect ratio required for efficient excitonic devices is currently accessible only through bottom-up synthetic approaches[4] and this is why hierarchical nanostructuring is becoming so important. Therefore, we are mainly going to discuss the applicability of hierarchical nanostructures to DSSCs and OPVs rather than other types of solar cells.

4.2.1 Hierarchically-branched Nanowires for Solar Cells

Various 1D nanostructures have been developed to reduce the recombination rate in excitonic solar cells. 1D nanostructures are usually crystalline materials that can serve to increase the electron diffusion length by providing a direct pathway for electron transport in the interior of the continuous crystal without grain boundary scattering from the point of electron injection to the substrate of the collection electrode.[1] This differs from the nanoporous materials or nanoparticles because the electrons take a random path among the nanoparticles and undergo many collisions at the grain boundary between nanoparticles or the semiconductor/electrolyte interface. In those cases, vertically aligned nanowire arrays (nanowires, nanotubes, nanotips) were developed to achieve functional 1D nanostructures and these were demonstrated by Law *et al.* in 2005 as shown in Figure 4.1.[6] However, the performance of the 1D nanostructured solar cell did not exceed expected values. A further enhancement of the solar cell efficiency could be realized by the development of 3D hierarchical nanostructures such as nanoflowers and nanoforests.

The vertically aligned 1D nanowire arrays were expected to show high photo-conversion efficiency due to a direct electron pathway with much reduced recombination loss by orthogonalizing the geometry. However, the insufficient surface area limited the conversion efficiency of that class of solar cells. The surface area could be increased dramatically by adding extra dimensions to the vertically aligned 1D nanowire arrays such as nanoflowers, nanotrees, nanoforests, nanodendrimers *etc.* In this regard, 3D branched nanostructures with larger surface areas in comparison to 1D nanowires have attracted increasing attention. Suh *et al.*[7] and Baxter *et al.*[8] have presented a dendritic ZnO NW DSSC. They grew ZnO NWs by expensive chemical vapor deposition (CVD), and they showed relatively low efficiency (0.5%) due to insufficient surface areas as shown in Figure 4.2. Jiang *et al.*[9] reported a ZnO nanoflower photoanode and Cheng *et al.*[10] reported hierarchical ZnO NWs produced *via* a hydrothermal method as shown in Figure 4.2. The nanoflower photoanodes consist of upstanding nanowires and outstretched branches. This is based on the consideration that the nanowires alone may not capture the photons completely due to the existence of intervals inherent in the morphology.[1] Nanoflower hierarchical nanostructures, however, have nanoscale branches that stretch to fill these intervals and, therefore, provide both a larger surface area and a direct pathway for electron transport along the channels from the branched

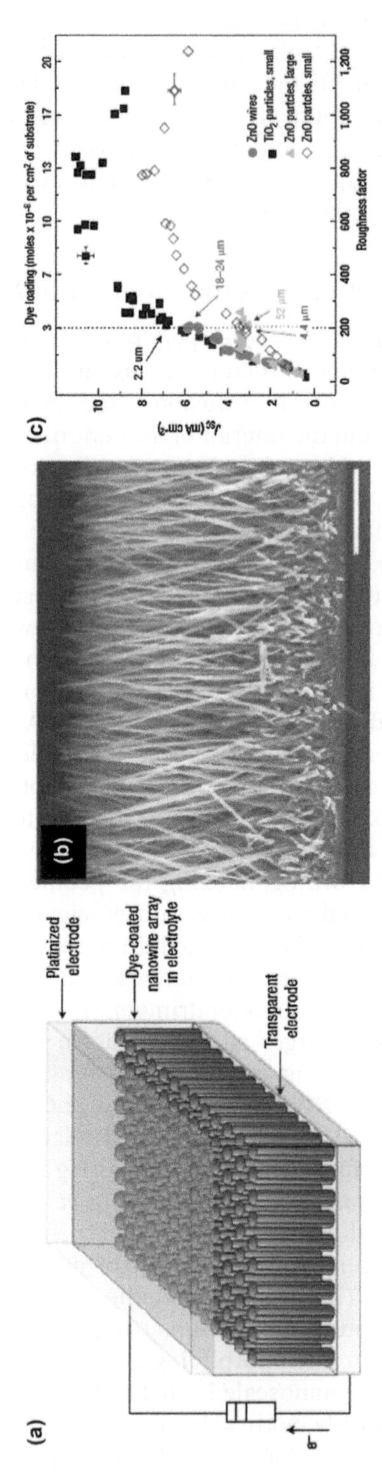

Figure 4.1 ZnO-nanowire dye-sensitized solar cells. (a) Schematic diagram of the cell with a photoelectrode comprised of the ZnO-nanowire array. (b) Cross-sectional SEM image of the ZnO-nanowire array. (c) Comparative performance of nanowire and nanoparticle cells.
Reproduced with permission from ref. 6. Nature Publishing Group.

Figure 4.2 Various hierarchical nanostructures for high efficiency DSSCs. (a) Nanoparticle decorated nanowire.[13] (b) Nanoflower.[14] (c) Nanotips.[15] (d) Branched nanowires.[10]
Reproduced with permission from ref. 10, 13–15.

"petals" to the nanowire backbone. Nanoflower films can be grown by a hydrothermal method at low temperatures, typically by employing a 5 mM zinc chloride aqueous solution with a small amount of ammonia.[10] These as-synthesized nanoflowers have dimensions of about 200 nm in diameter. The solar cell performance of ZnO nanoflower films was characterized by an overall conversion efficiency of 1.9%, a current density of 5.5 mA cm^2, and a fill factor of 0.53. These values are higher than the 1.0%, 4.5 mA cm^2, and 0.36 for films of nanorod arrays with comparable diameters and array densities that were also fabricated by the hydrothermal method. Nanotip arrays with different lengths could be synthesized to achieve high efficiency solar cells. The results confirmed that the energy-conversion efficiency of the cells increased with the length of the ZnO nanotips due to the increase in surface area of the photoelectrode film. An overall conversion efficiency of 0.55% was obtained for 3.2 mm long ZnO nanotips. It has been reported that ZnO nanotips present a maximum overall conversion efficiency at higher light intensities than in the case of TiO_2 nanoparticles. This implies a nontrap-limited electron transport in the respect that the nanotips provide a faster conduction pathway for electron transport. This feature allows for the use of ZnO nanotips in the fabrication of more stable and efficient DSSCs under high illumination. It has also been demonstrated that the overall conversion efficiency could be increased to 0.77% by combining the ZnO nanotips with a Ga-doped ZnO film as a transparent conducting layer. Dendritic ZnO nanowires, which possess a fractal structure more complicated than that of nanoflowers, are formed by a nanowire backbone with outstretched branches, on which the growth of smaller-sized nanowire backbones and branches is reduplicated.[1] Baxter *et al.*[8] described a MOCVD fabrication for dendritic ZnO nanowires by using a route of so-called multiple-generation growth. They first grew 100 nm diameter ZnO nanowires with 20 nm secondary nanowire branches that nucleated and grew from the primary nanowire backbone. It was indicated that each of the nanowires was crystalline with grain boundaries separating secondary nanowires from the primary nanowire. The substrate with both primary nanowires and secondary nanowire branches was then used to continue the nanowire growth, called "secondary generation" growth. During the growth of a second generation of nanowires, the outstretched nanowire branches act as new nucleation sites for next generation nanowire growth. The growth can also be continued for third and fourth generations for the attainment of a dendrite-like branched ZnO nanostructure. DSSC characterization showed that the short-circuit density increased with increasing growth generation due to the larger surface area, which in turn led to increased adsorption of dye molecules. A total improvement of over 250 times in current density and over 400 times in efficiency has been observed when the film morphology was changed from smooth nanowires to branched second-generation nanowires. The efficiencies obtained using fourth-generation dendritic nanowire films with a branched nanostructure and a thickness of 10 mm displayed an overall conversion efficiency of 0.5%, a more than 600-fold improvement over smooth nanowires. By integrating ZnO nanoparticles within the film of dendritic nanowires, the specific surface area was increased, leading to an improved conversion efficiency of 1.1% for these cells.[1] These hierarchical NWs were grown from seeds formed from $Zn(OAc)_2$ and still showed a relatively low efficiency of 1.5% due to insufficient surface area and lack of uniformity of the secondary branches that were produced by a randomly distributed seed layer.

To address these problems encountered during research into hierarchical nanostructured solar cells, Ko *et al.*[11] reported that "nanoforests" of high density, long branched "treelike" multi-generation hierarchical ZnO nanowire photoanodes grown *via* simple selective hierarchical growth by combining length-wise growth (LG) and branched growth (BG) can significantly increase the power conversion efficiency by 350–500% for the same backbone nanowire length as shown in Figure 4.3 and Table 4.1. The efficiency increase was due to a highly enhanced surface area for higher dye loading and light harvesting, and also due to reduced charge recombination by direct conduction pathway along the crystalline ZnO "nanotree"

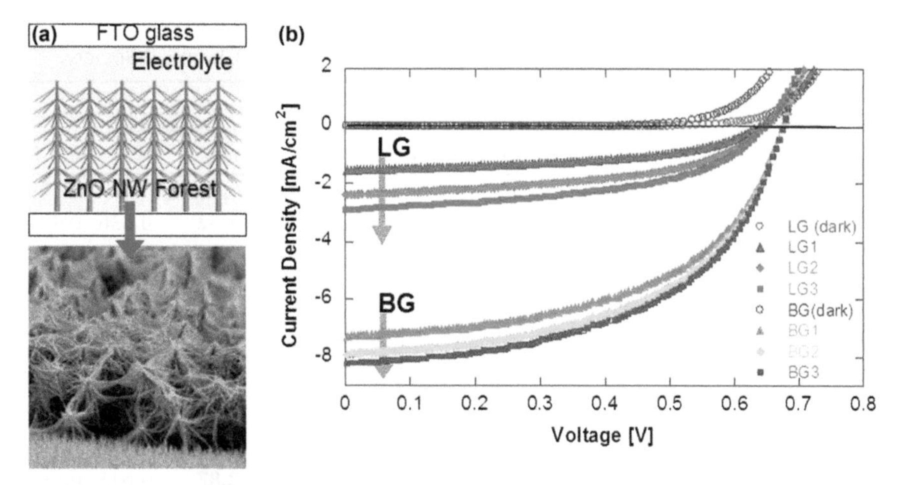

Figure 4.3 (a) Schematic structure and (b) IV curve of a dye-sensitized solar cell with a nanoforest ZnO nanowire.
Reproduced with permission from ref. 11. Copyright 2013, American Chemical Society.

multi-generation branches. This approach mimics branched plant structures with the objective to capture more sunlight.

The hierarchically branched nanowires grow in the normal direction on the vertically oriented 1st generation backbone nanowire surface. The diameter of the branched nanowire was usually 30–50 nm, which is much smaller than that of 1st generation backbone nanowires. The length of the branched nanowire was 2–10 μm after a single growth step. The diameter of the branched nanowires was always smaller than that of backbone nanowires. This may be attributed to the fact that the branched nanowires are based on one of the faces of the hexagonal backbone nanowires. Both backbone nanowires and branched nanowires have a hexagonal cross section and grow along the *c* axis of the wurtzite crystal. While the branched growth nanowire and the backbone nanowires have a monolithically single crystalline relationship for CVD grown comb-like ZnO nanostructures,[57,58] the hydrothermally grown secondary branched ZnO NW did not originate from the backbone NW but from the ZnO seed nanoparticles in the *c* axis direction with a single crystal wurtzite structure.[11] The efficiency increase can be explained by a combination of several effects. First, the enhanced photon absorption associated with the augmented surface area results in increased dye loading and corresponds to a large J_{sc} increase. Length-wise growth can grow only upstanding nanowires by adding extra length to the 1st generation backbone nanowire. However, branched growth can grow multi-branched hierarchical nanowires from just a single 1st generation backbone nanowire, thus surface area can be increased dramatically. The measured nanowire number density for upstanding conventional nanowires (10^9/cm^2) could be increased by 1–2 orders of magnitude ($10^{11,12}$/cm^2) by branched

Table 4.1 Characteristics of hierarchical "nanoforest" DSSCs. Reproduced with permission from ref. 11. Copyright 2013, American Chemical Society.

Symbol	Backbone NW length (μm)	Branching times	Configuration	Efficiency (%)	J_{sc} (mA cm^{-2})	V_{oc} (V)	FF
LG1	7			0.45	1.52	0.636	0.480
LG2	13	0	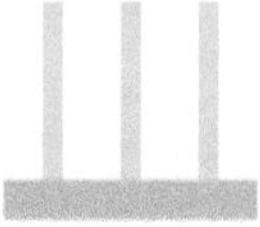	0.71	2.37	0.640	0.486
LG3	18			0.85	2.87	0.645	0.484
BG1	7	1	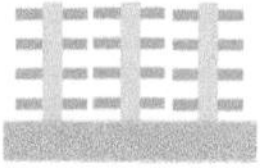	2.22	7.43	0.681	0.522
BG2	13		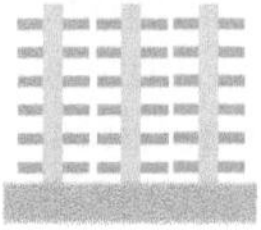	2.51	8.44	0.683	0.531
BG3		2	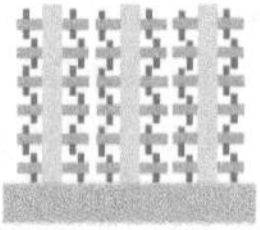	2.63	8.78	0.680	0.530

nanowire growth. Second, a dense network of crystalline ZnO nanowires can increase the electron diffusion length and electron collection because the nanowire morphology provides more direct conduction paths for electron transport from the point of injection to the collection electrode. Third, randomly branched nanowires exhibit enhanced light harvesting (light–dye interaction) without sacrificing efficient electron transport. The upstanding nanowires are not favorable for light harvesting because photons could travel between the vertical nanowires without being absorbed by the dye. Furthermore, branched nanowires can increase light harvesting efficiency by scattering enhancement and trapping.

Herman *et al.*[12] discuss the branching effect on solar cell efficiency further as shown in Figure 4.4. As the length of the branches increased, the branches became flaccid and the solar cell efficiency increase slowed down because the effective surface area increase was hindered by the branches bundling during the drying process and subsequently the dye loading decreased. The length of the branches of the ZnO nanowire tree grown once

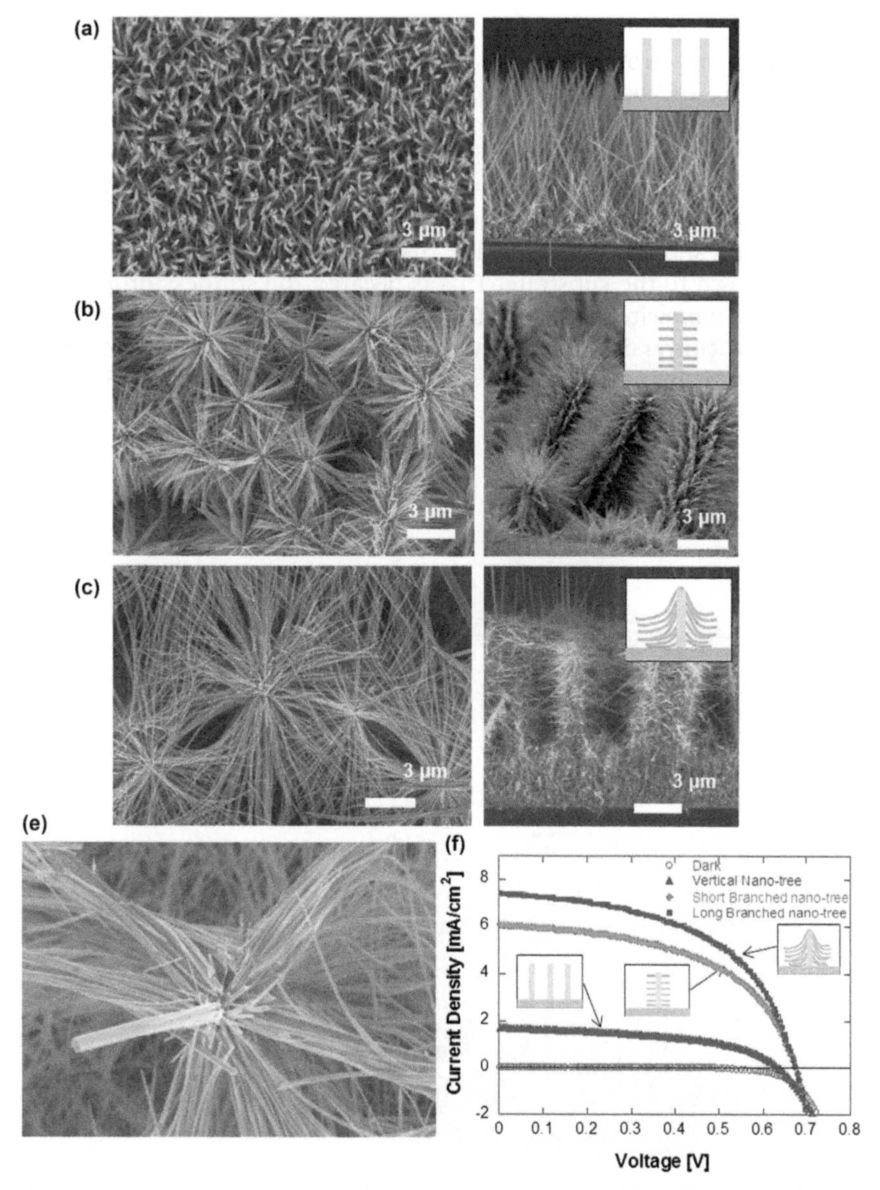

Figure 4.4 SEM pictures of the ZnO nanowire hierarchical nanostructures for dye sensitized solar cells (left column: top view, right column: cross sectional view). (a) Vertically aligned nanowire carpet. (b) Short branched nano weeping willow tree. (c) Long branched nano weeping willow tree. (e) Pseudo-colored SEM picture of the long branched nano weeping willow tree. Magenta colored nanowire represents backbone nanowire after first growth and purple colored nanowires represents branch nanowires after second growth. (f) I–V curve of dye-sensitized solar cell made from the long branched nano weeping willow trees of ZnO nanowires. Reproduced with permission from ref. 12.

(3–5 μm) was shorter than the backbone nanowire (5–9 μm). As the multiple growth steps can grow backbone ZnO nanowires 40–50 μm long, longer branches from the backbone nanowire with a weeping willow tree shape could be grown by repeating the hydrothermal growth steps without polymer layer removal and seed NP deposition. A weeping willow nanotree structure with a very long first generation branch can grow up to 12–15 μm. While the shorter nanotrees grown once show stiff spiny branches, nanotrees grown multiple times show a weeping willow tree shape with long flaccid branches that hang down to the substrate. This is because the stiffness drops as the length of the nanowire increases. The short circuit current density (J_{SC}) and overall light conversion efficiency (η) increase when the branches are introduced and also as the branches are getting longer by an effective surface area increase. J_{SC} and η could be significantly increased, just by adding

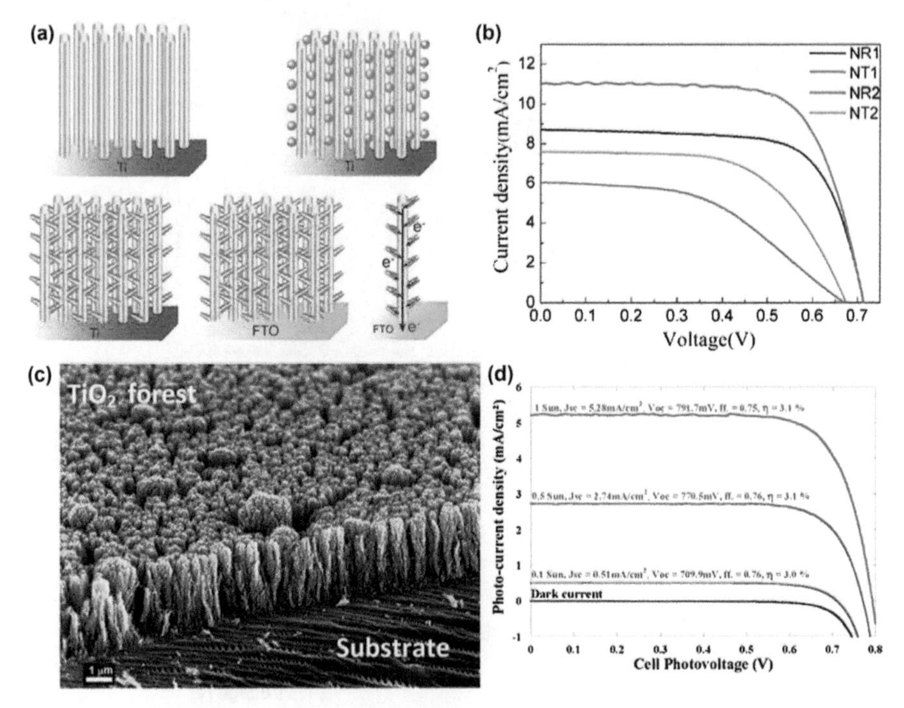

Figure 4.5 Branched TiO_2 nanostructures. (a) Schematic procedure for the formation of the forest-like hierarchical photoanodes and a cartoon of the presumed preferential electron pathway in the hierarchical photoanodes. (b) J–V characteristics of dye-sensitized solar cells assembled with different TiO_2 photoanodes.[16] (c) Scanning electron micrographs of PLD TiO_2 films: (d) J–V characteristics recorded at different light illumination for PLD films prepared at background oxygen pressures of 20 Pa with 2 μm thickness.[17]
Reproduced with permission from ref. 16. Copyright 2013, American Chemical Society. Reproduced with permission from ref. 17. Copyright 2013, the Royal Society of Chemistry.

branches on the backbone ZnO nanowires, up to 400% for the same backbone nanowire length. The overall light conversion efficiency increased as the length of the branches increased because the dense network of crystalline ZnO NWs can increase the electron diffusion length and electron collection. However, doubling or tripling the branch length did not double or triple the cell efficiency. When the branch length was over 10 μm, the cell efficiency increase with the branch length slowed down. This could be attributed to the recombination loss during the electron transport from the point of injection to the collection electrode and the partial collapse of the branches as the branches got longer. Branches could bundle during the drying process after the nanowire hydrothermal growth. The bundled branches could provide less surface area for dye attachment than the calculated expected values for a weeping willow nanotree structured solar cell with well separated branches.

TiO_2 and ZnO are the two most popular materials for hierarchical nanostructured solar cell fabrication. ZnO is a wide-band-gap semiconductor that possesses an energy-band structure and physical properties similar to those of TiO_2, but has higher electronic mobility that would be favorable for electron transport, with reduced recombination loss when used in DSSCs.[1] Many studies have already been reported on the use of ZnO materials for application in DSSCs. Although the conversion efficiencies of 0.4–5.8% obtained for ZnO are much lower than 11% for TiO_2, ZnO is still thought of as a distinguished alternative to TiO_2 due to its ease of crystallization and anisotropic growth. These properties allow ZnO to be produced in a wide variety of nanostructures. Very similar approaches to ZnO branched nanostructures were demonstrated for TiO_2 as shown in Figure 4.5.[16,17]

4.2.2 Hierarchically Structured Porous Materials for Solar Cells

The first generation of high efficiency DSSCs was demonstrated with uniform sized TiO_2 nanoparticle-based nanoporous structures. Since then, hierarchical nanoparticles at different scales have been studied to enhance the surface area and to capture the sunlight more efficiently by a light-scattering mechanism. The performance of solar cells with hierarchically structured ZnO films can be significantly affected by either the average size or the size distribution of the nanopores.

Natural materials have developed highly efficient hierarchical structures over a very long time to efficiently capture, convert and store sunlight energy. For example, green leaves and certain photosynthetic plants have hierarchical structures optimized for efficient light harvesting and sunlight conversion to chemical energy by photosynthesis. By learning from nature, materials with hierarchical porosity and structures have been heavily involved in newly developed energy storage and conversion systems. Owing to meticulous design and ingenious hierarchical structuration of porosities through mimicking natural systems, hierarchically structured porous materials can provide large surface areas for reactions, interfacial transport, or dispersion of active sites at different length scales of pores and they can shorten diffusion paths or reduce the diffusion effect.[18]

Zhou *et al.*[19] used natural leaves as a biotemplates to replicate all the fine hierarchical structures of leaves as shown in Figure 4.6. They used a pure inorganic structure of TiO_2 with the same hierarchy as leaves by a two-step procedure composed of the infiltration of inorganic precursors and the calcination of the biotemplates. All of the photosynthetic pigments were replaced by artificial catalysts such as Pt nanoparticles. The artificial leaf with catalyst components obtained was used for efficient light-harvesting and photochemical hydrogen production. The average absorbance intensities within the visible range increased 200–234% for the artificial leaves compared with conventional TiO_2 nanoparticles prepared without biotemplates. This should certainly contribute to hierarchical architectures with all the fine structures of leaves imprinted in artificial leaves. The photocatalytic activity is much higher than that of TiO_2 nanoparticles prepared without biotemplates and commercial nanoparticulate P25.[19] Similar structures were also demonstrate by Zhu *et al.* by sonochemical methods.[20] The important role of meso–macroporous structures in light harvesting photocatalysis has been revealed. In the macro–mesoporous TiO_2 photocatalyst, the macrochannels acted as a light-transfer path for introducing incident light onto the inner surface of mesoporous TiO_2. This allowed light waves to penetrate deep inside the photocatalyst, making it a more efficient light harvester. It is known that a wavelength of 320 nm is reduced to 10% of its original intensity after penetrating a distance of only 8.5 μm on thin film TiO_2. The presence of macrochannels, however, makes it possible to illuminate even the core TiO_2 particles with the emission from the four surrounding UV sources. Considering the light absorption, reflection, and scattering within such a hierarchical porous system, the effective light-activated surface area can be significantly enhanced. Moreover, the interconnected TiO_2 nanoparticle arrays embedded in the mesoporous wall may allow highly efficient photogenerated electron transport through the macrochannel network.[21]

There are other approaches to prepare hierarchically structured nanoparticle films with submicron sized nanoparticle aggregates to realize high efficiency solar cells as shown in Figure 4.7. The synthesis of hierarchical ZnO aggregates can be achieved by hydrolysis of zinc salt in a polyol medium at 160 °C.[22] ZnO aggregates with either a monodisperse or polydisperse size distribution can be prepared[23,24] by adjusting the heating rate during synthesis and using a stock solution containing ZnO nanoparticles of 5 nm in diameter. The hierarchically structured ZnO film is well packed by ZnO aggregates with a highly disordered stacking, while the spherical aggregates are formed by numerous interconnected nanocrystallites that have sizes ranging from several tens to several hundreds of nanometres.[22] The

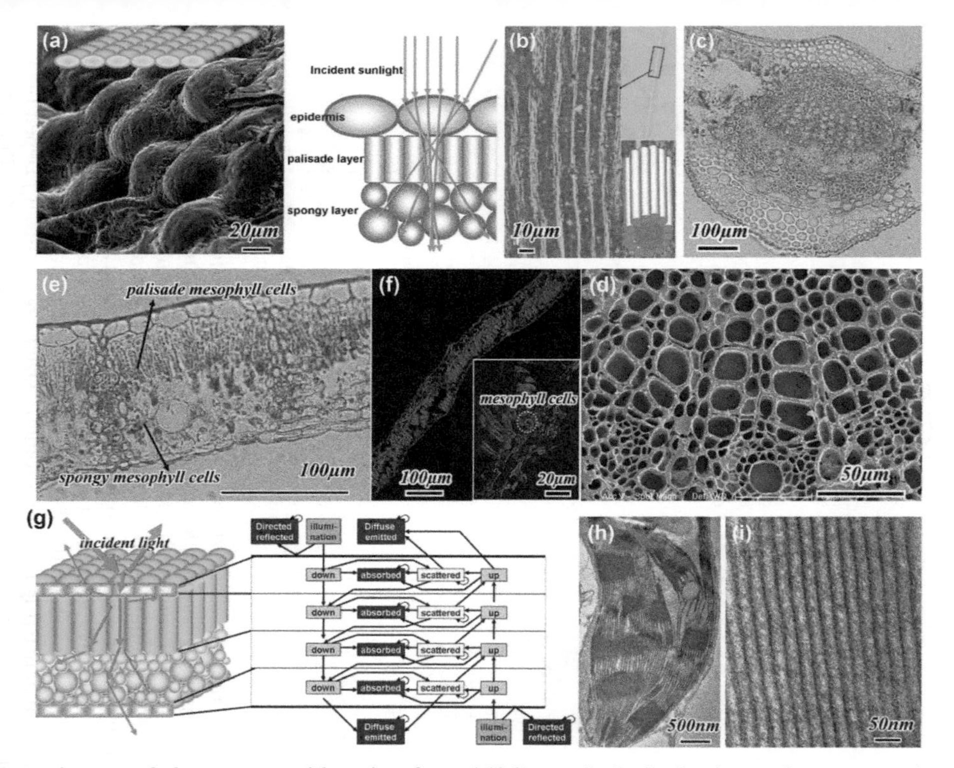

Figure 4.6 (a) Left: FESEM image of the upper epidermis of *A. vitifolia Buch.* leaf; the inset shows a 3D image of the same. Right: illustration of epidermal focusing by a lens mechanism and light distribution within the leaf. (b) Magnified surface image of bundle sheath extensions (top view) obtained by digital microscopy; the inset is a 3D illustration of the same. (c) Cross-section of the vein architecture observed with optical microscopy. (d) FESEM image of the cross-section of the vein architecture. (e) Optical microscopy image of the cross-section of *A. vitifolia Buch.* leaf, indicating the differentiation of leaf mesophyll into palisade and spongy layers. (f) Cross-section observed under CLSM; the inset is a magnified image, (g) pathway of light through a dicotyledon leaf as envisioned by stochastic theory. (h) TEM image of a chloroplast in mesophyll cells and (i) TEM image of a granum—the layered nanostructure of thylakoid membranes.
Reproduced with permission from ref. 19. Copyright 2013, Wiley-VCH.

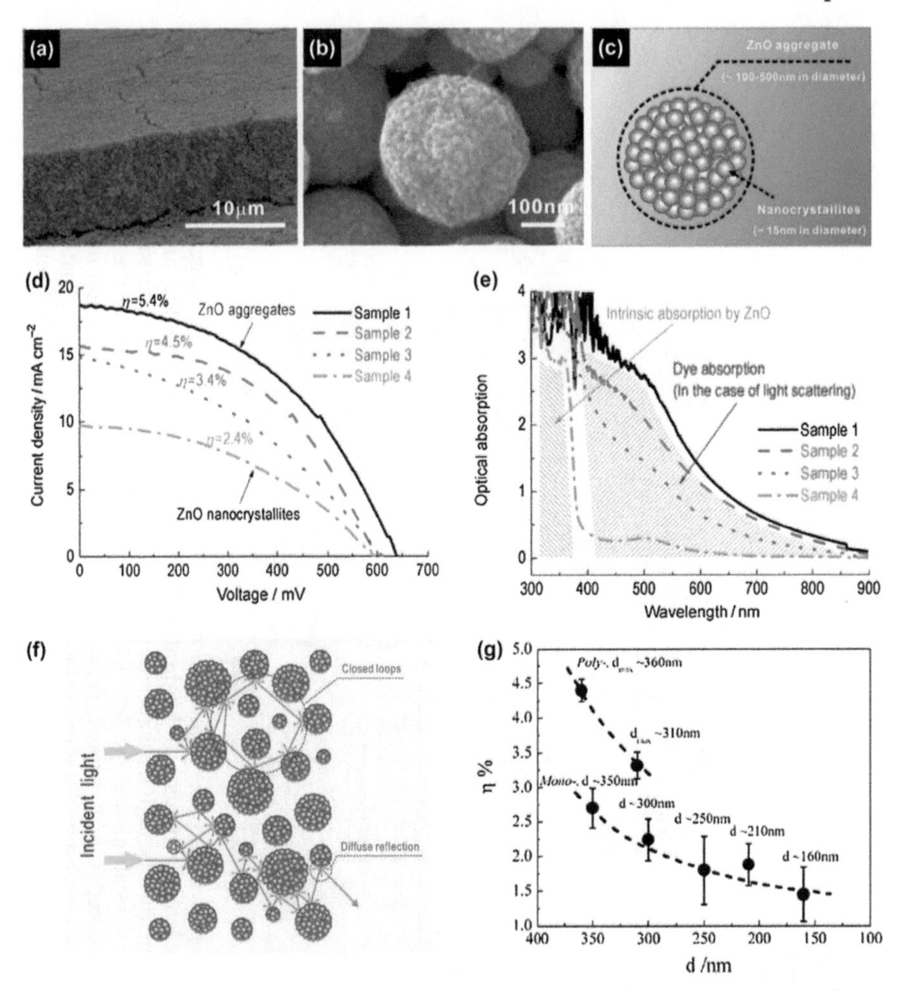

Figure 4.7 ZnO-aggregate dye-sensitized solar cells. (a) Cross-sectional SEM image of a ZnO-aggregate film. (b) Magnified SEM image of an individual ZnO aggregate. (c) Schematic diagram illustrating the microstructure of aggregated ZnO comprised of closely packed nanocrystallites. (d) Photovoltaic behavior of N3-dye-adsorbed ZnO-film samples with differences in the degree of aggregation of nanocrystallites. (e) Optical absorption spectra for (d). (f) Schematic of light scattering and photon localization within a film consisting of submicrometre-sized aggregates. (g) Dependence of the overall conversion efficiency on the size and size distribution of aggregates in dye-sensitized ZnO solar cells.
Reproduced with permission from ref. 1. Copyright 2013, Wiley-VCH.

structural features of the aggregates are their possession of a porosity and geometrical size comparable with the wavelengths of visible light. The degree of aggregation could be adjusted to get maximum DSSC performance. It has been demonstrated that all these samples present approximately the same crystallite size of about 15 nm and similar specific surface area of

80 m^2 g^{-1}. However, their photovoltaic behaviors exhibit a significant difference in the short-circuit current density, resulting in a difference of overall conversion efficiency. Typically, a maximum short-circuit current density of 19 mA cm^{-2} and conversion efficiency of 2.4–5.4% are observed. An obvious trend is that the overall conversion efficiency decreases as the degree of the spherical aggregation is gradually destroyed. In other words, the aggregation of ZnO nanocrystallites is favorable for achieving a DSSC with high performance.[1] Further studies reinforce the light-scattering mechanism. The performance of DSSCs with hierarchically structured ZnO films can be significantly affected by either the average size or the size distribution of aggregates.[22] Interestingly, the films with a more disordered polydisperse aggregate structure achieved better packing, establishing higher conversion efficiencies than those with monodisperse aggregates. The enhancement effect becomes more intense when the maximum size of the aggregates in polydisperse films, or the average size of the aggregates in monodisperse films, increases to be as large as, or comparable to, the wavelength of visible light. These results confirm the rationality of enhanced solar cell performance arising from light scattering, generated by hierarchically structured nanoporous films and promoting the light capturing ability of the efficient photoelectrode.

Another way to successfully enhance light harvesting is to introduce optical elements, such as highly scattering photonic crystal layers that consist of large particles to increase the photon path length in the cell as shown in Figure 4.8.[34] Photonic band gap materials in the form of 3D inverted TiO_2 opal or porous Bragg stacks have been applied to DSSCs to enhance light harvesting in specific parts of the spectrum while retaining cell transparency.[34–38] Guldin *et al.*[34] demonstrated a material assembly route for a double layer DSSC, integrating a high surface

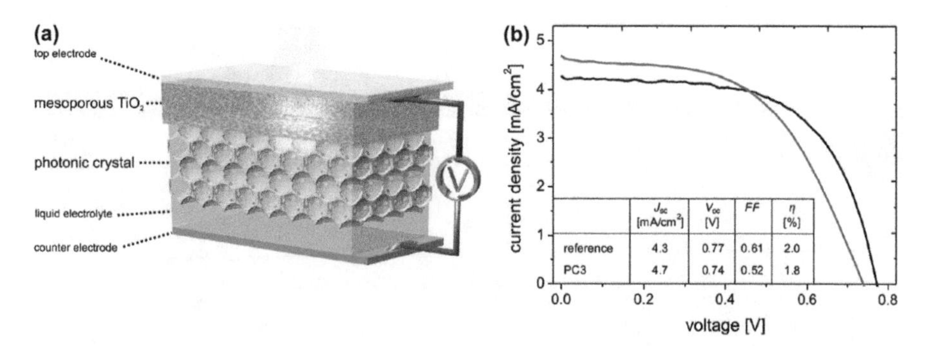

	J_{sc} [mA/cm²]	V_{oc} [V]	FF	η [%]
reference	4.3	0.77	0.61	2.0
PC3	4.7	0.74	0.52	1.8

Figure 4.8 (a) Schematic of a fully assembled DSSC based on the self-assembled double layer photoanode. (b) Current–voltage characteristics of the double layer device PC3 (blue) and a reference single layer cell of similar internal surface area (black) under standardized AM 1.5 illumination of 100 mW/cm^2 (active area 0.13 cm^2, spectral mismatch 22%). Reproduced with permission from ref. 34. Copyright 2013, American Chemical Society.

mesoporous underlayer with an optically active 3D photonic crystal overlayer. Self-assembly material synthesis on two length-scales enabled fabrication of a double layer DSSC with electric and pore connectivity at the mesoporous and the microporous level simultaneously. This construct allows effective dye sensitization, electrolyte infiltration, and charge collection from both the mesoporous and the photonic crystal layers. Due to the intimate physical contact between the layers, photonic crystal-induced resonances can significantly contribute to absorption enhancement in a specific part of the spectrum.[34]

4.2.3 Other Hierarchically Structured Solar Cells

Besides hierarchical nanoforest and nanoporous structures, nanostructures with other morphologies, such as nanosheets, nanobelts, and nanotetrapods as shown in Figure 4.9, have also been studied for high efficiency solar cell applications on account of the fact that they also have a large specific surface area.[1] However, for these nanostructures, the specific surface area is not the only factor that determines the photovoltaic efficiency of the DSSC.[1] Solar cell performance is also believed to be significantly affected by the geometrical structure of the photoelectrode films, which provides

Figure 4.9 SEM images of nanostructured ZnO films: (a) Dispersed nanosheets.[45] (b) Nanosheet-assembled spheres.[43] (c) Nanobelt array.[44] (d) A ZnO tetrapod formed in a three-dimensional structure with four arms extending from a common core.[41] (e) Networked film with interconnected ZnO tetrapods.[42]
Reproduced with permission from ref. 1. Copyright 2013, Wiley-VCH.

particular properties in terms of the electron transport and/or light propagation.[1]

4.2.3.1 Tetrapods

Tetrapod nanostructures for solar cell applications were first studied by Alivisatos's group at UC Berkeley.[39] Tetrapods possess a three-dimensional structure consisting of four arms extending from a common core. With the unique mechanical structures, tetrapods are shaped like schoolyard "jacks" since they come to rest on their three downward pointing legs. The tetrapod length of the arms can be adjusted within the range of 1–20 mm, while the diameter can be tuned from 100 nm to 2 mm by changing the substrate temperature and oxygen partial pressure during vapor deposition.[40] Multiple-layer deposition can result in tetrapods connected to each other so as to form a porous network with a large specific surface area. Tetrapod structures could be better for plastic solar cell applications, because the rods need to point in the same direction, and tetrapods will always do that. By simply scattering the tetrapods on a surface, they will all point upwards. The ZnO tetrapod-based DSSCs have achieved overall conversion efficiencies of 1.20–3.27%.[41,42] It was reported that the internal surface area of tetrapod films could be increased further by incorporating nanoparticles with these films, leading to significant improvement in the solar cell performance.[41]

4.2.3.2 Nanosheets

Nanosheets are thin film quasi-2D structures. The connected patches of ZnO nanosheet used in a DSSC have been shown to possess a relatively low conversion efficiency of 1.55%, possibly due to an insufficient internal surface area. It seems that ZnO nanosheet spheres prepared by hydrothermal treatment using oxalic acid as the capping agent may provide a significant enhancement in internal surface area, resulting in a conversion efficiency of up to 2.61%.[44] As for nanosheet spheres, the performance of the solar cell is also believed to benefit from a high degree of crystallinity and, therefore, low resistance with regards to electron transport.[1]

4.2.3.3 Nanobelts

Nanobelt array films prepared through an electrodeposition method were also studied for DSSC applications.[44] In fabricating these nanobelts, polyoxyethylene cetylether was added in the electrolyte as a surfactant. The ZnO nanobelt array obtained shows a highly porous striped structure with a nanobelt thickness of 5 nm, a typical surface area of 70 m^2/g, and a photovoltaic efficiency as high as 2.6%.

4.2.3.4 Mesh-based 3D Electrodes

Rustomji *et al.*[46] demonstrated a mesh-based 3D photoelectrode in an attempt to maximize the surface area of a photoelectrode comprised of TiO_2 nanotubes. By anodizing a commercial metal mesh of titanium (99.5 wt% Ti), TiO_2 nanotubes on the surface of titanium fibers were produced. The TiO_2 nanotubes grew in different orientations due to the use of a mesh as a substrate, and the mesh produced presented a surface area much larger than that of a flat titanium foil. A solar cell constructed with such a 3D mesh with TiO_2 nanotubes 37 microns in length and 110 nm in diameter showed 5.0% efficiency.

4.3 Fabrication/Synthesis Methods

Many studies on synthetic approaches towards hierarchical nanostructures have been reported for applications in high efficiency solar cells. The production of these structures can be achieved through sol–gel synthesis,[47] hydrothermal/solvothermal growth,[48] physical or chemical vapor deposition,[49,50] low-temperature aqueous growth,[51,52] chemical bath deposition,[53] and electrochemical deposition.[54–56] Interestingly, the majority of the hierarchical nanostructured solar cell studies apply wet chemistry instead of vacuum synthesis (for example, chemical vapor deposition) to realize a low cost solar cell fabrication process. This is very important because cost as well as efficiency is the most important factor in solar cell production. Solution processing becomes more important especially for DSSCs because DSSCs are low-cost at the cost of having a moderately lower efficiency compared with silicone-based crystalline solar cells, which have a high efficiency but are very expensive to produce.

One of the most notable hierarchical nanoforest synthesis methods was demonstrated by Ko *et al.*[11] and Herman *et al.*[12] by the hydrothermal growth of branched ZnO nanowires. A nanoforest of hierarchical ZnO nanowires is grown by a modified hydrothermal growth approach as illustrated in Figure 4.10. Depending on the growth conditions, there are two types of growth modes: length-wise growth (LG) and branched growth (BG). LG can yield nanowires of increased length by extending the growth at the tip of the backbone nanowire while BG produces highly branched nanowires by multiple generation hierarchical growth on nanowire side surfaces. The first generation (backbone) nanowires are grown from ZnO quantum dot seeds deposited on a substrate in an aqueous precursor solution. ZnO quantum dots (3–4 nm) in ethanol are drop casted on a substrate to form uniform seeds for nanowire growth. Nanowires were grown by immersing the seeded substrate in aqueous solutions containing 25 mM zinc nitrate hydrate [$Zn(NO_3)_2 \cdot 6H_2O$], 25 mM hexamethylenetetramine ($C_6H_{12}N_4$, HMTA) and 5–7 mM polyethylenimine (PEI) at 65–95 °C for 3–7 hours. After the reaction was complete, nanowires grown on the substrate were thoroughly rinsed with MilliQ water and dried in air to remove any residual polymer. Longer

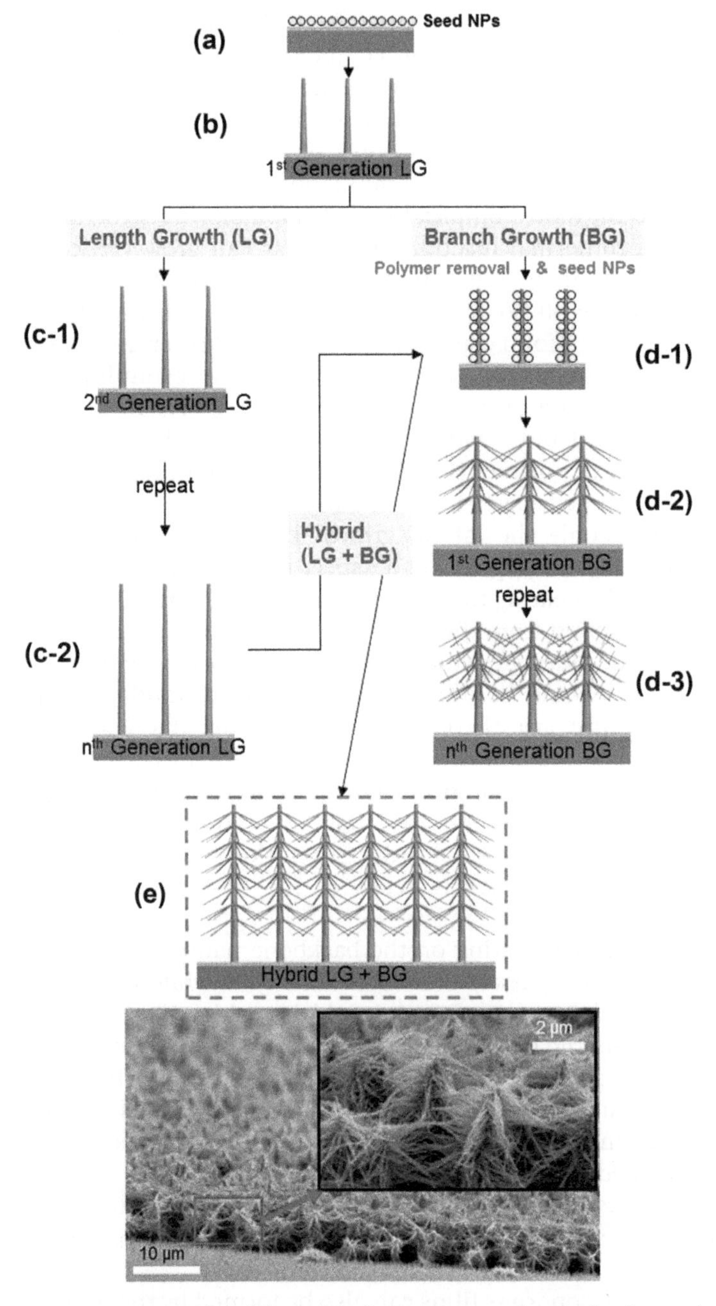

Figure 4.10 Two routes for hierarchical ZnO nanowire hydrothermal growth. Length-wise growth (LG) (a-b-c), branched growth (BG) (a-b-d), and hybrid (a-b-c-d-e). Notice the polymer removal and seed NPs for branched growth.
Reproduced with permission from ref. 11. Copyright 2013, American Chemical Society.

nanowires can be grown by repeating the hydrothermal growth in a fresh aqueous precursor solution by the LG mode. A dramatic change in nanowire structure could occur by heating the first generation nanowire at 350 °C (10 minutes), adding seed nanoparticles and then hydrothermal growth. Instead of LG, highly BG of nanowires on the first generation nanowire sidewall could be observed. The combination of multiple LG and BG steps can be applied for more complex hierarchical nanowire structures.

A single hydrothermal reaction LG process can grow vertically aligned nanowires 2–8 μm long (130–200 nm diameter). Multiple LG growth steps can grow nanowires 40–50 μm long of a high aspect ratio (>100). The length of the higher generation LG nanowires becomes shorter upon each growth step. A vertically aligned long nanowire forest grown by multiple LG can be used as the backbone of a hierarchical branched nanowire forest. High quality hierarchical branched nanowire forests can be grown only after both (1) removal of polymer (HMTA, PEI) by heating the nanowire and (2) coating with seed nanoparticles on the backbone nanowire surface. HMTA and PEI hinder only the lateral growth but allow axial growth of the nanowires in solution to yield high aspect ratio nanowires. The polymers can be removed by heating the nanowire at 350 °C for 10 minutes. Once the polymer was removed from the backbone nanowire, branched growth (BG) nanowires on the sidewall, which had been suppressed by HMTA and PEI for a regular LG, could be induced. In addition to random and sparsely branched nanowire growth on the side wall, the diameter of the first generation backbone nanowire also increased due to lateral growth after the removal of the HMTA and PEI polymer layer. After polymer removal, seed nanoparticle coating on the first generation backbone nanowire can grow densely packed higher order generation BG nanowires and could also suppress the diameter increase of the first generation backbone nanowire.

A seed nanoparticle coating on the backbone nanowire without polymer removal could grow sparsely branched nanowires while high quality hierarchical branched nanowire forests could be achieved with a seed nanoparticle coating on the backbone nanowire after polymer removal. Nanowire growth from the seed nanoparticle on HMTA and PEI polymer may be less favorable than on the nanowire surface without polymer. This signifies that both polymer removal and seed layer are important for high density hierarchical branched nanowire forest growth.

The "nanoforest" of the hierarchical branched ZnO nanowire could be easily grown on a large area by a low cost, all solution processed hydrothermal method.

Hierarchical mesoporous films can also be formed by the interconnected power of different dimensions by various self-assembly methods. A hierarchical porous material should exhibit at least a bimodal pore size distribution. The properties of the material and its capability of carrying out a specific function are directly dependent on the distribution of size, shape, and organization of the pores. Furthermore, while the pores can be randomly distributed within the material, controlling the order–disorder pore structure (Figure 4.11)[25] allows the opening of a new set of properties and material performances, especially when pore order is controlled at the nanoscale.[26] The key issue for the synthesis of porous materials is how to template the pores, and how to order and shape them. Several strategies have been proposed to produce porous materials *via* templates, which are removed after the material processing, or controlled phase separation.[27] A templating strategy is also at the base of self-assembly through evaporation, which allows the preparation of mesopores with an ordered and controlled organization *via* a combination of sol–gel and supramolecular chemistry. This process can be combined with other strategies to obtain hierarchical porous materials that exhibit organization at least in one length scale. Nakanishi *et al.*[28] and Huesing *et al.*[29] have synthesized hybrid organic–inorganic monoliths with continuous macropores formed by phase separation along with micro- and mesopores which are templated by micelles formed by an amphiphilic triblock copolymer. More recently, the formation of an organic polymer simultaneous to self-assembly has also been developed as a synthesis route toward macroporous–mesoporous monoliths.[30] The fabrication of hierarchical porous materials has also been extended to obtain films with different pore length scales.[31] Controlling the pore properties is, however, more difficult to achieve in films because the

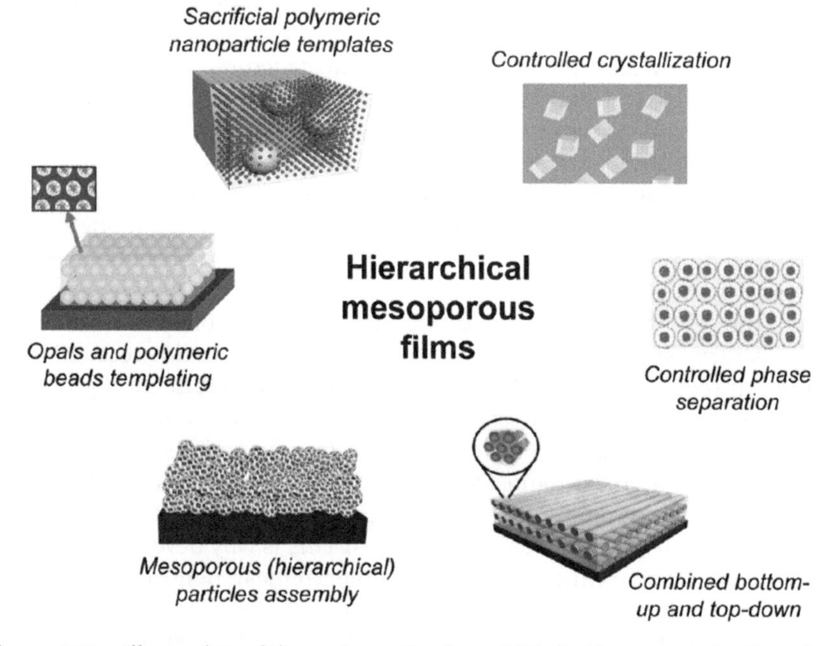

Figure 4.11 Illustration of the main synthesis and fabrication routes that have been used to obtain hierarchical porous films by combining self-assembly with other templating strategies.[25]
Reproduced with permission from ref. 25.

fast evaporation of the solvent makes the process kinetically controlled, and the chemistry has to be carefully modulated to get self-organization.[32,33] If hierarchical porous films are synthesized by self-assembly to obtain organized mesoporosity, the other templates should not interfere with self-organization.

4.4 Summary

Hierarchical nanostructures, such as branched nanoforest and nanoporous structures, have changed the recent research trend in developing high efficiency solar cells. Traditional research had focused on the development of new materials such as dyes, electrolytes, catalysts and so on. However, conversion efficiency enhancement by new material development has slowed down and new research trends to enhance solar cell efficiency by smart nanostructuring from the same material have started to garner tremendous attention. Nanowire-based solar cells have ignited this nanostructuring research and further progress with 2D and 3D hierarchical nanostructures has drawn noticeable solar cell efficiency enhancement. The major objectives of hierarchical nanostructuring in solar cells are: (1) high carrier mobility (mostly electron mobility in photoanodes) along the nanowire structures with less recombination, (2) a large surface area to capture more sunlight and adsorb more dye molecules, and (3) a light scattering layer to capture the sunlight more efficiently by multiple scattering. A large surface area and high carrier mobility are the requirements for most energy-related devices. Therefore, it is evident that hierarchical nanostructures can be applied to emerging energy conversion and storage fields, such as photocatalysis, photoelectrochemical water splitting, Li ion batteries, supercapacitors, fuel cells, thermoelectric devices, piezoelectric devices as well as solar cells. Furthermore, using 3D branched structures to harvest various types of ambient power, such as thermal, wind, vibration and electromagnetic energy, would also be very promising, which provides a potential endless source of energy.[5]

Even though there are a lot of published papers on 3D hierarchical nanostructured solar cell devices, further developments in this research field require improvements in synthetic methods and novel fabrication processes to provide better control of the structural complexity, composition uniformity, surface chemistry and interface electronics and last but not least, the yield, of hierarchical nanostructures.[5] These factors are directly related to the sustainability, high efficiency and production costs at an affordable level for the public for practical applications. This is why developing simple, economic and environmentally friendly hierarchical nanostructure mass production methods are of great interest. One of the most promising economical approaches usually consists of a solution process without using any expensive and complex vacuum-based vapour phase methods.

There are also drawbacks to using 3D branched nanostructures in energy applications, in that they bring challenges in quantifying the charge transport and recombination loss in terms of solar devices.[5] Understanding the light absorption property, interface electronic band structure, and photo-carrier dissociation and recombination are key issues to the overall performance.[5] To establish a reliable structure–performance correlation, a wide range of characterizations should be carried out in a consistent way. Overall, these are multidisciplinary topics, for which physics and chemistry experimentalists and theoreticians need to sit together and brainstorm innovative ideas to bring great enhancement to solar cells and energy devices with hierarchical nanostructures.

Acknowledgements

Parts of this work were supported by the Global Frontier R&D Program on Center for Multiscale Energy System funded by the National Research Foundation under the Ministry of Science, ICT & Future, Korea.

References

1. Q. Zhang, C. S. Dandeneau, X. Zhou and G. Cao, *Adv. Mater.*, 2009, **21**, 4087.
2. J. Weickert, R. B. Dunbar, H. C. Hesse, W. Widermann and L. Schmidt-Mende, *Adv. Mater.*, 2011, **23**, 1810.
3. B. O'Regan and M. Grätzel, *Nature*, 1991, **353**, 737–740.
4. A. I. Hochbaum and P. Yang, *Chem. Rev.*, 2010, **110**, 527.
5. C. Cheng and H. J. Fan, *Nano Today*, 2012, **7**, 327.
6. M. Law, L. E. Green, J. C. Johnson, R. Saykally and P. D. Yang, *Nat. Mater.*, 2005, **4**, 455.
7. D. I. Suh, S. Y. Lee, T. H. Kim, J. M. Chun, E. K. Suh, O. B. Yang and S. K. Lee, *Chem. Phys. Lett.*, 2007, **442**, 348.
8. J. B. Baxter and E. S. Aydil, *Sol. Cells*, 2006, **90**, 607.
9. C. Y. Jiang, X. W. Sun, G. Q. Lo, D. L. Kwong and J. X. Wang, *Appl. Phys. Lett.*, 2007, **90**, 263501.
10. H. Cheng, W. Chiu, C. Lee, S. Tsai and W. Hsieh, *J. Phys. Chem. C*, 2008, **112**, 16359.
11. S. H. Ko, D. H. Lee, H. W. Kang, K. H. Nam, J. Y. Yeo, S. J. Hong, C. P. Grigoropoulos and H. J. Sung, *Nano Lett.*, 2011, **11**, 666.
12. I. Herman, J. Yeo, S. Hong, D. Lee, K. H. Nam, J. Choi, W. Hong, D. Lee, C. P. Grigoropoulos and S. H. Ko, *Nanotechnology*, 2012, **23**, 194005.
13. M. Wang, C. Huang, Y. Cao, Q. Yu, Z. Deng, Y. Liu, Z. Huang, J. Huang, Q. Huang, W. Guo and J. Liang, *J. Phys. D: Appl. Phys.*, 2009, **42**, 155104.
14. C. Y. Jiang, X. W. Sun, G. Q. Lo, D. L. Kwong and J. X. Wang, *Appl. Phys. Lett.*, 2007, **90**, 263501.
15. Z. Yang, T. Xu, Y. Ito, U. Welp and W. K. Kwok, *J. Phys. Chem. C*, 2009, **113**, 20521.

16. F. Sauvage, F. D. Fonzo, A. L. Bassi, C. S. Casari, V. Russo, G. Divitini, C. Ducati, C. E. Bottani, P. Comte and M. Grätzel, *Nano Lett.*, 2010, **10**, 2562.
17. F. Shao, J. Sun, L. Gao, S. Yang and J. Luo, *J. Mater. Chem.*, 2012, **22**, 6824.
18. J. B. Baxter and E. S. Aydil, *Sol. Energy Mater. Sol. Cells*, 2006, **90**, 607.
19. H. Zhou, X. Li, T. Fan, F. E. Osterloh, J. Ding, E. M. Sabio, D. Zhang and Q. Guo, *Adv. Mater.*, 2010, **22**, 951.
20. S. M. Zhu, D. Zhang, Z. X. Chen, G. Zhou, H. B. Jiang and J. L. Li, *J. Nanopart. Res.*, 2010, **12**, 2445.
21. Y. Li, Z. Y. Fu and B. L. Su, *Adv. Funct. Mater.*, 2012, **22**, 4634.
22. T. P. Chou, Q. F. Zhang, G. E. Fryxell and G. Z. Cao, *Adv. Mater.*, 2007, **19**, 2588.
23. Q. F. Zhang, T. R. Chou, B. Russo, S. A. Jenekhe and G. Z. Cao, *Angew. Chem., Int. Ed.*, 2008, **47**, 2402.
24. Q. F. Zhang, T. P. Chou, B. Russo, S. A. Jenekhe and G. Z. Cao, *Adv. Funct. Mater.*, 2008, **18**, 1654.
25. P. Innocenzi, T. Kidchob, L. Malfatti and P. Falcaro, *Chem. Mater.*, 2009, **21**, 2501.
26. M. E. Davis, *Nature*, 2001, **417**, 813.
27. K. Nakanishi and N. Tanaka, *Acc. Chem. Res.*, 2007, **40**, 863.
28. K. Nakanishi, Y. Kobayashi, T. Amatani, K. Hirao and T. Kodaira, *Chem. Mater.*, 2004, **16**, 3652.
29. N. Huesing, C. Raab, V. Torma, A. Roig and H. Peterlik, *Chem. Mater.*, 2003, **15**, 2690.
30. G. L. Drisko, A. Zelcer, V. Luca, R. A. Caruso and G. J. A. A. Soler-Illia, *Chem. Mater.*, 2010, **22**, 4379.
31. G. J. A. Soler-Illia and P. Innocenzi, *Chem.–Eur. J.*, 2006, **12**, 4478.
32. D. Grosso, F. Cagnol, G. J. A. A. Soler-Illia, E. L. Crepaldi, H. Amenitsch, A. Brunet-Bruneau, A. Bourgeois and C. Sanchez, *Adv. Funct. Mater.*, 2007, **17**, 1247.
33. C. Sanchez, C. Boissiere, D. Grosso, C. Laberty and L. Nicole, *Chem. Mater.*, 2008, **20**, 682.
34. S. Guldin, S. Huttner, M. Kolle, M. E. Welland, P. Muller-Buschbaum, R. H. Friend, U. Steiner and N. Tetreault, *Nano Lett.*, 2010, **10**, 2303.
35. S. Nishimura, N. Abrams, B. Lewis and L. Halaoui, *J. Am. Chem. Soc.*, 2003, **125**, 6306.
36. L. Halaoui, N. Abrams and T. Mallouk, *J. Phys. Chem. B*, 2005, **109**, 6334.
37. A. Mihi, M. E. Calvo, J. A. Anta and H. Mıguez, *J. Phys. Chem. C*, 2008, **112**, 13.
38. S. Colodrero, A. Mihi, L. Haggman, M. Ocana, G. Boschloo, A. Hagfeldt and H. Mıguez, *Adv. Mater.*, 2009, **21**, 764.
39. L. Manna, D. J. Lilliron, A. Meisel, E. C. Scher and A. P. Alivisatos, *Nat. Mater.*, 2003, **2**, 382.
40. H. Q. Yan, R. R. He, J. Pham and P. D. Yang, *Adv. Mater.*, 2003, **15**, 402.

41. Y. F. Hsu, Y. Y. Xi, C. T. Yip, A. B. Djurisic and W. K. Chan, *J. Appl. Phys.*, 2008, **103**, 083114.
42. W. Chen, H. F. Zhang, I. M. Hsing and S. H. Yang, *Electrochem. Commun.*, 2009, **11**, 1057.
43. M. S. Akhtar, M. A. Khan, M. S. Jeon and O. B. Yang, *Electrochim. Acta*, 2008, **53**, 7869.
44. C. F. Lin, H. Lin, J. B. Li and X. Li, *J. Alloys Compd.*, 2008, **462**, 175.
45. A. E. Suliman, Y. W. Tang and L. Xu, *Sol. Energy Mater. Sol. Cells*, 2007, **91**, 1658.
46. C. Rustomji, C. Frandsen, S. Jin and M. Tauber, *J. Phys. Chem. B*, 2010, **114**, 14537.
47. M. Vafaee and M. S. Ghamsari, *Mater. Lett.*, 2007, **61**, 3265.
48. S. Kar, A. Dev and S. Chaudhuri, *J. Phys. Chem. B*, 2006, **110**, 17848.
49. P. D. Yang, H. Q. Yan, S. Mao, R. Russo, J. Johnson, R. Saykally, N. Morris, J. Pham, R. R. He and H. J. Choi, *Adv. Funct. Mater.*, 2002, **12**, 323.
50. X. D. Wang, Y. Ding, C. J. Summers and Z. L. Wang, *J. Phys. Chem. B*, 2004, **108**, 8773.
51. Q. C. Li, V. Kumar, Y. Li, H. T. Zhang, T. J. Marks and R. P. H. Chang, *Chem. Mater.*, 2005, **17**, 1001.
52. M. Wang, C. H. Ye, Y. Zhang, G. M. Hua, H. X. Wang, M. G. Kong and L. D. Zhang, *J. Cryst. Growth*, 2006, **291**, 334.
53. H. H. Wang and C. S. Xie, *J. Cryst. Growth*, 2006, **291**, 187.
54. M. Fu, J. Zhou, Q. F. Xiao, B. Li, R. L. Zong, W. Chen and J. Zhang, *Adv. Mater.*, 2006, **18**, 1001.
55. Y. Y. Xi, Y. F. Hsu, A. B. Djurisic and W. K. Chan, *J. Electrochem. Soc.*, 2008, **155**, D595.
56. K. Nonomura, T. Yoshida, D. Schlettwein and H. Minoura, *Electrochim. Acta*, 2003, **48**, 3071.
57. Z. L. Wang, X. Y. Kong and J. M. Zuo, *Phys. Rev. Lett.*, 2003, **91**, 185502.
58. H. Yan, R. He, J. Johnson, M. Law, R. Saykally and P. Yang, *J. Am. Chem. Soc.*, 2003, **125**, 4728.

CHAPTER 5

Hierarchical Nanostructures for Fuel Cells and Fuel Reforming

NICO HOTZ

Department of Mechanical Engineering and Materials Science,
Duke University, Box 90300, Durham, NC 27708, United States
Email: nico.hotz@duke.edu

5.1 Introduction

Fuel cells are a promising technology as an alternative to combustion engines for clean and efficient electricity generation. Since fuel cells can reduce our dependence on fossil fuels, the environmental and geopolitical advantages of fuel cells are tremendous. Fuel cells can be seen as a hybrid between combustion engines and batteries in terms of their functional principle: like combustion engines, fuel cells are fed by a chemical fuel as an energy source. However, like batteries, they directly convert chemical to electrical energy, avoiding the process of combustion, which creates pollution, noise, and inefficiencies. Fuel cells, on the other hand, offer high fuel-to-electricity efficiencies and low emissions without noise pollution.[1] Furthermore, fuel cell systems are perfect for distributed power generation due to their compact and modular design.[2–5]

Polymer electrolyte membranes (PEMFCs) and solid oxide fuel cells (SOFCs) are the focus of this chapter. These fuel cell types eliminate the usage of corrosive liquids, as used in phosphoric acid, alkali, and molten carbonate fuel cells. The main topic is the application of nanostructured

RSC Nanoscience & Nanotechnology No. 35
Hierarchical Nanostructures for Energy Devices
Edited by Seung Hwan Ko and Costas P Grigoropoulos

Published by the Royal Society of Chemistry, www.rsc.org

materials in these fuel cells and in fuel reforming systems. Fuel reforming is the process of converting primary fuels, typically alcoholic or higher hydrocarbons, to a hydrogen-rich gas mixture, which can be effectively converted by fuel cells.[6–9]

5.2 Types of Fuel Cells

5.2.1 Low-temperature Polymer Electrolyte Membrane Fuel Cells

For the last couple of decades, low-temperature polymer electrolyte membrane or proton exchange membrane fuel cells (PEMFCs) have been identified as the most promising type of fuel cell, particularly for automotive applications. Some of the several examples of research initiatives for automotive fuel cell systems are the New Generation of Vehicles program and the FreedomCAR by the U. S. Department of Energy and the U. S. Council for Automotive Research (USCAR) in the United States, the Fuel Cells and Hydrogen Joint Technology Initiative (FCH) launched in 2008 by the European Union, and the Next-Generation Automobile Fuel Initiative in Japan. The major goal of these initiatives was, and still is, to enable the mass production of efficient and affordable hydrogen-powered vehicles and to establish a hydrogen-supply infrastructure.[10] PEMFCs are typically favored for automotive applications due to their quick start-up thanks to their low operating temperature, high power density, simple thermal management at low operating temperature, and fast response to transient load dynamics.

5.2.1.1 Set-up of Low-temperature Polymer Electrolyte Membrane Fuel Cells

In PEMFCs, so-called bipolar plates impermeable to the reactants separate the oxidative and the reductive half reaction. As pictured in Figure 5.1, the set-up of a basic PEMFC comprises three main components: two flow channels (often referred to as bipolar or separator plates) and a membrane–electrode assembly (MEA). The MEA is an assembly of a polymer membrane, two catalyst layers or electrodes, and two gas diffusion layers (GDLs). On the anode side of the electrolyte membrane, the half reaction

$$H_2 \rightarrow 2\,H^+ + 2\,e^- \tag{5.1}$$

takes place and on the cathode side, the half reaction

$$0.5\,O_2 + 2\,H^+ + 2\,e^- \rightarrow H_2O \tag{5.2}$$

occurs. The membrane is impermeable to gas species, however, conductive for protons. Since it is not conductive for electrons, the electrons created on the anode side are forced to flow through an external circuit to the cathode

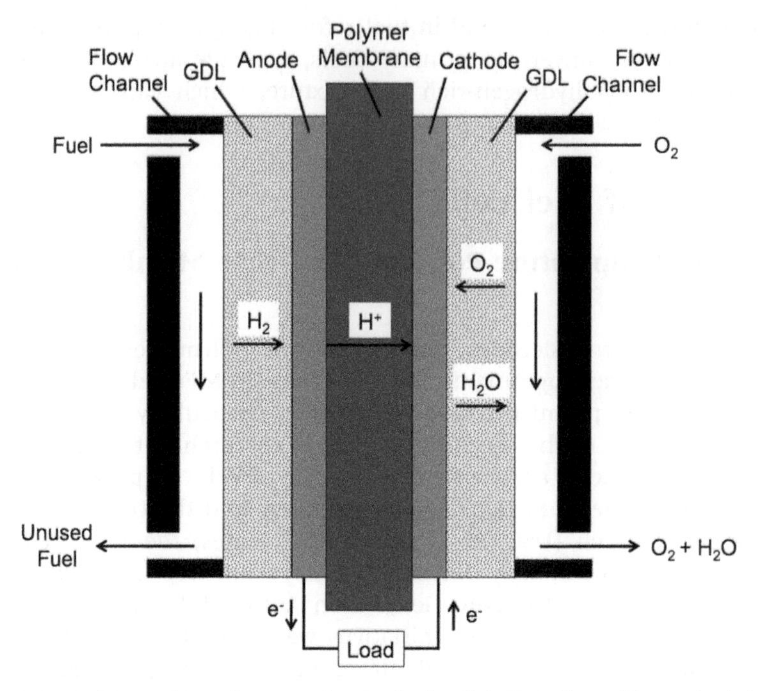

Figure 5.1 Polymer electrolyte membrane fuel cell (PEMFC) working principle.

side, thus creating an electric current and closing the circuit. The complete global reaction reads

$$H_2 + 0.5\,O_2 \rightarrow H_2O \tag{5.3}$$

The fuel (hydrogen in this case) and oxidant (typically pure oxygen or oxygen from air) is provided to the catalyst layer through the GDLs, ensuring distributed and uniform supply of both reactants. The bipolar plates typically have four functions:[11] distribution of the fuel and oxidant within the cell, facilitating water management within the cell, separation of the individual cells in the stack, and transporting current away from the cell. They facilitate the heat management as well, often assisted by additional cooling plates. By combining individual cells to stacks, the electric power generation increases, and each bipolar plate confines the anode flow channel and the cathode flow channel of two adjacent cells, respectively.

The cost of the platinum catalyst in conventional low-temperature, hydrogen-fed PEMFCs is a major obstacle for the widespread use of mass-produced fuel cells. A tremendous amount of research has been conducted to reduce the platinum loading, resulting in platinum loadings in the membrane–electrode assembly in the order of 0.6–0.8 $mg_{Pt}\ cm^{-2}$, while increasing the power density up to 0.7 $W\ cm^{-2}$ at cell voltages as high as 0.7 V at 80 °C and near-ambient pressure,[12] indicating a hydrogen-to-electricity efficiency of 58%.[13] The Pt loading-specific power densities of 0.85–1.1 $g_{Pt}\ kW^{-1}$

correspond to 72–94 g of platinum for a 85 kW fuel cell stack in a 75 kW_{net} automotive fuel cell system.[14,15]

Despite this significant reduction compared to the PEM fuel cell technology of the 1990s, a further reduction is required for commercially successful PEMFCs in automotive applications. Estimates have shown that the amount of platinum has to be further reduced five-fold, mostly due to cost[16] and limitations of Pt supply.[17] It has been demonstrated that these limitations can be met by reducing the Pt loading-specific power density below 0.2 g_{Pt} kW^{-1} at cell voltages of ≥0.65 V, keeping hydrogen-to-electricity efficiencies above 55%.[12] To achieve this, the fuel cell power density has to be improved to 0.8–0.9 W cm^{-2} at ≥0.65 V by reducing mass transport limitations at high current densities and by simultaneously decreasing the Pt-loading to 0.15 mg_{Pt} cm^{-2} without compromising cell performance.

5.2.1.2 Nanostructures in Low-temperature Polymer Electrolyte Membrane Fuel Cells

Reducing the catalyst loading while improving or at least maintaining the fuel cell performance is a very demanding challenge, which can be achieved by increasing the Pt utilization and the catalytic activity. In PEMFCs, the reaction kinetics of the oxygen reduction on the Pt catalyst of the cathode side is the rate-limiting step, creating a large overpotential for the oxygen reduction reaction (ORR).[18,19] Decreasing the overpotential of ORR is widely seen as the most critical step towards increasing the PEM fuel cell efficiency. Three approaches can lead to this goal: improve the catalyst layer structure, increase the catalytic activity for ORR, and increase the utilization of Pt by a higher dispersion of catalyst in the electrode.[12,20]

The second approach, increasing the catalytic activity of Pt for ORR, has been elaborated by many studies.[20–23] One possibility is to substitute Pt with Pt-transition metal alloys for enhanced catalytic performance.

The utilization of Pt can be enhanced by increasing the reactive surface area.[20,24] The catalyst can be deposited on highly conductive, high-surface area substrates, for example micro- and nano-structured carbon particles. Carbon black has commonly been used as a catalyst support for MEAs in PEMFCs.[20,24] By uniformly incorporating highly dispersed Pt nanoparticles into the carbon electrodes, the catalytically active surface area per mass of catalyst can be dramatically increased.

The main focus of this section is the third approach, improving the catalyst layer structure by using hierarchical nanostructures. Since the discovery of carbon nanotubes in 1991,[25] carbon nanotubes (CNTs) as well as carbon nanofibers (CNFs) have created great excitement in the nanoscience and nanotechnology community due to their unique chemical, structural, mechanical, and electromechanical properties.[26]

Compared to Vulcan XC-72R, a widely used carbon support with a conductivity of 4.0 S cm^{-1} and a specific surface area of 237 m^2 g^{-1},[27] CNTs and

CNFs have conductivities almost two orders of magnitude higher: 104 and 103–104 S cm^{-1}, respectively,[28,29] and significantly higher specific surface areas of 200–900 m^2 g^{-1}.[28] The high electric conductivity of CNTs and CNFs promises low ohmic losses when implemented in a PEMFC as a catalyst support, while the large surface area offers the opportunity to fabricate highly and uniformly dispersed catalytic sites.[28,30] Many studies have shown that the synthesis methods of Pt/CNT and Pt/CNF composites have a critical effect on morphology and catalytic activity.

Several synthesis methods for Pt on a CNT support such as impregnation deposition,[27,31] the sonochemical technique,[32] the microwave-heated polyol process,[33–37] electrodeposition,[38,39] sputter-deposition,[40,41] the γ-irradiation technique,[42] the self-regulated reduction technique of surfactants,[43,44] and the colloid method[45] have been reviewed by Lee *et al.*[46]

Especially for impregnation deposition, the CNTs require a surface oxidation treatment before the deposition of Pt. By placing the CNTs in nitric or sulfuric acid, functional groups such as hydroxyl, carboxyl, and carbonyl groups are created and allow for the successful integration of Pt nanoparticles on the CNT surfaces, acting as anchoring sites. Using this approach, it has been shown that smaller Pt nanoparticles with a more uniform distribution on the CNTs achieve higher catalytic activity than Pt/CNT without oxidation treatment.[31] A two-step sensitization-activation method appears to create even higher catalytic activity and nanoparticle dispersion.[47]

The sonochemical method further increases the number of surface functional groups by an oxidation treatment in acidic solution, leading to 30 wt% Pt in the Pt/CNT electrode.[32] γ-Irradiation can be used to establish surface functional groups without any oxidation treatment or chemical agents. This method can achieve Pt particles with an average size of 1.6–1.87 nm.[42] The self-regulated reduction of surfactants is another technique that avoids the danger of damaging the CNTs by acids.[43] Sputtering can be used for thinner layers with low Pt loading.[48]

The application of CNFs as a support for Pt catalysts in PEMFCs has been investigated less than CNTs, however, there is tremendous potential for CNFs as well. Platelet-shaped CNFs can support Pt nanoparticles of a smaller size and more homogeneously than herringbone or tubular CNFs.[49] Like CNTs, CNFs can decrease the required Pt loading.[50] Physical treatment by grinding of CNFs can lead to higher numbers of Pt ion anchoring sites, leading again to highly dispersed nanoparticles and higher catalytic activity.[21]

A first practical demonstration of a CNT-supported catalyst in a PEMFC was conducted by NEC Corporation in 2001.[51] It was said that the micro fuel cell containing a Pt/CNT catalyst had a fuel cell performance 20% higher than a conventional Pt/C catalyst.

While Pt has the high catalytic activity required for efficient fuel cell operation, it is very susceptible to CO poisoning. Small amounts of CO, usually starting at 10 ppm, will compromise and eventually deactivate the Pt catalyst. Since the vast majority of hydrogen for fuel cells is generated today by reforming alcoholic or higher hydrocarbon fuels and CO is an intrinsic by-product of any fuel reforming reaction, some CO, at least trace amounts of it, will always be present in the anode fuel gas. To decrease the Pt loading without losing catalytic performance and enhance the CO tolerance of the catalyst, especially for reformate gas mixtures containing 10–100 ppm CO, the possibility of binary and ternary catalysts has been investigated.[11]

One successful binary catalyst is Pt-Ru/C, which shows catalytic activity in the presence of 100 ppm CO comparable to that of Pt/C exposed to pure hydrogen.[52] Additionally, Pt-Ru/C is able to absorb more water, which supports the oxidation of CO. The optimum ratio of Pt and Ru was determined as approximately 50 : 50.[52] Pt-Mo/C catalysts, on the other hand, experience high tolerance towards low levels of CO (10–20 ppm) and can perform better than Pt-Ru/C catalysts.[53] Unfortunately, this beneficial effect is weakened at higher CO levels. Some binary catalysts completely avoid Pt. For example, Au-Pd catalysts have shown a three-fold increase in catalytic activity in the presence of CO compared to Pt-Ru catalysts.[54]

Many ternary catalysts are based on Pt-Ru with the addition of elements such as Ni, Pd, Co, Rh, Ir, Mn, Cr, W, Zr, and Nb. Pt-Ru-W has demonstrated better performance than pure Pt, Pt-Ru, and Pt-W individually.[55] Similarly, Pt-Ru-Mo can outperform Pt-Ru and Pt-Mo.[56] Some ternary catalysts perform well without a carbon support. Unsupported Pt-Ru-Al_4 achieves similar catalytic activity in the presence of 100 ppm CO compared to Pt-Ru/C and unsupported Pt-Re-MgH_2 performs better than Pt-Ru/C under the same conditions, with both ternary catalysts being fabricated by high-energy ball milling.[57,58]

5.2.2 High-temperature Polymer Electrolyte Membrane Fuel Cells

The great attraction of low-temperature PEMFCs during the past decades has been mainly due to their high efficiency and low environmental impact.[59] Traditionally, these PEMFCs contain polymeric membranes made of perfluorosulfonic acid-type materials such as DuPont's Nafion, which can be easily produced and are mechanically robust. Their drawbacks, however, include their humidification requirement due to water-based proton conductivity, relatively high cost, and proneness to catalyst poisoning by CO and other impurities. The humidification requirement basically limits the operating temperature to a maximum of 80 °C to avoid dry-out and power loss. Novel membrane materials are needed to elevate the operating temperature of PEMFCs, leading to faster reaction kinetics, higher tolerance to CO and other impurities, and simpler system design due to the lack of humidification needs.[60,61]

Promising alternative membrane materials are sulfonated aromatic polymers such as polyimides,[62] polysulfones,[63] polybenzoxazoles,[64] poly(ether ketones),[65–67] poly(arylene ether),[68,69] and poly(benzobisthiazole).[70] These sulfonated polymers are cheaper than fluorinated ones and recent successful studies have shown similar or even improved membrane properties, such as mechanical strength, conductivity, and water retention.[71]

The first breakthrough was achieved with high-temperature PEM fuel cell membranes made of phosphoric acid-doped polybenzimidazole (PBI), a highly stable aromatic polymer.[72] PBI membranes are usually produced by dissolving the material in a mixture of an organic solvent (*e.g.* dimethylacetamide) and inorganic salts (*e.g.* lithium chloride), applying the solution as a thin film, drying and washing the film (to remove solvent, salts, and water), and placing the film in phosphoric acid to generate a phosphoric acid-doped, robust, and highly conductive membrane.[72–77] These membranes exhibit perfect properties for PEM fuel cell membranes: high ionic conductivity at high temperature and low humidity (in the order of 0.1 S cm^{-1} at 160 °C), low electro-osmotic drag, high CO tolerance, and low gas permeability. This leads to good fuel cell performance, for example 0.35 W cm^{-2} for H_2/air and 0.55 W cm^{-2} for H_2/O_2, both at 160 °C and 0.55 V.[71] An extensive review of polybenzimidazole derivatives for high-temperature PEMFCs has been published recently.[78] Challenges related to PBI are the low phosphoric acid loading and its retention during operation, membrane durability, and low molecular weight.[71]

A novel method to overcome these issues related to PBI is the creation of a highly robust and stable, high molecular weight PBI film by using polyphosphoric acid as the polycondensation agent, polymerization solvent, and casting solvent.[79–82] However, an effective alternative replacing Nafion as the standard electrolyte for PEMFCs is still missing.

An interesting approach besides sulfonated polymers is the synthesis of organic–inorganic membranes consisting of Nafion and hygroscopic oxides, such as TiO_2,[83,84] SiO_2,[85,86] and ZrO_2[87] for PEMFCs operating at 100–150 °C.[88] The hygroscopic nanoparticles enhance water retention in the membrane and improve transport properties, thus increasing the fuel cell performance at higher temperatures.[86] In principle, there are two methods to incorporate inorganic nanoparticles in organic electrolytes: on the one hand, a mixture of the oxide nanoparticles and Nafion solution in a solvent is applied by casting as a so-called composite membrane. On the other hand, oxides can be formed *in situ* inside the polymer membrane by sol-gelation, creating a hybrid membrane.[89]

The synthesis of casting composites allows for the production of membranes with well-defined composition, volume fraction, and oxide distribution, which strongly affect the water retention and electrochemical performance of the membranes.[90]

In hybrid membranes, the oxides are mostly created by hydrolysis or condensation reactions on hydrophilic sites of the Nafion during the sol-gelation process, due to the affinity of the inorganic precursor with water.[61] It is important that the synthesis takes place at a low temperature (<100 °C) and that the oxide particle size is compatible with the ionic clusters, to ensure that the intrinsic properties of the polymer remain unaltered.

Titanium dioxide created by conventional sol-gelation, for example, has to be crystallized at higher temperatures, which affects the polymer and increases the particle size.[91,92] Another issue of hybrid membranes is their low chemical stability in acids, which can lead to the leaching out of the inorganic material due to insufficient condensation during the hydrolysis.[84,93]

Nafion-TiO_2 hybrid[94] and composite[95–99] membranes have demonstrated enhanced fuel cell performance, which is typically attributed to their higher water retention capacity thanks to the inorganic phase.[94,97–99] A significant improvement at high temperature and low humidity conditions despite an ohmic drop has been explained by more efficient water management at the electrode/electrolyte interface due to the hygroscopic properties of the oxide nanoparticles.[61] The exact mechanism leading to the improvement of the membranes, however, is not fully understood yet and the properties of organic–inorganic fuel cell membranes have to be studied further in the future.

5.2.3 Direct Alcohol Fuel Cells

The direct alcohol fuel cell (DAFC) has been used in several commercial applications, taking advantage of its ability to directly use alcohol as a fuel instead of hydrogen. In other words, the conversion of a primary fuel to hydrogen can be avoided, thus leading to potentially higher efficiencies and energy densities. Due to the membrane and catalyst materials used, DAFCs are counted as a sub-type of PEMFCs. Methanol and ethanol are the most widely used alcohols for DAFC applications[100–107] and can be generated in a renewable way from biomass. Ethanol is sometimes favored over methanol due to its low toxicity, safety, higher energy density, and simple mass production from biomass.[108–110]

The major drawbacks of DAFCs are the low catalytic activity of electrocatalysts at low operating temperatures, especially on the anode side, and alcohol crossover to the cathode, which can lead to poisoning of the cathode catalyst.[111–113] The only active and stable noble metal for alcohol oxidation, particular in an acidic environment, is Pt, however, pure Pt is prone to being poisoned by CO-like intermediates from electro-oxidation of permeated alcohol.[114] Similar to PEMFCs, a major issue of DAFCs is the high cost of Pt and an essential goal is the reduction of the required Pt loading. As seen in an earlier section of this chapter, Pt-based alloys can help reduce the amount of Pt. Additionally, support materials with a high surface area such as carbon particles increase the utilization of Pt and lower the required Pt loading.

To oxidize methanol or ethanol on Pt, the cell potential has to be in a range where adsorbed intermediates such as CO are effectively oxidized. This creates a substantial overpotential and reduces the efficiency of DAFCs. Pt-Ru binary electrocatalysts have a more negative potential and lead to a more efficient cell operation compared to Pt.[115–118] This is generally thought to be caused by their CO tolerance due to a bifunctional effect, where OH species on Ru surface atoms oxidize CO. Besides Pt-Ru, several other binary Pt alloys including Pt-Sn,[119–121] Pt-Pd,[122,123] Pt-Rh,[124] and Pt-Mo[125] have been investigated, as well as ternary catalysts.[126,127]

Palladium is a very interesting alternative for Pt, since it is an excellent catalyst for organic fuel electro-oxidation[128] and at least 50 times more abundant on the Earth than Pt.[129] The release of hydrogen contained in Pd can lead to a lower surface concentration of adsorbed CO.[130] Binary Pt-Pd catalysts experience a strong resistance towards CO poisoning from the oxidation of formic acid[131] and an enhancement of the catalytic activity of Pt-Pd compared to pure Pt has been confirmed by cyclic voltammetry and complete direct ethanol fuel cell measurements.[132] Carbon-supported Pt-Pd catalysts also have a reduced ethanol adsorption compared to Pt, leading to an oxygen reduction catalyst with higher ethanol tolerance and reducing the effect of performance loss due to fuel crossover.[133] Uniformly distributed Pt-Pd nanoparticles with a 2–4 nm diameter on a carbon support can be prepared by co-reduction of mixed oxides by a single-step alcohol reduction without stabilizers.[129] The oxidation current densities of both methanol and ethanol on Pt-Pd/C increased 2 or 3 times compared to Pt/C electrodes.

5.2.4 Solid Oxide Fuel Cells

Next to PEMFCs, solid oxide fuel cells (SOFCs) have attracted tremendous attention, especially due to their potential for stationary power generation. Applications of micro-SOFCs for portable applications have been investigated as well.[6] With their high operating temperature, SOFCs achieve high conversion efficiencies due to increased reaction kinetics, avoid the use of expensive materials, particularly for the catalyst, and can be directly operated with hydrocarbon fuels instead of hydrogen, reducing the need for fuel reforming and increasing fuel flexibility. While SOFCs offer unique advantages over PEMFCs, the high operating temperature and corresponding lengthy heat-up process are major drawbacks of SOFC technology.

5.2.4.1 Set-up of Solid Oxide Fuel Cells

The basic set-up of an SOFC MEA consists of the same three key components as PEMFCs: anode, cathode, and electrolyte.[134] However, instead of a polymeric membrane, SOFCs utilize ceramic materials, thus reducing issues of corrosion and water management. The electrolyte membrane is required to be compact, dense, and exhibit very low electrical conductivity, but allow for efficient ion transport. Comparable to the MEA of PEMFCs, the ceramic membrane is sandwiched by two porous and catalytic electrodes, where the fuel oxidation and the oxygen reduction take place.

The fundamental working principle of an SOFC is shown in Figure 5.2. Oxygen is dissociated and ionized on the cathode and these oxygen ions diffuse to the cathode/electrolyte interface and transfer through the electrolyte to the anode/electrolyte interface. From there, they diffuse to the catalytic sites of the anode, where they react with hydrogen from the fuel

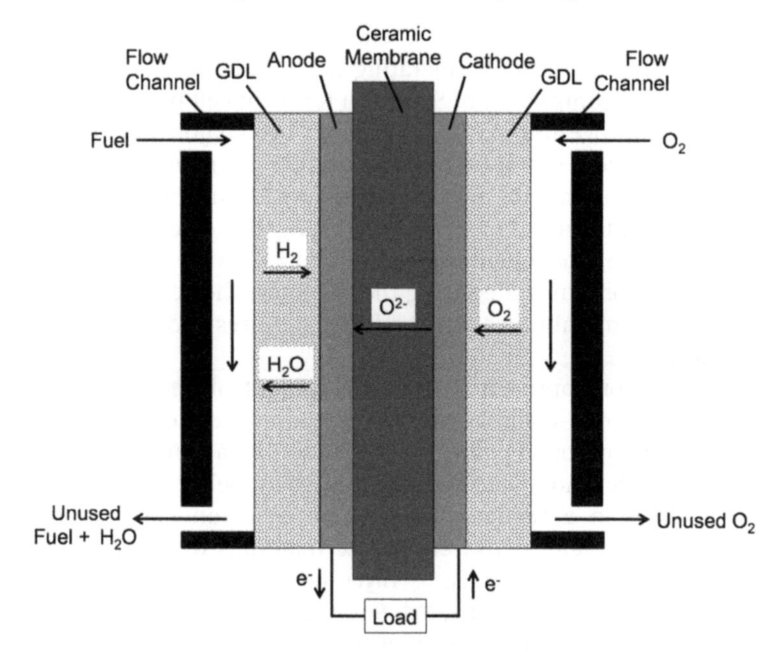

Figure 5.2 Solid oxide fuel cell (SOFC) working principle.

gas flow. As a product of this oxidation reaction, electrons are created, which travel through an external circuit to the cathode, where they are required for the oxygen dissociation reaction. The water vapor generated during the oxidation reaction diffuses through the porous anode into the gas stream and is removed from the fuel cell. Besides the generated electric power, a significant amount of heat is produced due to ionic and electronic resistance of the materials and reaction polarization. Since SOFCs are operated at high temperature, the exergy content of this heat generation is high, enabling efficient combined heat and power (CHP) generation, in contrast to low-temperature fuel cells.

5.2.4.2 Electrolyte Materials in Solid Oxide Fuel Cells

The goal of increased fuel cell performance and efficiency can be achieved by improving the catalytic activity of the electrodes and the ionic conductivity of the electrolyte. Taking advantage of the fast development of better micro- and nanofabrication methods and functional materials of the past decades, satisfactory cell performance can be achieved at lower temperature, sometimes referred to as an intermediate to high temperature range (600–1000 °C).[135] Further reduction of the SOFC temperature to intermediate temperatures (500–600 °C) or lower is required for successful commercialization of realistic SOFC applications thanks to reduced operational and installation cost and increased system reliability.[136–141] The faster start-up and shutdown for intermediate-temperature SOFCs extends the range of applications, like portable devices, thus further stimulating research endeavors targeting lower SOFC operating temperatures.[6,142,143]

The most critical component of SOFCs in terms of cell performance is the electrolyte. To improve the oxide membrane, two major approaches have been pursued in recent years: first, by decreasing the membrane thickness, the ohmic loss of the membrane is reduced, since this loss is linearly proportional to the thickness.[138,140,144] By limiting the electrolyte thickness, even to micro-scale, the operating temperature can be reduced without losing electric performance. Reducing the thickness, however, decreases the robustness of the membranes and typically requires more difficult fabrication methods.

The second major approach is the development of new electrolyte materials with high ionic conductivity to replace conventionally used yttria stabilized zirconia or doped ceria. These state-of-the-art materials are cubic fluorite oxides with inadequate ionic conductivity below 700 °C or partial mixed conduction in a reduced atmosphere or at an elevated temperature.[145,146] Promising alternative electrolyte materials discovered for SOFC applications include $Ca_{12}Al_{14}O_{33}$,[147] Bi_2O_3-based materials,[148] perovskite $LaGaO_3$,[149] $Ln_{10}(SiO_4)_6O_3$ (Ln = La, Nd, Sm, Gd and Dy),[150] and La_2Mo_2O,[151] to name only a few. The conductivity of several ion conductive materials is summarized by S. M. Haile.[1]

Unfortunately, many of these newly found materials exhibit properties hindering or even inhibiting successful application in realistic fuel cells. The refractory and highly reactive $LaGaO_3$ is difficult to be applied as thin films.[152–154] Bi_2O_3 can fail due to its phase structure change.[148] To achieve satisfactory fuel cell efficiency, the ionic conductivity should be above a critical level of 0.1 S cm^{-1}.[155] None of these novel materials has yet been able to fulfill this requirement for temperatures of 600 °C and less.[134] More research is necessary to improve the stability and ionic conductivity of known materials or to find entirely new materials.

5.2.4.3 Nanotechnology in Composite Electrolytes

Nanostructures have emerged as highly promising materials applied as functional components of advanced energy conversion and storage devices, such as lithium-ion batteries, supercapacitors, solar cells, low-temperature fuel cells, and fuel reformers.[6,9,111,156–158] Nanostructured materials have a very large surface-to-volume ratio and experience a strong effect of grain boundaries, leading to unique material characteristics that can be exploited in novel energy technologies. Like low-temperature fuel cells, high-temperature fuel cells including SOFCs can benefit significantly from the use of nano-scale structures and materials,[111,144,159–162] for example in the form of nanostructured solid conductors, sometimes referred to as 'nanoionics'.[163]

The high cell fabrication temperature of SOFCs can cause difficulties when applying nanostructures due to their instability at high temperatures. Nano-sized particles, for example, may form microstructures during high-temperature fabrication above 1000 °C.[111] By aiming for SOFC operating temperatures far below 1000 °C, the fabrication temperature can be reduced as well and, therefore, the application of nanomaterials in SOFCs is very likely to expand in the future.

Ceria nanoparticles are one interesting example of novel nano-scale materials for SOFC electrolytes. The grain growth of these nanoparticles can be suppressed by a second phase, such as in core-shelled ceria-carbonate particles, which can tolerate severe heat treatment without undesired particle growth due to the carbonate shell.[164] It has been demonstrated that such composite electrolytes consisting of two phases are able to reach impressive ionic conductivity due to unique transport phenomena at the interface between the phases.[134] By reducing the structure size to the nano-scale, this interfacial area is vastly increased, thus increasing the ionic conductivity. A larger interfacial surface area and more grain boundaries in nanostructured materials create a high density of mobile defects in the space-charge region.[111]

5.3 Fuel Reforming and Fuel Treatment for Fuel Cell Systems

5.3.1 Fuel Reforming

A very promising approach to resolve the energy situation which mankind is currently facing is the use of fossil and renewable fuel in combination with novel and more efficient hydrogen-based technologies such as fuel cells.[3,4] This benefits from the continued use of the existing fuel supply infrastructure, the exceptionally high energy density per volume and mass of liquid fuels, and at the same time the reduction of consumed fuel due to the tremendously high efficiency of fuel cells compared to combustion engines. These fuels include fossil hydrocarbon fuel as well as renewable fuels such as alcoholic and hydrocarbon biofuels (*e.g.* methanol or biodiesel). A major challenge that must be solved in order to achieve this goal is the energetically and economically efficient conversion of conventional fuels to hydrogen.

The processing of butane on catalytic nanostructures, as an example of liquid hydrocarbon fuel reforming, has been undertaken with Rh/ceria/zirconia nanoparticles by using packed beds with catalytic nanoparticles,[9] by using an improved reactor design,[8] and by applying a novel sol-gelation method.[7] Conventional ceramic foams coated with Rh have already been used for the processing of different hydrocarbons.[165–168] The idea of many of these studies is to use a small-scale fuel-to-syngas processor as part of an entire fuel cell system.[6]

Multiple studies available in the literature investigate the reforming of hydrocarbon fuels on micro- and nano-scale catalysts, both by numerical modeling and by experiments. For hydrocarbon reforming, steam reforming (SR),[169–175] partial oxidation (POX),[7–9,166,176–183] autothermal reforming (ATR),[169,180,181,184–186] and thermal cracking or pyrolysis[187,188] have been identified as possible reaction pathways to reformate fossil fuels. In a previous study, the reaction kinetics of butane POX on catalytic nanoparticles has been investigated and a simple quantitative reaction model has been presented.[189]

Steam reforming is often seen as the most promising route to reform alcoholic and higher hydrocarbon fuels because it achieves the highest hydrogen yield since large amounts of molecular hydrogen are produced both from the fuel and the water itself.[169–175] However, steam reforming is highly endothermic and a large amount of heat has to be provided to compensate for the reaction enthalpy, the heat losses due to imperfect thermal insulation, and the latent heat during evaporation of water and (possibly) fuel. In conventional reactors, the required heat is generated by partial combustion of the fuel, typically consuming between a third and half of the original fuel.

On average, SR-based reformers show the highest efficiency and hydrogen yield. Nevertheless, the endothermic nature of POX and, even more, the thermally balanced ATR lead to advantages due to a lower system complexity and simpler thermal management. Practically all fossil fuels can be considered for hydrocarbon reforming, starting from kerosene,[188] gasoline,[166,171,180,185] diesel,[174,179,188] biodiesel,[184,187,188] butane,[9,176,178] propane,[170,172,176,178,181] and compressed natural gas.[176–178,182] In general, the overall efficiency of the reformer increases with higher hydrogen-to-carbon ratio: kerosene $\approx$ diesel $<$ gasoline $<$ butane $<$ propane $<$ natural gas.

5.3.2 Removal of CO from Fuel

As described earlier, CO tolerance is a major issue for fuel cells, especially low-temperature PEMFCs. Hydrogen-rich gas mixtures generated from hydrocarbon fuels by fuel reforming reactions such as SR, POX, and ATR intrinsically contain a significant amount of CO. Reformate gas resulting from the SR of methanol, for example, contains a mole fraction of CO in the order of 1%.[3,5,157] Conventional PEMFC catalysts exhibit a significant decrease in performance when in contact with CO mole fractions as low as only several parts per million, due to the strong chemisorption of CO on the Pt catalyst.[52] Efficient long-term operation can be guaranteed for most PEMFCs with CO mole fractions below 10 or 20 ppm. As long as the CO tolerance of PEMFC catalysts has not been significantly increased, effective solutions have to be found to reduce the amount of CO in the hydrogen fuel mixture before it can be fed to the fuel cell.[158] Various approaches have been investigated in the past, such as selective or preferential CO oxidation, water–gas shift reaction (WGS), methanation, and selective diffusion. Preferential oxidation (PROX) of CO offers the advantage of the lowest cost and highest efficiency compared to other methods.[190,191]

Preferential oxidation of CO can be performed to successfully reduce the amount of residual CO at temperatures below 100 °C, which is identical to the typical operating temperature of low-temperature PEMFCs. Popular catalysts are based on gold or copper. Gold is widely known as a catalytically inert material due its highly stable structure, however, Haruta *et al.* have shown that gold is extremely active at sizes below 15 nm and in the presence of 3d transition metal oxides.[192] It has been demonstrated that there exists a $1/d^3$ relationship between particle size and CO oxidation.[193] Metal oxide supports with high oxygen mobility such as TiO_2, Fe_2O_3, NiO_x, and CoO_x enhance the catalytic activity of Au.

Conventional Au-based catalysts exhibit reduced catalytic activity in the presence of CO_2 and H_2O, typically attributed to a blockage of catalytic sites by these species. In realistic fuel reforming scenarios, significant amounts of CO_2 and H_2O are part of the reformate gas. For methanol steam reforming, as an example, approximately 25% of the gas is CO_2 and a substantial part of the water steam remains unreacted. A major challenge of CO-PROX is the creation of novel catalysts enabling close to complete CO conversion at temperatures below 100 °C with potentially large amounts of CO_2 and H_2O in the gas mixture. Furthermore, this conversion must be achieved without the addition of excess O_2 to minimize competitive oxidation of H_2 fuel.

Au/Fe_2O_3 catalyst nanoparticles calcined at 400 °C achieved 99.5% CO conversion at 100 °C reaction temperature and with excess O_2, in the presence of CO_2 and H_2O. These nanoparticles were synthesized by wet impregnation, co-precipitation, and sol-gelation.[194] Au/Fe_2O_3 nanoparticles have been shown to reach higher catalytic activity after calcination than Au/TiO_2 and Au/Al_2O_3 in the presence of H_2O at 80 °C.[195,196] Traces of water were found to increase the catalytic activity of Au/Fe_2O_3 nanostructures, but larger amounts of water required significantly higher reaction temperatures.[197] Gold nanostructures can be catalytically active at low temperatures, even down to 0 °C.[198,199] The size and dispersion of these nanocrystalline gold particles are essential parameters determining the catalytic activity towards CO oxidation.[200] A recent study demonstrated that the negative effect of CO_2 and H_2O on the catalytic activity can be compensated by creating an inverse catalyst with smaller Fe_2O_3 support particles (5–7 nm) and relatively large active Au particles (15–25 nm), achieving 99.85% CO conversion at 80 °C and in the presence of up to 10% H_2O and 25% CO_2.[158] The increased Fe_2O_3 surface area compared to conventional larger support structures significantly enhances O_2 adsorption and, therefore, CO oxidation.

References

1. S. M. Haile, *Acta Mater.*, 2003, **51**, 5981.
2. N. Hotz, *J. Sol. Energy Eng.*, 2012, **134**, 041010.
3. N. Hotz, M. T. Lee, C. P. Grigoropoulos, S. M. Senn and D. Poulikakos, *Int. J. Heat Mass Transfer*, 2006, **49**, 2397.
4. N. Hotz, S. M. Senn and D. Poulikakos, *J. Power Sources*, 2006, **158**, 333.
5. N. Hotz, R. Zimmerman, C. Weinmueller, M.-T. Lee, C. P. Grigoropoulos, G. Rosengarten and D. Poulikakos, *J. Power Sources*, 2010, **195**, 1676.

6. A. Bieberle-Huetter, D. Beckel, A. Infortuna, U. P. Muecke, J. L. M. Rupp, L. J. Gauckler, S. Rey-Mermet, P. Muralt, N. R. Bieri, N. Hotz, M. J. Stutz, D. Poulikakos, P. Heeb, P. Mueller, A. Bernard, R. Gmuer and T. Hocker, *J. Power Sources*, 2008, **177**, 123.
7. N. Hotz, N. Koc, T. Schwamb, N. C. Schirmer and D. Poulikakos, *AIChE J.*, 2009, **55**, 1849.
8. N. Hotz, N. Osterwalder, W. J. Stark, N. R. Bieri and D. Poulikakos, *Chem. Eng. Sci.*, 2008, **63**, 5193.
9. N. Hotz, M. J. Stutz, S. Loher, W. J. Stark and D. Poulikakos, *Appl. Catal., B*, 2007, **73**, 336.
10. S. G. Chalk and J. E. Miller, *J. Power Sources*, 2006, **159**, 73.
11. V. Mehta and J. S. Cooper, *J. Power Sources*, 2003, **114**, 32.
12. H. A. Gasteiger, S. S. Kocha, B. Sompalli and F. T. Wagner, *Appl. Catal., B*, 2005, **56**, 9.
13. S. S. Kocha, in *Handbook of Fuel Cells – Fundamentals, Technology and Applications*, ed. W. Vielstich, A. Lamm and H. Gasteiger, Wiley, Chichester, UK, 2003, vol. 3.
14. D. A. Masten and A. D. Bosco, in *Handbook of Fuel Cells – Fundamentals, Technology and Applications*, ed. W. Vielstich, A. Lamm and H. Gasteiger, Wiley, Chichester, UK, 2003, vol. 4.
15. A. Rodrigues, M. Fronk and B. McCormick, in *Handbook of Fuel Cells – Fundamentals, Technology and Applications*, ed. W. Vielstich, A. Lamm and H. Gasteiger, Wiley, Chichester, UK, 2003, vol. 4.
16. M. F. Mathias and H. A. Gasteiger, Proton Conducting Membrane Fuel Cells III, in *Proceedings 3rd International Symposium Proton Conducting Membrane Fuel Cells, Salt Lake City, US, 2002*, ed. M. Murthy, T. F. Fuller, J. W. Van Zee and S. Gottesfeld, Electrochemical Society, 2005, vol. 2002–31, pp. 1–24.
17. C. Jaffray and G. A. Hards, in *Handbook of Fuel Cells – Fundamentals, Technology and Applications*, ed. W. Vielstich, A. Lamm and H. Gasteiger, Wiley, Chichester, UK, 2003, vol. 3.
18. K. Kinoshita, *Electrochemical Oxygen Technology*, John Wiley & Sons, New York, US, 1992.
19. K. Kordesch and G. Simader, *Fuel Cells and Their Applications*, VCH, New York, US, 1996.
20. T. R. Ralph and M. P. Hogarth, *Platinum Met. Rev.*, 2002, **46**, 3.
21. K. Ota, A. Ishihara, S. Mitsushima, K. Lee, Y. Suzuki, N. Horibe, T. Nakagawa and N. Kamiya, *J. New Mater. Electrochem. Syst.*, 2005, **8**, 25.
22. O. Savadogo and P. Beck, *J. Electrochem. Soc.*, 1996, **143**, 3842.
23. J. Shim, C. R. Lee, H. K. Lee, J. S. Lee and E. J. Cairns, *J. Power Sources*, 2001, **102**, 172.
24. T. Tada, in *Handbook of Fuel Cells – Fundamentals, Technology and Applications*, ed. W. Vielstich, A. Lamm and H. Gasteiger, Wiley, Chichester, UK, 2003, vol. 3.
25. S. Iijima, *Nature*, 1991, **354**, 56.
26. H. J. Dai, *Surf. Sci.*, 2002, **500**, 218.

27. W. Z. Li, C. H. Liang, W. J. Zhou, J. S. Qiu, Z. H. Zhou, G. Q. Sun and Q. Xin, *J. Phys. Chem. B*, 2003, **107**, 6292.
28. P. Serp, M. Corrias and P. Kalck, *Appl. Catal., A*, 2003, **253**, 337.
29. A. Thess, R. Lee, P. Nikolaev, H. J. Dai, P. Petit, J. Robert, C. H. Xu, Y. H. Lee, S. G. Kim, A. G. Rinzler, D. T. Colbert, G. E. Scuseria, D. Tomanek, J. E. Fischer and R. E. Smalley, *Science*, 1996, **273**, 483.
30. N. M. Rodriguez, *J. Mater. Res.*, 1993, **8**, 3233.
31. N. Rajalakshmi, H. Ryu, M. M. Shaijumon and S. Ramaprabhu, *J. Power Sources*, 2005, **140**, 250.
32. Y. C. Xing, *J. Phys. Chem. B*, 2004, **108**, 19255.
33. W. X. Chen, J. Y. Lee and Z. L. Liu, *Chem. Commun.*, 2002, **21**, 2588.
34. W. X. Chen, J. Zhao, J. Y. Lee and Z. L. Liu, *Mater. Chem. Phys.*, 2005, **91**, 124.
35. Z. L. Liu, L. M. Gan, L. Hong, W. X. Chen and J. Y. Lee, *J. Power Sources*, 2005, **139**, 73.
36. Z. L. Liu, J. Y. Lee, W. X. Chen, M. Han and L. M. Gan, *Langmuir*, 2004, **20**, 181.
37. B. Yang, Q. Y. Lu, Y. Wang, L. Zhuang, J. T. Lu and P. F. Liu, *Chem. Mater.*, 2003, **15**, 3552.
38. H. S. Kim, N. P. Subramanian and B. N. Popov, *J. Power Sources*, 2004, **138**, 14.
39. S. D. Thompson, L. R. Jordan and M. Forsyth, *Electrochim. Acta*, 2001, **46**, 1657.
40. S. Hirano, J. Kim and S. Srinivasan, *Electrochim. Acta*, 1997, **42**, 1587.
41. S. Mukerjee, S. Srinivasan and A. J. Appleby, *Electrochim. Acta*, 1993, **38**, 1661.
42. S. D. Oh, B. K. So, S. H. Choi, A. Gopalan, K. P. Lee, K. R. Yoon and I. S. Choi, *Mater. Lett.*, 2005, **59**, 1121.
43. C. L. Lee, Y. C. Ju, P. T. Chou, Y. C. Huang, L. C. Kuo and J. C. Oung, *Electrochem. Commun.*, 2005, 7, 453.
44. C. L. Lee, C. C. Wan and Y. Y. Wang, *Adv. Funct. Mater.*, 2001, **11**, 344.
45. T. Yoshitake, Y. Shimakawa, S. Kuroshima, H. Kimura, T. Ichihashi, Y. Kubo, D. Kasuya, K. Takahashi, F. Kokai, M. Yudasaka and S. Iijima, *Physica B*, 2002, **323**, 124.
46. K. Lee, J. J. Zhang, H. J. Wang and D. P. Wilkinson, *J. Appl. Electrochem.*, 2006, **36**, 507.
47. Z. L. Liu, X. H. Lin, J. Y. Lee, W. Zhang, M. Han and L. M. Gan, *Langmuir*, 2002, **18**, 4054.
48. C. C. Chen, C. F. Chen, C. H. Hsu and I. H. Li, *Diamond Relat. Mater.*, 2005, **14**, 770.
49. K. Sasaki, K. Shinya, S. Tanaka, A. Furukawa, K. Ando, T. Kuroki, H. Kusaba and Y. Teraoka, 11th FCDIC Fuel Cell Symposium, Tokyo, Japan, Fuel Cell Development Information Center, 2004, pp. 239–242.
50. K. Sasaki, K. Shinya, S. Tanaka, A. Furukawa, K. Ando, T. Kuroki, H. Kusaba and Y. Teraoka, 206th Electrochemical Society Meeting, Honolulu, US, 2004, in *Proceedings 4th International Symposium Proton*

Conducting Membrane Fuel Cells, Honolulu, US, 2004, ed. M. Murthy, K. Ota, J. W. Van Zee, S. R. Narayanan and E. S. Takeuchi, Electrochemical Society, 2006, vol. 2004–21, pp. 159–170.
51. G. Weaver, *World Fuel Cells – An Industry Profile with Market Prospects to 2010*, Elsevier Science, Oxford, UK, 2002.
52. M. Iwase and S. Kawatsu, Proton Conducting Membrane Fuel Cells I, in *Proceedings 1st International Symposium Proton Conducting Membrane Fuel Cells, Chicago, US, 1995*, ed. S. Gottesfeld, G. Halpert and A. Landgrebe, Electrochemical Society, 1995, vol. 95–23, pp. 12–23.
53. J. Bauman, T. Zawodzinski, T. Rockward, P. Haridoss, F. Uribe and S. Gottesfeld, Proton Conducting Membrane Fuel Cells II, in *Proceedings 2nd International Symposium Proton Conducting Membrane Fuel Cells, Boston, US, 1998*, ed. S. Gottesfeld and T. F. Fuller, Electrochemical Society, 1999, vol. 98–27, pp. 200–208.
54. P. N. Ross, N. M. Markovic, T. J. Schmidt and V. Stamenkovic, *New Electrocatalysts for Fuel Cells*, Transportation Fuel Cell Power System, 2000 Annual Progress Report, US Department of Energy, 2000.
55. G. L. Holleck, D. M. Pasquariello and S. L. Clauson, Proton Conducting Membrane Fuel Cells II, in *Proceedings 2nd International Symposium Proton Conducting Membrane Fuel Cells, Boston, US, 1998*, ed. S. Gottesfeld and T. F. Fuller, Electrochemical Society, 1999, vol. 98–27, pp. 150–161.
56. A. L. N. Pinheiro, A. Oliveira-Neto, E. C. de Souza, J. Perez, V. A. Paganin, E. A. Ticianelli and E. R. Gonzalez, *J. New Mater. Electrochem. Syst.*, 2003, **6**, 1.
57. M. C. Denis, P. Gouerec, D. Guay, J. P. Dodelet, G. Lalande and R. Schulz, *J. Appl. Electrochem.*, 2000, **30**, 1243.
58. M. C. Denis, G. Lalande, D. Guay, J. P. Dodelet and R. Schulz, *J. Appl. Electrochem.*, 1999, **29**, 951.
59. L. Carrette, K. A. Friedrich and U. Stimming, *ChemPhysChem*, 2000, **1**, 162.
60. J. Lobato, P. Canizares, M. A. Rodrigo, J. J. Linares and J. A. Aguilar, *J. Membr. Sci.*, 2007, **306**, 47.
61. E. I. Santiago, R. A. Isidoro, M. A. Dresch, B. R. Matos, M. Linardi and F. C. Fonseca, *Electrochim. Acta*, 2009, **54**, 4111.
62. T. Watari, J. H. Fang, K. Tanaka, H. Kita, K. Okamoto and T. Hirano, *J. Membr. Sci.*, 2004, **230**, 111.
63. F. Lufrano, I. Gatto, P. Staiti, V. Antonucci and E. Passalacqua, *Solid State Ionics*, 2001, **145**, 47.
64. Y. J. Kim, B. R. Einsla, C. N. Tchatchoua and J. E. McGrath, *High Perform. Polym.*, 2005, **17**, 377.
65. Y. Gao, G. P. Robertson, M. D. Guiver, S. D. Mikhailenko, X. Li and S. Kaliaguine, *Macromolecules*, 2004, **37**, 6748.
66. M. Gil, X. L. Ji, X. F. Li, H. Na, J. E. Hampsey and Y. F. Lu, *J. Membr. Sci.*, 2004, **234**, 75.

67. X. G. Jin, M. T. Bishop, T. S. Ellis and F. E. Karasz, *Br. Polym. J.*, 1985, **17**, 4.
68. M. A. Hickner, H. Ghassemi, Y. S. Kim, B. R. Einsla and J. E. McGrath, *Chem. Rev.*, 2004, **104**, 4587.
69. G. Y. Xiao, G. M. Sun, D. Y. Yan, P. F. Zhu and P. Tao, *Polymer*, 2002, **43**, 5335.
70. S. G. Kim, D. A. Cameron, Y. K. Lee, J. R. Reynolds and C. R. Savage, *J. Polym. Sci., Part A: Polym. Chem.*, 1996, **34**, 481.
71. J. A. Mader and B. C. Benicewicz, *Macromolecules*, 2010, **43**, 6706.
72. J. S. Wainright, J. T. Wang, D. Weng, R. F. Savinell and M. Litt, *J. Electrochem. Soc.*, 1995, **142**, L121.
73. R. Bouchet and E. Siebert, *Solid State Ionics*, 1999, **118**, 287.
74. J. Kerres, A. Ullrich, F. Meier and T. Haring, *Solid State Ionics*, 1999, **125**, 243.
75. Q. F. Li, R. H. He, J. A. Gao, J. O. Jensen and N. J. Bjerrum, *J. Electrochem. Soc.*, 2003, **150**, A1599.
76. Y. L. Ma, J. S. Wainright, M. H. Litt and R. F. Savinell, *J. Electrochem. Soc.*, 2004, **151**, A8.
77. S. R. Samms, S. Wasmus and R. F. Savinell, *J. Electrochem. Soc.*, 1996, **143**, 1225.
78. J. Mader, L. Xiao, T. J. Schmidt and B. C. Benicewicz, in *Fuel Cells II*, ed. G. G. Scherer, Springer, Berlin, Germany, 2008, vol. 216, pp. 63–124.
79. G. Qian and B. C. Benicewicz, *J. Polym. Sci., Part A: Polym. Chem.*, 2009, **47**, 4064.
80. G. Qian, D. W. Smith, Jr. and B. C. Benicewicz, *Polymer*, 2009, **50**, 3911.
81. L. Xiao, H. Zhang, T. Jana, E. Scanlon, R. Chen, E. W. Choe, L. S. Ramanathan, S. Yu and B. C. Benicewicz, *Fuel Cells*, 2005, **5**, 287.
82. S. Yu, L. Xiao and B. C. Benicewicz, *Fuel Cells*, 2008, **8**, 165.
83. A. Sacca, A. Carbone, E. Passalacqua, A. D'Epifanio, S. Licoccia, E. Traversa, E. Sala, F. Traini and R. Ornelas, *J. Power Sources*, 2005, **152**, 16.
84. H. Uchida, Y. Ueno, H. Hagihara and M. Watanabe, *J. Electrochem. Soc.*, 2003, **150**, A57.
85. K. T. Adjemian, S. J. Lee, S. Srinivasan, J. Benziger and A. B. Bocarsly, *J. Electrochem. Soc.*, 2002, **149**, A256.
86. A. K. Sahu, G. Selvarani, S. Pitchumani, P. Sridhar and A. K. Shukla, *J. Electrochem. Soc.*, 2007, **154**, B123.
87. N. H. Jalani, K. Dunn and R. Datta, *Electrochim. Acta*, 2005, **51**, 553.
88. Q. F. Li, R. H. He, J. O. Jensen and N. J. Bjerrum, *Chem. Mater.*, 2003, **15**, 4896.
89. D. J. Jones and J. Roziere, in *Handbook of Fuel Cells – Fundamentals, Technology and Applications*, ed. W. Vielstich, H. A. Gasteiger and A. Lamm, Wiley, Chichester, UK, 2003, vol. 3.
90. G. Alberti and M. Casciola, *Annu. Rev. Mater. Res.*, 2003, **33**, 129.
91. J. P. Hsu and A. Nacu, *Langmuir*, 2003, **19**, 4448.

92. J. Livage, M. Henry and C. Sanchez, *Prog. Solid State Chem.*, 1988, **18**, 259.
93. H. Hagihara, H. Uchida and M. Watanabe, *Electrochim. Acta*, 2006, **51**, 3979.
94. K. T. Adjemian, R. Dominey, L. Krishnan, H. Ota, P. Majsztrik, T. Zhang, J. Mann, B. Kirby, L. Gatto, M. Velo-Simpson, J. Leahy, S. Srinivasant, J. B. Benziger and A. B. Bocarsly, *Chem. Mater.*, 2006, **18**, 2238.
95. V. Baglio, A. S. Arico, A. Di Blasi, V. Antonucci, P. L. Antonucci, S. Licoccia, E. Traversa and F. S. Fiory, *Electrochim. Acta*, 2005, **50**, 1241.
96. V. Baglio, A. Di Blasi, A. S. Arico, V. Antonucci, P. L. Antonucci, F. S. Fiory, S. Licoccia and E. Traversa, *J. New Mater. Electrochem. Syst.*, 2004, 7, 275.
97. E. Chalkova, M. V. Fedkin, D. J. Wesolowski and S. N. Lvov, *J. Electrochem. Soc.*, 2005, **152**, A1742.
98. E. Chalkova, M. B. Pague, M. V. Fedkin, D. J. Wesolowski and S. N. Lvov, *J. Electrochem. Soc.*, 2005, **152**, A1035.
99. B. R. Matos, E. I. Santiago, F. C. Fonseca, M. Linardi, V. Lavayen, R. G. Lacerda, L. O. Ladeira and A. S. Ferlauto, *J. Electrochem. Soc.*, 2007, **154**, B1358.
100. E. A. Batista, G. R. P. Malpass, A. J. Motheo and T. Iwasita, *J. Electroanal. Chem.*, 2004, **571**, 273.
101. C. Lamy, S. Rousseau, E. M. Belgsir, C. Coutanceau and J. M. Leger, *Electrochim. Acta*, 2004, **49**, 3901.
102. H. S. Liu, C. J. Song, L. Zhang, J. J. Zhang, H. J. Wang and D. P. Wilkinson, *J. Power Sources*, 2006, **155**, 95.
103. F. Vigier, C. Coutanceau, F. Hahn, E. M. Belgsir and C. Lamy, *J. Electroanal. Chem.*, 2004, **563**, 81.
104. F. Vigier, C. Coutanceau, A. Perrard, E. M. Belgsir and C. Lamy, *J. Appl. Electrochem.*, 2004, **34**, 439.
105. S. Wasmus and A. Kuver, *J. Electroanal. Chem.*, 1999, **461**, 14.
106. X. H. Xia, H. D. Liess and T. Iwasita, *J. Electroanal. Chem.*, 1997, **437**, 233.
107. C. Weinmueller, G. Tautschnig, N. Hotz and D. Poulikakos, *J. Power Sources*, 2010, **195**, 3849.
108. C. Coutanceau, S. Brimaud, C. Lamy, J. M. Leger, L. Dubau, S. Rousseau and F. Vigier, *Electrochim. Acta*, 2008, **53**, 6865.
109. F. Delime, J. M. Leger and C. Lamy, *J. Appl. Electrochem.*, 1999, **29**, 1249.
110. J. T. Wang, S. Wasmus and R. F. Savinell, *J. Electrochem. Soc.*, 1995, **142**, 4218.
111. A. S. Arico, P. Bruce, B. Scrosati, J. M. Tarascon and W. Van Schalkwijk, *Nat. Mater.*, 2005, **4**, 366.
112. V. Baglio, A. Stassi, A. Di Blasi, C. D'Urso, V. Antonucci and A. S. Arico, *Electrochim. Acta*, 2007, **53**, 1360.
113. C. Lamy, E. M. Belgsir and J. M. Leger, *J. Appl. Electrochem.*, 2001, **31**, 799.

114. C. Lamy, A. Lima, V. LeRhun, F. Delime, C. Coutanceau and J. M. Leger, *J. Power Sources*, 2002, **105**, 283.
115. N. Jha, A. L. M. Reddy, M. M. Shaijumon, N. Rajalakshmi and S. Ramaprabhu, *Int. J. Hydrogen Energy*, 2008, **33**, 427.
116. J.-J. Jow, S.-W. Yang, H.-R. Chen, M.-S. Wu, T.-R. Ling and T.-Y. Wei, *Int. J. Hydrogen Energy*, 2009, **34**, 665.
117. S. G. Lemos, R. T. S. Oliveira, M. C. Santos, P. A. P. Nascente, L. O. S. Bulhoes and E. C. Pereira, *J. Power Sources*, 2007, **163**, 695.
118. E. Spinace, A. O. Neto and M. Linardi, *J. Power Sources*, 2004, **129**, 121.
119. P. Ferrin, A. U. Nilekar, J. Greeley, M. Mavrikakis and J. Rossmeisl, *Surf. Sci.*, 2008, **602**, 3424.
120. T. Frelink, W. Visscher and J. A. R. Vanveen, *Electrochim. Acta*, 1994, **39**, 1871.
121. D. M. Han, Z. P. Guo, R. Zeng, C. J. Kim, Y. Z. Meng and H. K. Liu, *Int. J. Hydrogen Energy*, 2009, **34**, 2426.
122. F. Kadirgan, B. Beden, J. M. Leger and C. Lamy, *J. Electroanal. Chem.*, 1981, **125**, 89.
123. H. Wang, C. Xu, F. Cheng, M. Zhang, S. Wang and S. P. Jiang, *Electrochem. Commun.*, 2008, **10**, 1575.
124. D. F. A. Koch, D. A. J. Rand and R. Woods, *J. Electroanal. Chem.*, 1976, **70**, 73.
125. L. C. Ordonez, P. Roquero, P. J. Sebastian and J. Ramirez, *Int. J. Hydrogen Energy*, 2007, **32**, 3147.
126. F. Kadirgan, A. M. Kannan, T. Atilan, S. Beyhan, S. S. Ozenler, S. Suzer and A. Yorur, *Int. J. Hydrogen Energy*, 2009, **34**, 9450.
127. N. M. Markovic and P. N. Ross, *Surf. Sci. Rep.*, 2002, **45**, 121.
128. K. Machida and M. Enyo, *J. Electrochem. Soc.*, 1987, **134**, 1472.
129. F. Kadirgan, S. Beyhan and T. Atilan, *Int. J. Hydrogen Energy*, 2009, **34**, 4312.
130. O. Yepez and B. R. Scharifker, *J. Appl. Electrochem.*, 1999, **29**, 1185.
131. G. Q. Lu, A. Crown and A. Wieckowski, *J. Phys. Chem. B*, 1999, **103**, 9700.
132. W. J. Zhou, Z. H. Zhou, S. Q. Song, W. Z. Li, G. Q. Sun, P. Tsiakaras and Q. Xin, *Appl. Catal., B*, 2003, **46**, 273.
133. T. Lopes, E. Antolini and E. R. Gonzalez, *Int. J. Hydrogen Energy*, 2008, **33**, 5563.
134. L. D. Fan, C. Y. Wang, M. M. Chen and B. Zhu, *J. Power Sources*, 2013, **234**, 154.
135. Z. P. Shao and S. M. Haile, *Nature*, 2004, **431**, 170.
136. S. deSouza, S. J. Visco and L. C. DeJonghe, *Solid State Ionics*, 1997, **98**, 57.
137. R. Doshi, V. L. Richards, J. D. Carter, X. P. Wang and M. Krumpelt, *J. Electrochem. Soc.*, 1999, **146**, 1273.
138. H. Huang, M. Nakamura, P. Su, R. Fasching, Y. Saito and F. B. Prinz, *J. Electrochem. Soc.*, 2007, **154**, B20.
139. B. C. H. Steele, *Solid State Ionics*, 2000, **129**, 95.

140. M. Tsuchiya, B.-K. Lai and S. Ramanathan, *Nat. Nanotechnol.*, 2011, **6**, 282.
141. E. D. Wachsman and K. T. Lee, *Science*, 2011, **334**, 935.
142. B.-K. Lai, K. Kerman and S. Ramanathan, *J. Power Sources*, 2011, **196**, 6299.
143. Y. Takagi, B.-K. Lai, K. Kerman and S. Ramanathan, *Energy Environ. Sci.*, 2011, **4**, 3473.
144. P.-C. Su, C.-C. Chao, J. H. Shim, R. Fasching and F. B. Prinz, *Nano Lett.*, 2008, **8**, 2289.
145. A. Atkinson, *Solid State Ionics*, 1997, **95**, 249.
146. S. P. S. Badwal, F. T. Ciacchi and J. Drennan, *Solid State Ionics*, 1999, **121**, 253.
147. M. Lacerda, J. T. S. Irvine, F. P. Glasser and A. R. West, *Nature*, 1988, **332**, 525.
148. T. Takahashi, T. Esaka and H. Iwahara, *J. Appl. Electrochem.*, 1977, 7, 303.
149. T. Ishihara, H. Matsuda and Y. Takita, *J. Am. Chem. Soc.*, 1994, **116**, 3801.
150. S. Nakayama, H. Aono and Y. Sadaoka, *Chem. Lett.*, 1995, **6**, 431.
151. P. Lacorre, F. Goutenoire, O. Bohnke, R. Retoux and Y. Laligant, *Nature*, 2000, **404**, 856.
152. P. N. Huang, A. Horky and A. Petric, *J. Am. Ceram. Soc.*, 1999, **82**, 2402.
153. A. Matraszek, L. Singheiser, D. Kobertz, K. Hilpert, M. Miller, O. Schulz and M. Martin, *Solid State Ionics*, 2004, **166**, 343.
154. A. L. Shaula, V. V. Kharton and F. M. B. Marques, *J. Eur. Ceram. Soc.*, 2004, **24**, 2631.
155. T. H. Etsell and S. N. Flengas, *Chem. Rev.*, 1970, **70**, 339.
156. C. Xia, L. Li, Y. Tian, Q. Liu, Y. Zhao, L. Jia and Y. Li, *J. Power Sources*, 2009, **188**, 156.
157. M.-T. Lee, M. Werhahn, D. J. Hwang, N. Hotz, R. Greif, D. Poulikakos and C. P. Grigoropoulos, *Int. J. Hydrogen Energy*, 2010, **35**, 118.
158. T. Shodiya, O. Schmidt, W. Peng and N. Hotz, *J. Catal.*, 2013, **300**, 63.
159. M. G. Bellino, D. G. Lamas and N. E. W. de Reca, *Adv. Funct. Mater.*, 2006, **16**, 107.
160. C.-C. Chao, Y. B. Kim and F. B. Prinz, *Nano Lett.*, 2009, **9**, 3626.
161. Y. Liu, S. W. Zha and M. L. Liu, *Adv. Mater.*, 2004, **16**, 256.
162. T. Z. Sholklapper, H. Kurokawa, C. P. Jacobson, S. J. Visco and L. C. De Jonghe, *Nano Lett.*, 2007, 7, 2136.
163. J. Maier, *Nat. Mater.*, 2005, **4**, 805.
164. X. Wang, Y. Ma, R. Raza, M. Muhammed and B. Zhu, *Electrochem. Commun.*, 2008, **10**, 1617.
165. G. J. Panuccio, B. J. Dreyer and L. D. Schmidt, *AIChE J.*, 2007, **53**, 187.
166. G. J. Panuccio, K. A. Williams and L. D. Schmidt, *Chem. Eng. Sci.*, 2006, **61**, 4207.
167. K. A. Williams, R. Horn and L. D. Schmidt, *AIChE J.*, 2007, **53**, 2097.

168. R. Horn, K. A. Williams, N. J. Degenstein, A. Bitsch-Larsen, D. D. Nogare, S. A. Tupy and L. D. Schmidt, *J. Catal.*, 2007, **249**, 380.
169. A. Cutillo, S. Specchia, M. Antonini, G. Saracco and V. Specchia, *J. Power Sources*, 2006, **154**, 379.
170. M. Dokupil, C. Spitta, J. Mathiak, P. Beckhaus and A. Heinzel, *J. Power Sources*, 2006, **157**, 906.
171. G. Kolb, T. Baier, J. Schuerer, D. Tiemann, A. Ziogas, H. Ehwald and P. Alphonse, *Chem. Eng. J.*, 2008, **137**, 653.
172. G. Kolb, R. Zapf, V. Hessel and H. Lowe, *Appl. Catal., A*, 2004, **277**, 155.
173. T. Rampe, A. Heinzel and B. Vogel, *J. Power Sources*, 2000, **86**, 536.
174. J. Thormann, P. Pfeifer, K. Schubert and U. Kunz, *Chem. Eng. J.*, 2008, **135**, S74.
175. X. Wang and R. J. Gorte, *Appl. Catal., A*, 2002, **224**, 209.
176. A. S. Bodke, S. S. Bharadwaj and L. D. Schmidt, *J. Catal.*, 1998, **179**, 138.
177. D. A. Hickman and L. D. Schmidt, *Science*, 1993, **259**, 343.
178. M. Huff, P. M. Torniainen and L. D. Schmidt, *Catal. Today*, 1994, **21**, 113.
179. B. Lindstrom, J. A. J. Karlsson, P. Ekdunge, L. De Verdier, B. Haggendal, J. Dawody, M. Nilsson and L. J. Pettersson, *Int. J. Hydrogen Energy*, 2009, **34**, 3367.
180. C. Severin, S. Pischinger and J. Ogrzewalla, *J. Power Sources*, 2005, **145**, 675.
181. B. Silberova, H. J. Venvik, J. C. Walmsley and A. Holmen, *Catal. Today*, 2005, **100**, 457.
182. S. Specchia, G. Negro, G. Saracco and V. Specchia, *Appl. Catal., B*, 2007, **70**, 525.
183. K. A. Williams and L. D. Schmidt, *Appl. Catal., A*, 2006, **299**, 30.
184. G. J. Kraaij, S. Specchia, G. Bollito, L. Mutri and D. Wails, *Int. J. Hydrogen Energy*, 2009, **34**, 4495.
185. A. D. Qi, S. D. Wang, G. Z. Fu and D. Y. Wu, *Appl. Catal., A*, 2005, **293**, 71.
186. V. Recupero, L. Pino, A. Vita, F. Cipiti, M. Cordaro and M. Lagana, *Int. J. Hydrogen Energy*, 2005, **30**, 963.
187. E. E. Iojoiu, M. E. Domine, T. Davidian, N. Guilhaume and C. Mirodatos, *Appl. Catal., A*, 2007, **323**, 147.
188. A. Pastore and E. Mastorakos, *Fuel*, 2011, **90**, 64.
189. J. von Rickenbach, M. Nabavi, I. Zinovik, N. Hotz and D. Poulikakos, *Int. J. Hydrogen Energy*, 2011, **36**, 12238.
190. M. J. Kahlich, H. A. Gasteiger and R. J. Behm, *J. Catal.*, 1999, **182**, 430.
191. E. C. Njagi, H. C. Genuino, C. K. King'ondu, C. H. Chen, D. Horvath and S. L. Suib, *Int. J. Hydrogen Energy*, 2011, **36**, 6768.
192. M. Haruta, N. Yamada, T. Kobayashi and S. Iijima, *J. Catal.*, 1989, **115**, 301.
193. N. Lopez, T. V. W. Janssens, B. S. Clausen, Y. Xu, M. Mavrikakis, T. Bligaard and J. K. Norskov, *J. Catal.*, 2004, **223**, 232.

194. G. Avgouropoulos, T. Ioannides, C. Papadopoulou, J. Batista, S. Hocevar and H. K. Matralis, *Catal. Today*, 2002, **75**, 157.
195. P. Landon, J. Ferguson, B. E. Solsona, T. Garcia, S. Al-Sayari, A. F. Carley, A. A. Herzing, C. J. Kiely, M. Makkee, J. A. Moulijn, A. Overweg, S. E. Golunski and G. J. Hutchings, *J. Mater. Chem.*, 2006, **16**, 199.
196. P. Landon, J. Ferguson, B. E. Solsona, T. Garcia, A. F. Carley, A. A. Herzing, C. J. Kiely, S. E. Golunski and G. J. Hutchings, *Chem. Commun.*, 2005, 3385.
197. S. T. Daniells, M. Makkee and J. A. Moulijn, *Catal. Lett.*, 2005, **100**, 39.
198. M. Okumura, S. Nakamura, S. Tsubota, T. Nakamura, M. Azuma and M. Haruta, *Catal. Lett.*, 1998, **51**, 53.
199. R. M. T. Sanchez, A. Ueda, K. Tanaka and M. Haruta, *J. Catal.*, 1997, **168**, 125.
200. S. Carrettin, P. Concepcion, A. Corma, J. M. L. Nieto and V. F. Puntes, *Angew. Chem., Int. Ed.*, 2004, **43**, 2538.

CHAPTER 6

Thermoelectric Materials and Devices

CHANYOUNG KANG,[a] HONGCHAO WANG,[a] JE-HYEONG BAHK,[b] HOON KIM[a] AND WOOCHUL KIM*[a]

[a] School of Mechanical Engineering, Yonsei University, Seoul, Republic of Korea; [b] Birck Nanotechnology Center, Purdue University, West Lafayette, Indiana 47907, U.S.A.
*Email: woochul@yonsei.ac.kr

6.1 Introduction

Thermoelectric effects are the physical principles that are used to directly convert heat into electricity or *vice versa* based on charge carrier and phonon transport phenomena in a solid.[1–8] When a temperature gradient is imposed on a solid, an electric potential is created due to the redistribution of charge carriers in the material. This phenomenon is called the Seebeck effect. The reverse phenomenon, in which a temperature difference is created across a material by an electric current due to the lattice cooling or heating at the interfaces, is called the Peltier effect. These two effects can be used in various power generation and refrigeration applications *via* the thermal-to-electric (or *vice versa*) energy conversion.

Thermoelectric effects were first discovered in the early 1820s and developed quickly through the 1850s. Their applications in thermoelectric devices (power generation and refrigeration) were recognized during this time period.[9] In the 1950s, there were many research activities on the use of thermoelectric power generation for spacecraft. Prominent achievements among these were the projects led by NASA to power spacecraft using

RSC Nanoscience & Nanotechnology No. 35
Hierarchical Nanostructures for Energy Devices
Edited by Seung Hwan Ko and Costas P Grigoropoulos

Published by the Royal Society of Chemistry, www.rsc.org

radioisotope thermoelectric generators (RTG).[1] NASA devoted its efforts to developing highly reliable and robust RTGs, and successfully used a series of RTGs in their spacecraft without any failures for more than 10 years. A thermoelectric generator is highly robust because it is a solid-state device that has no moving parts. The pioneering efforts by NASA and others in the 1950s led to the discoveries of many important thermoelectric materials such as Bi_2Te_3, PbTe and SiGe, all of which are still widely used.[9] These bulk thermoelectric materials are semiconductor alloys that have high electrical conductivities by sufficient doping, and possess inherently low lattice thermal conductivities. Typically, semiconductors are found to be more efficient thermoelectric materials than insulators or metals.[6] We will discuss the reason for this later in this chapter.

Thermoelectric refrigeration drew renewed attention as an alternative method of refrigeration in the 1990s, when the eco-toxicity of refrigerant fluids used in conventional refrigerators was revealed.[9] Thermoelectric devices have many advantages as alternative refrigerators; they have high power density and do not have moving parts, so their operation is quiet and robust. A critical advantage is that they do not require a toxic or environmentally harmful refrigerant.[10]

In the past few decades, both thermoelectric refrigeration and power generation have been explored extensively for various potential applications both in academia and industry. The most promising application is waste heat recovery. Thermoelectric devices can generate electrical power even from a small temperature gradient, whereas conventional energy conversion devices such as turbines or combustion engines are not suitable to produce electricity under a low temperature gradient. Therefore, thermoelectric generators can be used to harvest energy from low-quality heat sources such as waste heat.[11] Thermoelectric refrigerators are widely used for cooling small solid-state devices as well these days. Controlling the hot spots in a CPU and cooling optoelectronic laser modules are representative applications for thermoelectric refrigerators.[12]

Although thermoelectric devices have many advantages over conventional energy conversion systems, their main weakness comes from their low energy conversion efficiency.[10] Significant efforts have been made throughout the 20th century to enhance the thermoelectric efficiency.[13] Figure 6.1(a) shows several important studies conducted on thermoelectric materials with high figures of merit in each year for various temperature ranges. The dimensionless thermoelectric figure of merit *ZT*, which represents the efficiency of a thermoelectric material, is defined as

$$ZT = \frac{S^2 \sigma T}{k} \tag{6.1}$$

where S, σ, and k are the Seebeck coefficient, electrical conductivity, thermal conductivity of the material used, respectively, and T is the absolute temperature. The Seebeck coefficient is defined as the ratio of the induced electric, E, field to the temperature gradient within the solid, which is equal to the induced voltage, V, across the material divided by the temperature difference when there is no current flow, such that

$$S = \frac{E}{\nabla T} = -\frac{dV}{dx}\frac{dx}{dT} = -\frac{dV}{dT} \tag{6.2}$$

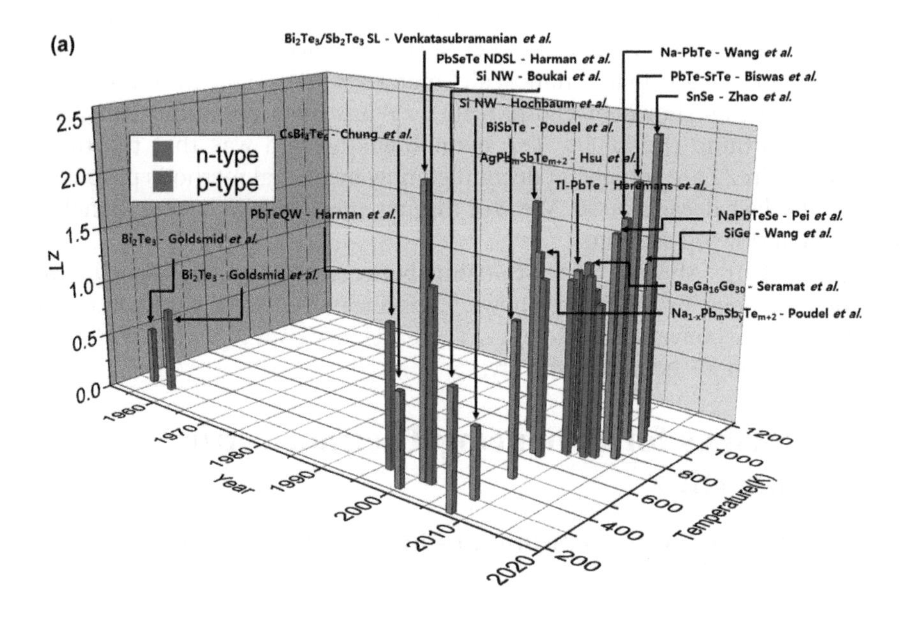

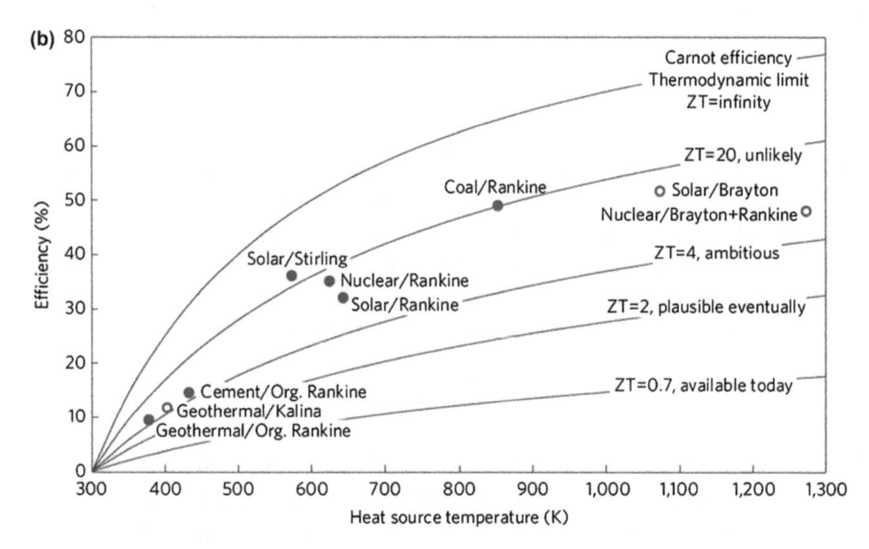

Figure 6.1 (a) Some remarkable studies of thermoelectric materials. Reprinted with permission from Ref. 13. Copyright © 2010 WILEY-VCH Verlag GmbH & Co. KGaA, Weinheim. (b) Efficiency of mechanical heat engines compared with thermoelectric figures of merit. Reprinted with permission from Ref. 11. Copyright © 2009 Nature Publishing Group.[11,13]

As shown in Figure 6.1(a), there had been no significant improvement in *ZT* for a long time until the last few decades. Improving ZT is challenging, because the individual properties in the definition of *ZT* are mutually coupled, so that it is hard to enhance one property without affecting the others unfavourably. In 1993, however, Hicks and Dresselhaus reported the possibility of enhancing the thermoelectric effect by using low-dimensional materials.[14,15] Since then, a great number of studies have been reported for the enhancement of *ZT* using nanoscale materials and nanostructures.[16–21] Consequently, as shown in Figure 6.1(a), *ZT* has been enhanced considerably since around the 2000s. The main breakthrough was that the three parameters in *ZT* are somewhat decoupled now by several methods proposed during the last decade or so. We will discuss these methods of enhancing *ZT* in the following sections.

Figure 6.1(b)[11] shows the relationship between *ZT* and the energy conversion efficiency. As the figure shows, *ZT* is a representative parameter that determines the conversion efficiency. The efficiency converges to the Carnot efficiency when *ZT* becomes infinity.[11]

6.2 Schemes for Enhancing Thermoelectric Properties

6.2.1 Three Coupled Parameters in the Thermoelectric Figure of Merit

Before describing the enhancement of *ZT* by using nanostructures, we will explain why improving *ZT* can be a difficult task. The electrical conductivity, σ, is given as

$$\sigma = nq\mu \tag{6.3}$$

where n, q and μ are carrier concentration, electron charge unit and carrier mobility, respectively. According to this relation, electrical conductivity is proportional to carrier concentration. Electron mobility changes only slightly with carrier concentration in most semiconductors. Therefore electrical conductivity generally increases with carrier concentration. However, the Seebeck coefficient generally follows the rule given by[22]

$$S \approx -\frac{1}{qT}(E_{\text{ave}} - E_{\text{F}}) \tag{6.4}$$

where E_{ave} and E_{F} are the average energy of free charge carriers and Fermi energy, respectively. Figure 6.2(a) illustrates the band structures and electron densities in metals, semiconductors and insulators. The electrical conductivity is proportional to the carrier concentration (orange area) and decreases exponentially as the Fermi energy moves away from the band edge. At the same time, the Seebeck coefficient increases because $E_{\text{ave}} - E_{\text{F}}$ increases (eqn (6.4)). Thus, the Seebeck coefficient and electrical conductivity have opposite tendencies with carrier concentration.[23] Figure 6.2(b) shows the relationship

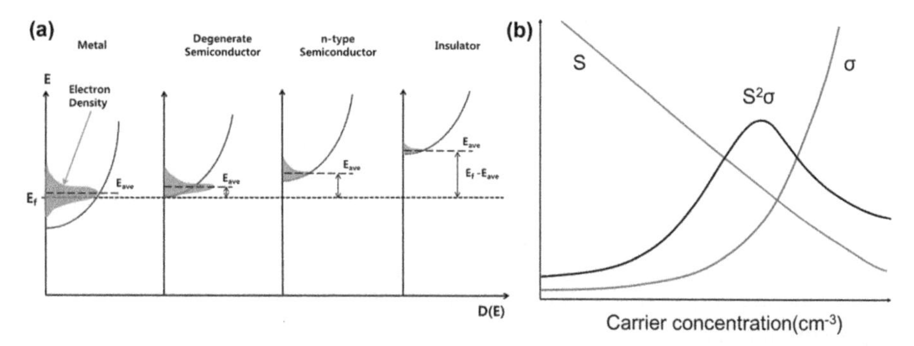

Figure 6.2 (a) Band diagrams of metal, semiconductor and insulator. Electrical conductivity increases as electron density increases. Seebeck coefficient increases when $E_{ave} - E_f$ increases.[22,23] (b) Relationship between Seebeck coefficient and electrical conductivity, given as carrier concentration. Reprinted by permission from Ref. 23. Copyright © 2008 Nature Publishing Group.

between the power factor (which is defined as $S^2\sigma$) and carrier concentration in a typical bulk semiconductor. As the carrier concentration increases, the electrical conductivity generally increases. On the other hand, the Seebeck coefficient decreases as the carrier concentration increases. As one can see in Figure 6.2(b), there is an optimum carrier concentration that maximizes the power factor.

The thermal conductivity consists of two contributions from electrons, *i.e.*, electronic thermal conductivity (k_e) and from phonons or lattice vibrations, *i.e.*, lattice thermal conductivity (k_l):[24]

$$k = k_e + k_l \tag{6.5}$$

The lattice contribution of thermal conductivity is more or less independent of the carrier concentration. The electron contribution is typically negligibly small at low carrier concentrations, but increases as the carrier concentration increases, although it is known that typically the lattice thermal conductivity dominates over the electronic thermal conductivity in semiconductors.[25] Thus, the total thermal conductivity increases as the carrier concentration increases. Therefore, there exists an optimum carrier concentration that maximizes *ZT* in bulk materials.

6.2.2 Power Factor Enhancement

From the 1950s to 1960s, the only way to control the power factor was doping.[1,8] As one can see in Figure 6.2(b), there is an optimum *ZT* value because the Seebeck coefficient and electrical conductivity have different trends with carrier concentration, so it is important to control the carrier concentration close to the optimal value, which is a function of temperature, and varies among different materials.

6.2.2.1 Quantum Confinement Effect

As we discussed in the previous section, the three parameters, particularly the power factor, determining ZT are interdependent. To uncouple the three parameters, it has been proposed[14,15] that as a dimensionality of a material changes, so does the electron density of states, which offers excellent potential for increasing the Seebeck coefficient without affecting the electrical conductivity too much (see Figure 6.3). This is due to the confinement of electron waves in the direction of reduced dimensionality. Thus, the electron states become discrete in that direction.[3] The Seebeck coefficient can be approximated as[26]

$$\begin{aligned} S &= \frac{\pi^2}{3}\frac{k_B}{q}k_BT\left\{\frac{d[\ln(\sigma(E))]}{dE}\right\}_{E=E_F} \\ &= \frac{\pi^2}{3}\frac{k_B}{q}k_BT\left\{\frac{1}{n}\frac{dn(E)}{dE}+\frac{1}{\mu}\frac{d\mu(E)}{dE}\right\}_{E=E_F} \end{aligned} \tag{6.6}$$

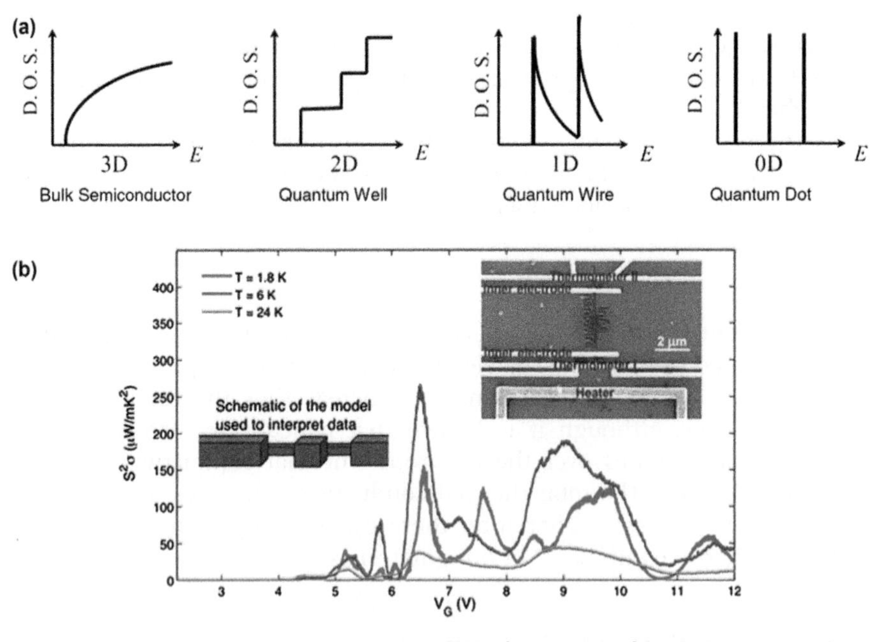

Figure 6.3 (a) Electronic density of states for (left to right) a bulk 3D crystalline semiconductor, a 2D quantum well, a 1D nanowire or nanotube and a 0D quantum dot. Reprinted with permission from Ref. 3. Copyright © 2007 WILEY-VCH Verlag GmbH & Co. KGaA, Weinheim. (b) Power factor of the InAs nanowire at various temperatures *versus* gate voltage. Inset shows SEM image of experimental setup of the demonstration of the quantum confinement effect in an InAs nanowire. Reprinted with permission from Ref. 27. Copyright © 2013 American Chemical Society.

where k_B is the Boltzmann constant. As we can see in eqn (6.6), the Seebeck coefficient is proportional to the energy derivative of the differential conductivity, $\sigma(E)$, at the Fermi energy E_F. The differential conductivity is defined as[2]

$$\sigma(E) = q^2\tau(E)\iint v_x^2 \mathrm{d}k_y \mathrm{d}k_z \cong q^2\tau(E)\bar{v}_x^2(E)D(E) \tag{6.7}$$

where τ is the momentum dependent relaxation time, k is the wavevector, v is the carrier velocity and $D(E)$ is the density of state. Then electrical conductivity can be written as

$$\sigma = \int_E \sigma(E)dE \tag{6.8}$$

Here, $n(E)$ is the energy-dependent electron density, *i.e.* $n(E) = D(E)\,f(E)$. Further, $D(E)$ is the density of states, $f(E)$ is the Fermi–Dirac distribution, q is the electron charge and $\mu(E)$ is the carrier mobility. We can rewrite eqn (6.6) using $D(E)$ and $f(E)$ as

$$S = \frac{\pi^2}{3}\frac{k_B}{q}k_BT\left\{\frac{1}{n}\left(D(E)\frac{df(E)}{dE} + f(E)\frac{dD(E)}{dE}\right) + \frac{1}{\mu}\frac{d\mu(E)}{dE}\right\}_{E=E_F} \tag{6.9}$$

We can confirm from eqn (6.9) that the Seebeck coefficient increases when the energy derivative of the electron density of states ($dD(E)/dE$) at the Fermi level increases. Low-dimensional materials have discrete, sharp densities of states due to the so-called quantum confinement effect (see Figure 6.3(a)). Thus, if we control the Fermi level so as to locate it near a sharp peak of the density of states, we can enhance the Seebeck coefficient dramatically without reducing electrical conductivity too much. In order to achieve this quantum confinement effect in 1D or 2D materials, however, a very small confinement direction size, typically less than 20 nm, and a very low temperature are required. Careful doping control is also necessary to locate the Fermi level near a sharp DOS peak, which could be realized by gate-induced charge density modulation. Wu *et al.*[27] demonstrated large power factor enhancement in relatively thick InAs nanowires with 50–70 nm diameters at temperatures below 20 K, which they attributed to the resonance effect by quantum dot states in the nanowires, not to the 1D quantum confinement effect.

6.2.2.2 *Electron Filtering*

Several strategies can be used to enhance a material's thermoelectric properties, including the energy filtering effect.[28–31] An energy barrier appears at a hetero-junction between different semiconductors, and between a semiconductor and a metal (Figure 6.4(a)).[32] This barrier acts as a filter for charge carriers. While high-energy electrons can cross this barrier, electrons with energies lower than the barrier height cannot pass. The high-energy electron transport is required to be a thermionic or quasi-thermionic emission in

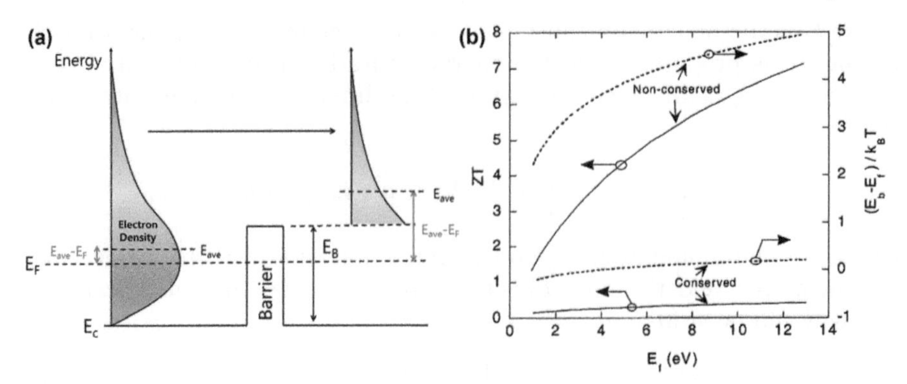

Figure 6.4 (a) Schematic of energy filtering effect. Energy filtering increases $E_{ave} - E_F$, so it can enhance the Seebeck coefficient dramatically. Reprinted with permission from Ref. 100. Copyright © 2002 American Institute of Physics. (b) Figure of merit (left axis) and optimal barrier height (right axis) *versus* Fermi level for a metallic superlattice with conserved and non-conserved lateral momentum.[31,32] Reprinted with permission from Ref. 31. Copyright © 2004 American Institute of Physics.

order to achieve the filtering effect. A large number of scatterings within the layers may force the filtered electron distribution back to the equilibrium, so that the filtering effect is relaxed. This filtering effect can enhance the Seebeck coefficient dramatically without decreasing the electrical conductivity too much because it can yield a large $E_{ave} - E_F$ (Figure 6.4(a)). Using multiple layers of thermionic emission was suggested to achieve this electron filtering effect.[31,33] Figure 6.4(b) shows figure of merit (left axis) and optimal barrier height (right axis) *versus* Fermi level for a metallic superlattice with conserved and non-conserved lateral (toward the barrier) momentum. It has been reported[23] that if the Fermi energy is aligned at an appropriate level with a sufficiently large carrier density, and if the lateral momentum conservation is broken at the interfaces, *ZT* could be enhanced beyond 7.

However, Kim *et al.*[34] recently pointed out that the enhancement due to the lateral momentum non-conservation could be modest because the smallest number of modes in the well and barrier layers limits the emission current over the barrier. Later, Bahk *et al.*[35] proposed the use of distributed resonant scatterings to partially realize the non-planar electron energy filtering effect in bulk materials, which is not limited by the lateral momentum conservation as in the planar filtering.

6.2.2.3 Band Engineering (Resonant State, Band Convergence etc*)*

Recently, there have been many remarkable studies on the enhancement of the power factor.[26,36] Band engineering by resonant impurities is one of the most remarkable results.[26] Figure 6.5(a) shows a schematic diagram of the distortion of the electronic density of states by Tl impurities in PbTe.

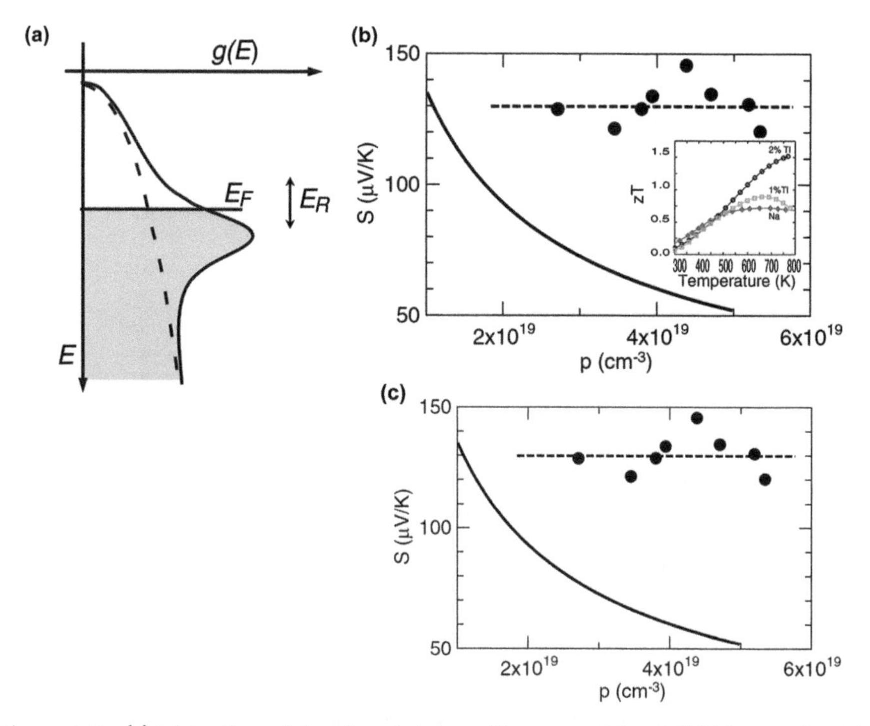

Figure 6.5 (a) Distortion of density of states with resonant level. (b) Figure of merit for 2 wt% Tl-PbTe. (c) Pisarenko plot of Tl-PbTe. Solid black line is calculated; solid black dots are experimental data.

Thallium produces resonant levels inside the valence band of PbTe. Special impurities, *e.g.*, thallium in PbTe, can create resonant states inside the valence or conduction bands in the host material. Then, these resonant states can distort the electronic density of states to achieve a large differential density of states $[dD(E)/dE]$ at the Fermi level. If the Fermi level is located on the rapidly rising side of the distorted density of states, according to eqn (6.9) we can increase the Seebeck coefficient without significantly decreasing the electrical conductivity.[26] Figure 6.5 (b) and (c) show the figure of merit and Pisarenko plot for this material, respectively. Doping PbTe with 2 wt% Tl yielded a ZT of ~1.5 at 773 K.[26] The Pisarenko plot shows the relationship between the carrier concentration and the Seebeck coefficient. As mentioned above, the Seebeck coefficient generally decreases with increasing carrier concentration (Figure 6.2). However, in Tl-PbTe, the Seebeck coefficient does not decrease as the carrier concentration increases, but instead stays high at ~150 $\mu V\ K^{-1}$ even at very high carrier concentrations because of the resonant level.

In 2011, a ZT of ~1.8 was reported for Na-doped $PbTe_{1-x}Se_x$.[36] In PbTe systems, there are two valence bands, the L band and Σ band. In the alloys with Se, $PbTe_{1-x}Se_x$, when the temperature increases, the band maximum of

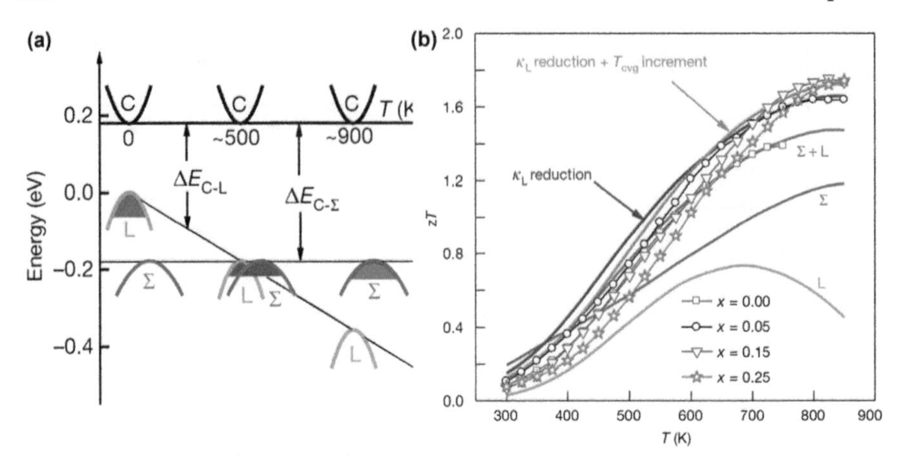

Figure 6.6 (a) Schematic of band convergence in $Na_{0.02}Pb_{0.98}Se_{0.15}Te_{0.85}$. (b) Figure of merit of $Na_{0.02}Pb_{0.98}Se_xTe_{1-x}$ *versus* selenium concentration x. Reprinted with permission from Ref. 36. Copyright © 2008 Nature Publishing Group.

the L band increases, but that of the Σ band remains the same because it is temperature-independent and depends only on the selenium concentration. Therefore, the two bands become close and finally converge with each other at certain Se compositions at 500–600 K (Figure 6.6(a)). The resulting converged band has a high valley degeneracy of 16, which stays high at higher temperatures with a 12 valley degeneracy of the Σ valley itself. This material has yielded the ZT at ~1.8 at 850 K without any nanostructures (Figure 6.6(b)).[36] The lattice thermal conductivity was also reduced due to the increased point defect scattering and alloy scattering of phonons. Many other studies have also shown that a large degeneracy in thermoelectric materials is desirable for power factor enhancement.[1,4,10]

6.2.3 Conventional Approaches to Lower Thermal Conductivity (Alloys)

Traditionally, alloying is a way of achieving low thermal conductivity in crystalline solids.[37] The typical thermal conductivity of an alloy is 5–10 W/m-K at room temperature.[37] Figure 6.7(a) shows the thermal conductivity of Si_xGe_{1-x} as a function of Si content in Ge.[38] As one can see, the thermal conductivity decreases dramatically when Si and Ge form an alloy compared to the values for pure Si and Ge. It has been difficult to reduce the thermal conductivity below the alloy while maintaining crystallinity of a material—this is called the alloy limit. However, it has been demonstrated that the alloy limit can be beaten using nanostructures.[16,18,39,40] Alloys possess low thermal conductivity because phonon scattering occurs due to atomic substitutions. This is known as alloy scattering (Figure 6.7(b)). In particular, atomic-scale defects scatter high-frequency phonons effectively.[41–43]

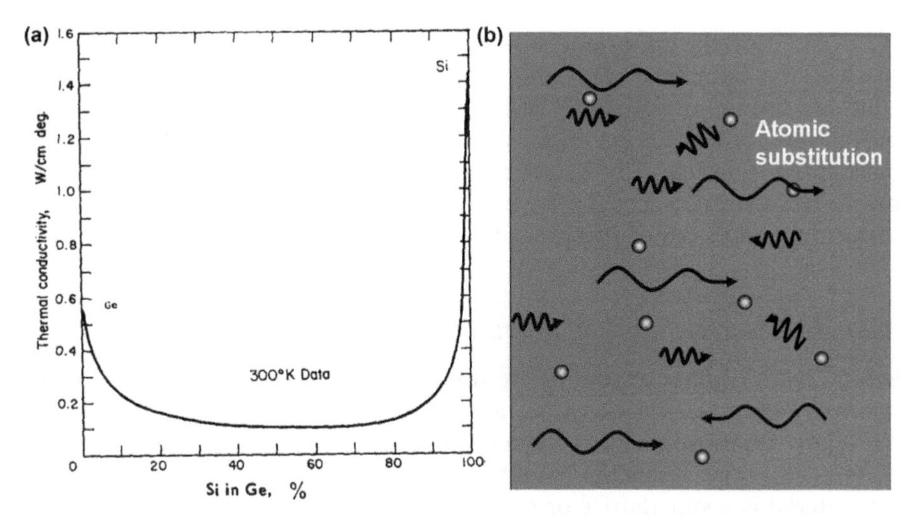

Figure 6.7 (a) Thermal conductivity of an Si_xGe_{1-x} alloy. Thermal conductivity dramatically decreases when Si and Ge form an alloy. Reprinted with permission from Ref. 38. Copyright © 1967 Elsevier. (b) Schematic of alloy scattering. Atomic substitutions scatter short-wavelength phonons.[37]

6.2.4 Thermal Conductivity Reduction by Nanostructures

The earlier experimental demonstration of nanostructured materials for thermoelectrics exhibited both reduced thermal conductivity and reduced power factor.[16] The increased surface-to-volume ratio by nanostructures increases the phonon scattering at the interfaces, but it also increases the electron scattering at the same time. However, it is still possible to reduce the lattice thermal conductivity without affecting the electral transport too much, if one could find a material such that the electron mean free path is shorter than the phonon mean free path.[6] In general, the phonon mean free path is generally on the order of ~100 nm, but the electron mean free path is on the order of ~10 nm.[37,44] Therefore, if the system dimensions are maintained in between the electron and the phonon mean free paths, phonons can be effectively scattered to reduce the thermal conductivity, while the transport of electrons is not significantly affected.

The phonon (lattice) thermal conductivity is written as[2]

$$k_p = \frac{1}{3}C(\omega)v_p(\omega)l(\omega)d\omega \tag{6.10}$$

where C, v_p and l are the specific heat, phonon group velocity and mean free path, respectively, all of which are functions of the phonon frequency, ω. The effective mean free path is determined by various scattering mechanisms and is written as[45,46]

$$l_{eff}^{-1} = l_B^{-1} + l_U^{-1} + l_N^{-1} + l_A^{-1} + l_D^{-1} + l_{e-ph}^{-1} \tag{6.11}$$

The terms on the right-hand side of eqn (6.11) represent, respectively from the left, the boundary scattering, Umklapp scattering, normal scattering, alloy scattering, nanoparticle scattering and electron–phonon scattering mean free path. This equation shows that the shortest mean free path mechanism dominates over other mean free paths. Nanostructured thermoelectric materials usually have reduced mean free paths by boundary scattering and/or nanoparticle scattering.

6.2.4.1 *Interface and/or Boundary Scattering*

Several types of nanostructures are used to reduce the lattice thermal conductivity. They can be classified as either 1-dimensional (1D), 2D or 3D materials. A nanowire is a 1D material with a nanometre-scale diameter, so phonons are scattered at the boundaries of the nanowires.[17] A representative 2D material is a superlattice or multi-layers. When material A is wetted and adheres to material B epitaxially, a two-dimensional (2D) film growth results and a traditional superlattice (SL) may be created if films A and B are alternated in a periodic manner.[47] Each layer is a few nanometres thick, and the layers are stacked periodically. Phonons are scattered at the interfaces between the two layers, which results in a dramatic decrease of the thermal conductivity.[16,18,39,48]

Li *et al.*[40] reported reduced thermal conductivities in individual Si nanowires grown by the vapor–liquid–solid method. The thermal conductivities of Si nanowires decrease as the diameter is reduced (Figure 6.8, black squares). The reduction in thermal conductivity is caused by the decreasing boundary size. The phonon mean free path for boundary scattering becomes

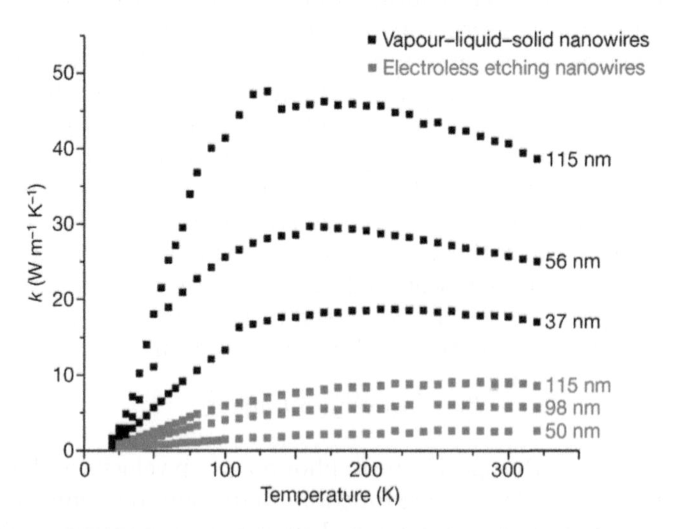

Figure 6.8 Thermal conductivities of rough and smooth Si nanowires. Reprinted with permission from Ref. 19. Copyright © 2008 Nature Publishing Group.

shorter as the diameter decreases. In 2008, Hochbaum *et al.*[19] observed extremely low thermal conductivities in silicon nanowires with rough surfaces, where they used the electroless etching (EE) method for the synthesis of the rough nanowires. The thermal conductivity was reduced almost fourfold in the wires of the same diameter at room temperature, compared to the vapor–liquid–solid nanowires[19] (Figure 6.8, red squares).

Two-dimensional superlattices can also have low thermal conductivities due to the increased interface scattering. Venkatasubramanian *et al.*[49] reported a ZT of $\sim$2.4 at room temperature with Bi_2Te_3/Sb_2Te_3 superlattices. They obtained a cross-plane lattice thermal conductivity of 0.22 W/m-K, which is only 50% of that of a $Bi_{2-x}Sb_xTe_3$ bulk alloy. To minimize reduction in power factor while achieving more reduction in thermal conductivity, the superlattice period was carefully chosen to be larger than the electron mean free path, but smaller than the phonon mean free path.

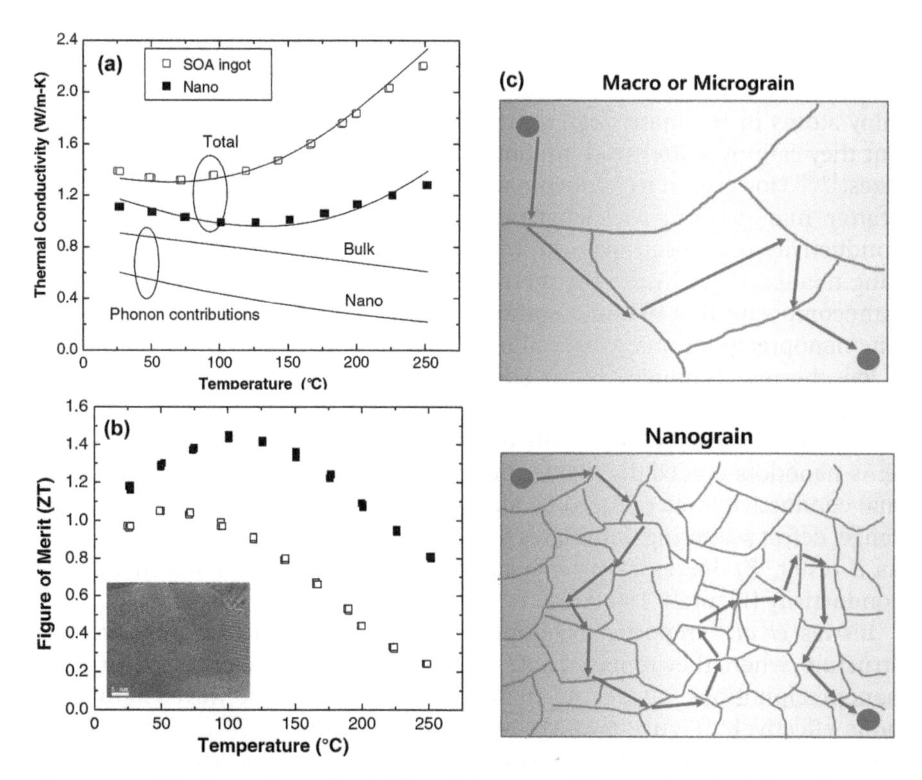

Figure 6.9 (a) Thermal conductivities of the state-of-the-art (SOA) BiSbTe alloy and nanograined BiSbTe. Solid lines are calculated phonon contributions of bulk and nanograin composite. (b) Figures of merit of SOA (white squares) and nanograined BiSbTe alloy (black squares). Inset shows a scanning electron microscope image of nanograin BiSbTe. Reprinted with permission from Ref. 21. Copyright © 2008 American Association for the Advancement of Science. (c) Schematic of phonon grain boundary scattering in normal polycrystal and nanograins.[21]

In 3D materials, the thermal conductivity can be reduced by increasing grain boundary scattering. To achieve this, the grain size should be kept small. In Figure 6.9(a), the thermal conductivity of a BiSbTe nanograin composite was reduced over the entire temperature range compared to that of a normal crystalline state-of-the-art (SOA) BiSbTe alloy. Black lines below the symbols in Figure 6.9(a) represent the calculated lattice thermal conductivities of nano and bulk BiSbTe. As a result of this, Poudel *et al.*[21] reported the BiSbTe nanograin composite that yielded a ZT of $\sim$1.4 at 373 K (Figure 6.9(b)). Nanostructured BiSbTe has a lower thermal conductivity than the bulk material due to the fact that nanograins increased the phonon scattering (Figure 6.9(c)). Small grains reduce the phonon mean free path by increasing grain-boundary scattering, which results in an enhancement of the thermoelectric figure of merit.[21]

6.2.4.2 Nanosized Defect Scattering

Figure 6.10(a) shows the mechanism by which nanoparticles reduce the thermal conductivity. The aforementioned atomic substitutions such as alloy atoms in the matrix can scatter short-wavelength phonons effectively, but they cannot scatter mid- to long-wavelength phonons due to their small sizes.[45,50] However, if we embed nanosize defects inside the matrix, they can scatter mid- to long-wavelength phonons effectively. Thus, the thermal conductivity is reduced over the wide phonon wavelength range by embedding nanoparticles inside the matrix. Hsu *et al.*[20] reported a $AgPb_mSbTe_{2+m}$ nanocomposite that exhibits epitaxial nanoprecipitation inside the matrix. The nanoprecipitations scatter phonons effectively, so that the material has a low thermal conductivity. In addition, it also has a large power factor, which resulted in a maximum ZT of $\sim$2.2 at 800 K. Kim *et al.*[46] demonstrated the reduced thermal conductivity for InGaAs semiconductors by embedding ErAs nanodots epitaxially inside the matrix (Figure 6.10(b) inset). The thermal conductivity was reduced by almost a factor of two. In this temperature range, defect scattering dominates the thermal conductivity (Figure 6.10 (b)).[43] As a result, ZT increases by a factor of two owing to the reduced thermal conductivity (Figure 6.10(c)).

Biswas *et al.*[51] reported Na-doped PbTe-SrTe with all-scale hierarchical structure, where they reported a ZT of $\sim$2.2 at 950 K. PbTe-SrTe has intrinsic nanoprecipitations inside the matrix, which scatter mid-wavelength phonons effectively (Figure 6.11(a)). Moreover, they produced grains of a few micrometres in size by spark plasma sintering (SPS). These small grains can scatter the long-wavelength phonons, and intrinsic atomic substitutions in the alloy scatter short-wavelength phonons. By using this all-scale hierarchical structure, phonons in a wide range of wavelengths can be effectively scattered. Interestingly, Biswas *et al.*[51] observed a rich concentration of Na at the interfaces between the matrix and the precipitation (Figure 6.11(b)). As temperature increases, sodium atoms that were located at the interface diffuse into the bulk region and become electrically active, donating charge carriers to the matrix.

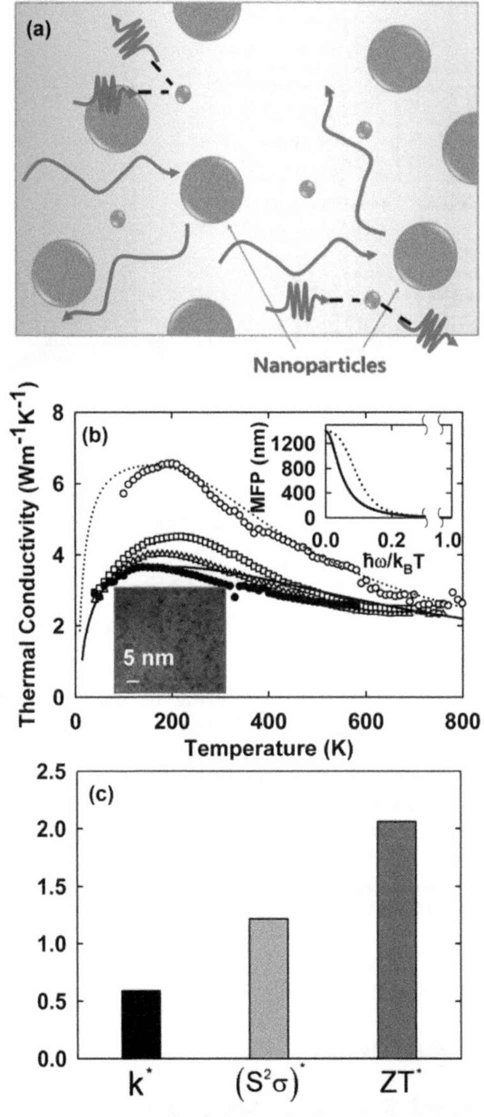

Figure 6.10 (a) Schematic of thermal conductivity reduction by nanoparticle scattering. (b) Thermal conductivity of $In_{0.53}Ga_{0.47}As$ with randomly distributed embedded ErAs nanoparticles (solid circles). Thermal conductivity of an $In_{0.53}Ga_{0.47}As$ alloy (open circles), 0.4 ML ErAs/$In_{0.53}Ga_{0.47}As$ superlattice with a 40 nm period thickness (open squares) and 0.1 ML ErAs/$In_{0.53}Ga_{0.47}As$ superlattice with a 10 nm period thickness (open triangles). Dotted and solid lines represent theoretical calculations. The lower inset shows a transmission electron microscope (TEM) image of randomly distributed ErAs/$In_{0.53}Ga_{0.47}As$. The upper inset shows the phonon mean free path (MFP) *versus* normalized frequency at 300 K. (c) Thermoelectric figure of merit at 300 K compared with that of $In_{0.53}Ga_{0.47}As$ alloy.[46]

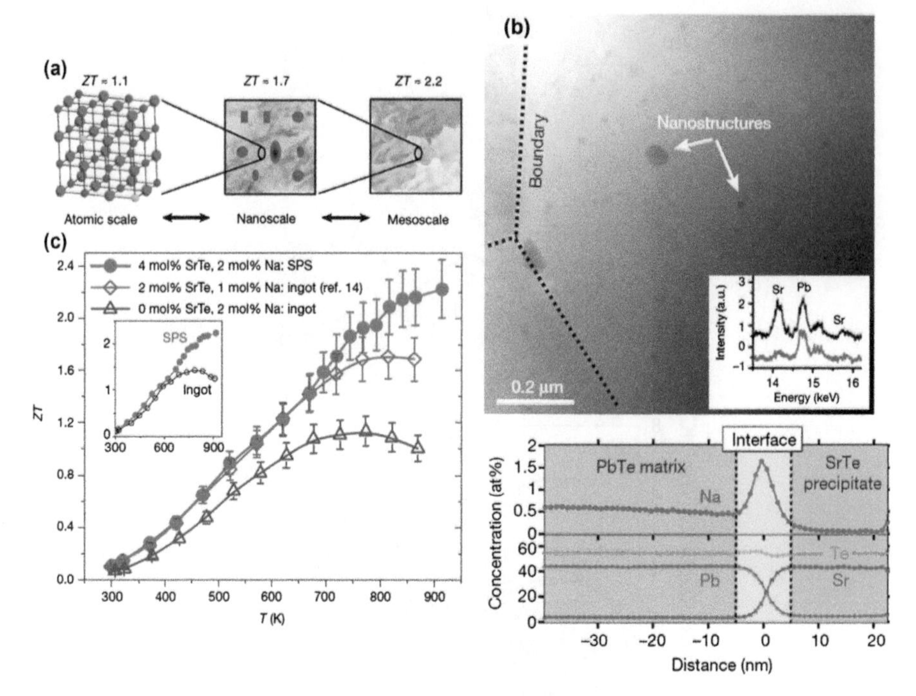

Figure 6.11 (a) Schematic of phonon scattering by various scales of hierarchical structures. (b) Scanning TEM image showing the presence of SrTe nanostructures in the PbTe matrix. The inset is the energy dispersion X-ray spectrum (black, precipitate; blue, matrix). The histogram shows the concentration profiles of Pb, Te, Sr and Na across the interface of the large SrTe precipitate. (c) Figure of merit of Na-doped PbTe-SrTe with various SrTe concentrations.

6.3 Material Synthesis

6.3.1 Conventional Synthesis Schemes

Traditional bulk thermoelectric materials such as Bi_2Te_3, PbTe and SiGe have been intensively studied in the last few decades. All of them are alloys because alloys have low thermal conductivity, as we mentioned in the previous sections. Researchers obtain superior thermoelectric performance with these materials by adjusting the synthesis methods, optimizing the composition, and so on. These traditional thermoelectric materials have been made by conventional alloy synthesis methods, which are very useful because of their simple processes and ease of control. Here we introduce two commonly used synthesis methods: 'melting and quenching' and 'mechanical alloying' (MA).

6.3.1.1 Melting Quenching

Melting quenching is a structure formation and crystallisation process that begins with mixing of the elements constituting the raw materials. For melting, the mixture of raw materials is heated beyond the melting temperatures. When the temperature exceeds the melting temperature of each element, the elements start to change phase to a homogeneous liquid. Then the temperature is held above the highest melting point of the elements for several hours until full liquefaction occurs. Quenching means 'rapid cooling' and is generally performed in water, oil or air. The melted liquid mixture is placed in the cooling medium; it is cooled rapidly and will crystallize quickly. The mixture does not have sufficient time to grow large crystals, so one can obtain alloys with a very small grain size and homogeneous structure. Although this melting quenching method is very simple, it is difficult to control the secondary phase of the quenching process. Thus, it is easy to obtain non-crystalline phases.

6.3.1.2 Mechanical Alloying

Mechanical alloying (MA) uses collisional energy for alloying. It is usually performed using a ball mill. First, the powdered elements are prepared. Each elemental powder is placed in a jar with an alumina or steel ball, and ball milling is performed. During the ball milling, the powders collide with each other, and fracturing and cold welding among them occur. As these processes are repeated during ball milling, the mixture powder becomes an alloy powder. The milled powder is then consolidated into a bulk shape and heat-treated to obtain the desired microstructure and properties. The results of this MA process depend on the milling speed, type of mill, type of protective gas and so on. MA can avoid the high temperatures used in the melting process, and it can reduce the synthesis time.

6.3.2 Nanocomposites Based on Conventional Thermoelectric Materials

The two synthesis methods, *i.e.* melting quenching and MA, have yielded good results in bulk alloy thermoelectric materials. Many studies have reported further improvement in the thermoelectric performance by using special microstructures and nanostructures. In this section, we will introduce several special synthesis methods used for obtaining nanostructures and compare them to the conventional synthesis methods.

6.3.2.1 Bismuth Telluride/Antimony Telluride

Bi_2Te_3 and related compounds are the most popular thermoelectric materials for refrigeration applications because they have a maximum ZT range at around room temperature. There are many compounds based on Bi_2Te_3, such as p-type $Bi_{0.5}Sb_{1.5}Te_3$, n-type $Bi_2Te_{2.4}Se_{0.6}$ and n-type $Bi_2Te_{2.7}Se_{0.3}$. The highest figure of merit for the Bi_2Te_3 system as a pure alloy is $ZT \sim 1$ of $Bi_{0.5}Sb_{1.5}Te_3$ alloys.[52] This system was developed in the 1950s and now is widely used in commercial thermoelectric devices. Recent research work demonstrated enhanced thermoelectric properties for the nanograined composites of the BiSbTe system.[21] The first step in making nanograins is to make a nanopowder. A BiSbTe ingot is made by a conventional melting method and loaded into a jar with balls. The process is similar to that of MA, but the alloying is performed not by mechanical action but by melting. Secondly, high-energy ball milling is performed for several hours. This yields a nanosized BiSbTe powder. The next step is to form the powder by hot pressing or spark plasma sintering (SPS). Figure 6.12 shows the schematics of hot pressing and SPS processes.[1,21,51–56] Hot pressing is a type of sintering with applied pressure and temperature. Hot pressing has many advantages. Firstly, it can reduce the sintering time and temperature. Therefore, grain growth can be controlled. Further, it can eliminate the pores inside the sample because it applies a high pressure to the sample. SPS is a special type of hot pressing that uses a DC current supply instead of a furnace. The DC current is applied inside the powder during the process. The current flow produces considerable joule heating at the powder surface. The temperature of the powders rapidly increases, so their surfaces are melted in a moment.

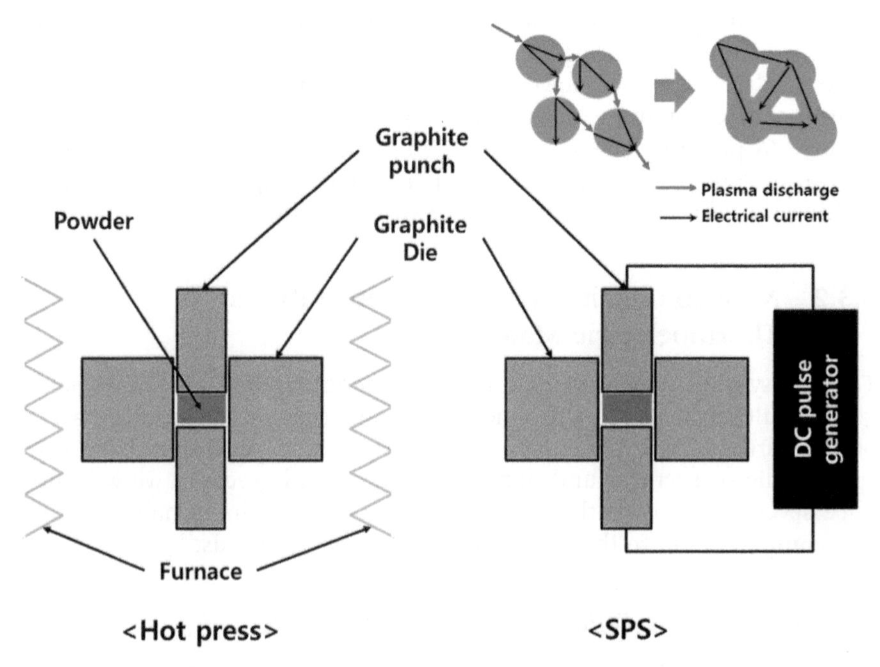

Figure 6.12 Schematics of hot pressing and spark plasma sintering (SPS) processes. Reprinted with permission from Ref. 101. Copyright © 2004, John Wiley and Sons.

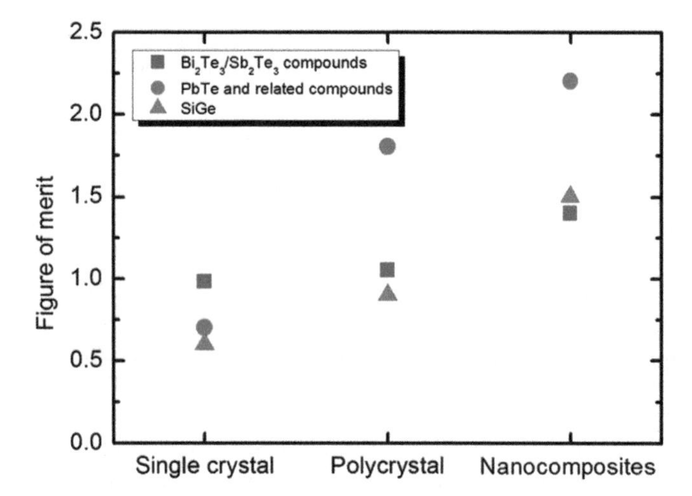

Figure 6.13 Thermoelectric figures of merit of BiSbTe, PbTe and SiGe systems in various crystal structures.[1,21,51–56]

Finally, the melted surfaces adhere to each other. SPS has the great advantage of providing extreme confinement of grain growth. Grain growth occurs by diffusion, and the diffusion rate is determined by the diffusivity of the material and time during the process. In SPS, the process is completed in about 5–10 minutes. Because the time is short, the degree of diffusion is limited.

Figure 6.13 shows the thermoelectric figures of merit of the BiSbTe system in various crystal structures: a single crystal made by the zone melting method,[52] a polycrystal made by melting,[21] and nanograins made by ball milling and hot pressing. *ZT* increases as the grain size decreases. One of the reasons could be the thermal conductivity reduction owing to a decrease in the grain size.

6.3.2.2 *Silicon Germanium*

Silicon germanium is ideal for high temperature applications. It was first used in radioisotope thermoelectric generators for spacecraft.[1] SiGe devices can operate at up to 1300 K without performance degradation. The optimal composition is generally considered to be $Si_{0.8}Ge_{0.2}$.[1] It can be made by conventional methods such as zone melting, ball milling and hot pressing. It can also produce nanograined composites although a high operating temperature is required. A nanograined SiGe composite[54] yields a *ZT* of ~1.5 at 1173 K, which is almost twice that of a SiGe bulk alloy (*ZT* ~ 0.9 at 1173 K).[1] Figure 6.13 shows the thermoelectric figures of merit of the $Si_{0.8}Ge_{0.2}$ system in various crystal structures: a single crystal made by the zone levelling method,[53] a polycrystal made by a conventional sintering method[53] and a nanocomposite with nanosized grains produced by SPS.[54] The thermoelectric figure of merit increases as the average grain size decreases owing to the increased phonon scattering at the grain boundaries.

6.3.2.3 *Lead Telluride and Related Compounds*

Lead telluride is widely used for a temperature range of 600–900 K, *i.e.* mid-temperature thermoelectric material, which corresponds to a use in waste heat recovery of exhaust gas in vehicles. Therefore, the lead telluride system could be used to harvest waste heat from vehicles. There are many reports of lead telluride synthesis. It is known that it is difficult to make small PbTe grains because the grains grow very rapidly.[57] However, the PbTe system has its own advantages. Lead antimony silver telluride (LAST) has intrinsic nanodots inside the matrix.[20] These nanodots scatter phonons, so LAST has low thermal conductivity. Similarly, the PbTe-SrTe system also has intrinsic nanostructures.[55] In the following subsections, we will discuss this synthesis method.

6.3.2.3.1 Melting, Quenching, Annealing and Hot Pressing. Annealing is a heat motional dispersion process. It operates at a relatively high temperature below the melting point and includes a recrystallization process. The actual annealing is conducted in a furnace for over a long period of time. After annealing, a homogeneous composition can be obtained in the ingot. Hot pressing is a typical method to consolidate into a bulk shape, and heat treatment is a method of obtaining a desired microstructure and improving the density of materials. Therefore, annealing and hot pressing have often been added to the synthesis process for PbTe-based thermoelectric materials made by the conventional melting quenching method. For example, Na-doped PbTe[58] and Na-doped $PbTe_{1-x}Se_x$ alloys[36] have been synthesized in this way. For these two examples, homogeneous samples with high relative density have been synthesized successfully. Higher thermoelectric figures of merit (ZT ~1.4 and ~1.8) were obtained for Na-doped PbTe and Na-doped $PbTe_{1-x}Se_x$, respectively. Therefore, the synthesis method of melting, quenching, annealing and hot pressing is a good choice for successfully synthesizing PbTe-based materials with a high thermoelectric performance.

6.3.2.3.2 Melting, Annealing, Hot Pressing and Post Annealing. Quenching is a process of solidification and crystallisation. The desired composition should appear after quenching. However, some defects, such as non-crystalline phases, nanoparticles and nanopores also occur during this process because the temperature drops in a very short time.[59] Therefore, to obtain a pure phase without defects, some researchers omit the quenching process. Instead, they set the temperature to decrease slowly from the melting point to the annealing temperature, and post-annealing is added after hot pressing. The main aim of post-annealing is to ensure stable properties at a high temperature. Because the operating temperature of thermoelectric materials, especially PbTe, is relatively high, the sample properties may change during operation. For example, stable K-doped $PbTe_{1-x}Se_x$ have been prepared in this way,[60] and a higher figure of merit (ZT ~ 1.7) was obtained. In addition, they show high stability at high temperatures above the hot pressing temperature. Therefore, the process of melting, annealing, hot pressing and post-annealing has the advantage of stability and extends the operating temperature range.

6.4 Nanodot Nanocomposites

Nanostructured materials are currently a focus of research to improve thermoelectric properties. As noted, it is difficult to obtain nanosized PbTe grains because of rapid grain growth during the annealing process. However, some PbTe systems have excellent intrinsic nanostructures that can enhance the thermoelectric properties. The melting annealing method is suitable for making nanostructures consisting of nanosized precipitations. Generally, precipitations appear at a very specific temperature range, so it is very important to control the temperature of the solid solution properly.[20] To control the temperature precisely, the solid solution should be cooled slowly from the melting temperature and be held at the temperature at which precipitation occurs. The PbTe-SrTe system was also made by the melting annealing method. These systems exhibited *ZT* values of 2.2 and 1.7,[51,55] respectively, at 800 K. Later, the Kanatzidis group achieved a *ZT* of ~2.2 at 950 K in the same material by using all-scale hierarchical structures.[51] They adopted the quenching and SPS processes in the PbTe-SrTe system. Despite the fast grain growth of the PbTe system, relatively small grains could be formed. In addition, quenching produces many atomic-scale defects. Thus, this system has intrinsic atomic-scale defects, nanodots and small grains (Figure 6.11(a)). Quenching and SPS are both rapid synthesis processes that may produce special microstructures and nanostructures. Figure 6.13 shows[1,21,51–56] the thermoelectric figures of merit of PbTe and related compounds in various crystal structures. The figure shows the highest figure of merit for each of the following three methods: the Bridgman method of producing single crystals,[56] the melting annealing method of producing polycrystals[55] and melting, quenching and SPS for producing a nanocomposite.[51] The PbTe system and related compounds could possess intrinsic nanostructures depending on synthesis schemes, so the polycrystals in particular could have high *ZT*s. In addition to this, SPS reduces the grain size, yielding the highest *ZT*. In summary, melting, quenching, annealing, hot pressing, post-annealing and SPS can be combined to prepare PbTe-based thermoelectric materials. However, the specific synthesis method for a given sample depends on the specific requirements, such as high density, stability and special microstructure. Wang *et al.*[99] reported that 2% Na doped PbTe can have an enhanced *zT* of 2.0 at 773 K by modifying synthesis conditions. A rapid quenching process makes nano- and microstructures inside the PbTe matrix. These structures scatter phonons so that the thermal conductivity decreases (Figure 6.14).

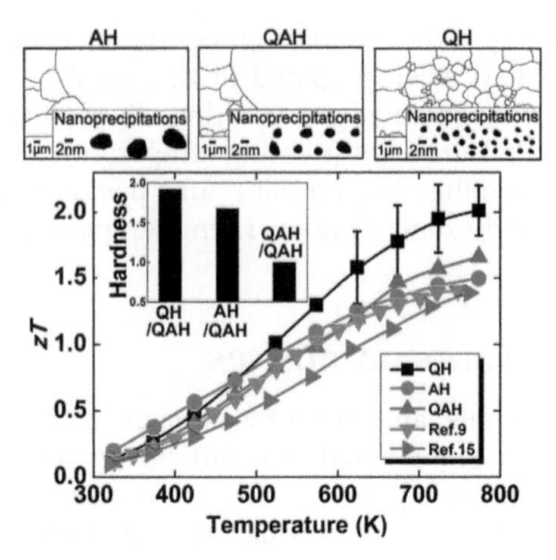

Figure 6.14 Thermoelectric figure of merit for 2% Na doped PbTe made using various synthesis processes and their structures.

6.4.1 Mg_2Si as a Thermoelectric Material

Mg_2Si-based materials have a high effective mass, high carrier mobility and low lattice thermal conductivity.[61] Thus, they are good for power generation in the 500–800 K temperature range. In addition, their low cost and non-toxicity are great advantages over other conventional thermoelectric materials based on telluride. Therefore, they are considered to be promising Te-free thermoelectric materials. It is hard to make Mg_2Si using conventional processes because Mg evaporates and oxidizes easily in air. Therefore, special synthesis processes should be used to prevent these problems. We will introduce two types of synthesis methods for Mg_2Si. One is a two-step solid-state reaction (SSR) method based on the processes of melting and quenching.[62] The other is a mechanochemical method that combines a chemical reaction and MA.[63] These two methods will be discussed in the following examples.

6.4.1.1 Two-step Solid-state Reaction Method and SPS

The two-step SSR method consists of two processes in a melting reaction. This method can supply sufficient reaction opportunities for particles of each element. It is easy to obtain a homogeneous sample. Because it is easy to lose Mg in the reaction, Mg is generally added during synthesis to compensate for its loss. The Sb-doped $Mg_2Si_{1-x}Sn_x$ alloy is chosen as an example to explain each step in the synthesis process in detail.[62] Sb-doped $Mg_2Si_{1-x}Sn_x$ compounds were prepared by the two-step SSR method and SPS. High-purity powders of the elements were weighed in stoichiometric quantities and Mg powder was added with an excess of 5–10 mol% over the stoichiometric amount. The raw materials were mixed in a glove box, shaped by cold pressing and sealed in quartz tubes under vacuum for the first-step SSR. The reaction temperature was 873–973 K. The reaction products were then used for a second-step reaction at 973 K to promote solid solution formation and increase the homogeneity. Finally, the products were ground to fine powders in the glove box, and SPS processing was performed at 823–1073 K. The two-step SSR method can effectively improve the degree of reaction and solve the problem of the raw elements having entirely different melting temperatures.

6.4.1.2 *Mechanochemical Synthesis, Incremental Milling Technique and Hot Pressing*

Mechanochemical synthesis was initially a synthetic method for preparing toxic materials.[64] Many researchers have introduced this technique to be applied to thermoelectric materials. It has proven to be an effective synthetic method for some materials. In particular, it is useful when the stoichiometric proportion is difficult to control owing to evaporation or other problems. In this process, the elemental powders are loaded into a milling vial with ball bearings; the vial is then sealed, and ball milling is performed. Some of the mechanical work among the powders and balls during ball milling is converted into chemical work, as mentioned above. Chemical reactions among the elemental powders occur during ball milling, and their reaction energy comes from the mechanical work. This method can be used to prepare materials at room temperature and avoid the vaporisation of Mg at high temperature. However, the elemental Mg is a ductile material with a low bulk modulus, so aggregation of Mg may occur in the vial. To solve this problem, the amount of Mg is often controlled in the mechanochemical synthesis process, so an incremental milling technique, *i.e.* adding more raw materials during milling, is chosen. Finally, the pure phase of a Mg_2Si-based thermoelectric material can be formed by mechanochemical synthesis and incremental milling.[63] We will consider Bi-doped Mg_2Si as an example. Mg metals are added incrementally to a stoichiometric amount of silicon and bismuth and ball-milled for intervals of less than 1 h. Additional increments of Mg are then added and ball-milled until the correct stoichiometry is achieved. This technique is called incremental milling. These processes can prevent the oxidation and evaporation of Mg and effectively prepare phase-pure thermoelectric materials. Figure 6.15 shows the highest figure of merit of Mg_2Si-based materials produced by this method and the two methods discussed above.

6.4.2 Oxide Materials

Since 1997, oxides have received attention as a new environmentally friendly thermoelectric material. $Ca_3Co_4O_9$-based and $SrTiO_3$-based ceramics are chosen as examples of typical p- and n-type oxide thermoelectric materials, respectively, in order to discuss the synthetic methods in this section. Conventional SSR technology and sol–gel methods are two typical synthesis methods.

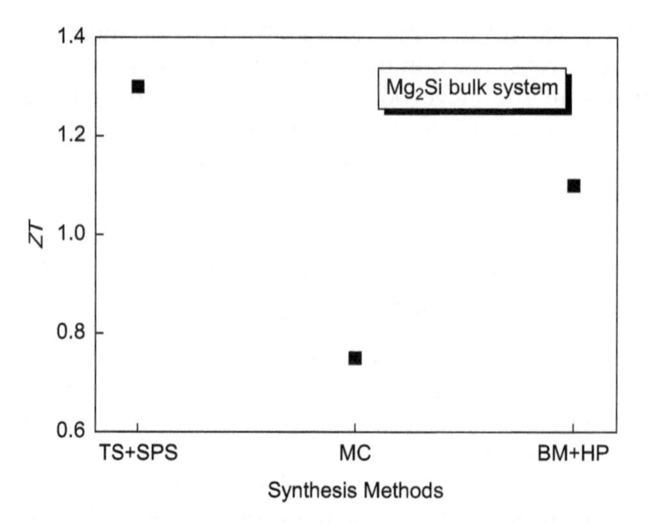

Figure 6.15 Figure of merit for Mg_2Si produced by different methods. TS + SPS: two-step solid-state reaction method and SPS; MC: mechanochemical technique; BM + HP: ball milling and hot pressing.[62,63]

6.4.2.1 *Conventional Solid-state Reaction Technology*

Conventional SSR technology is the most widely used method of synthesizing polycrystalline bulk oxide thermoelectric materials. This method can be applied to various starting materials such as oxides and carbonates. Because solids do not react with each other at room temperature, it is necessary to apply a high temperature for the proper reaction to occur at an appreciable rate. Thus, the thermodynamic and kinetic factors are important in the SSR. We select the reported La/Dy-doped $SrTiO_3$ as an example of how to perform each step in preparing the sample by conventional SSR.[65] The starting materials were La_2O_3, $SrCO_3$, TiO_2 and Dy_2O_3. These raw materials were weighed in stoichiometric proportions and mixed by ball milling for 12 h. After drying, they were pressed into pellets and calcined at 1350 °C for 6 h in air. The pellets were crushed and ball-milled for 12 h, yielding a fine powder. The powder was re-pressed into pellets and sintered at 1460 °C for 4 h under an argon atmosphere with 5 mol% hydrogen. A stable sample was obtained because of the relatively high sintering temperature; in addition, the synthesis conditions can be optimized to obtain an oxide nanocomposite. This method is simple, inexpensive and suitable for large-scale production.

6.4.2.2 *Sol–gel and SPS*

Sol–gel is a chemical solution process. It has been used to synthesize various materials, especially mixed oxides. The sol–gel method has been widely used not only for inorganic ceramics but also organics. It works with many types

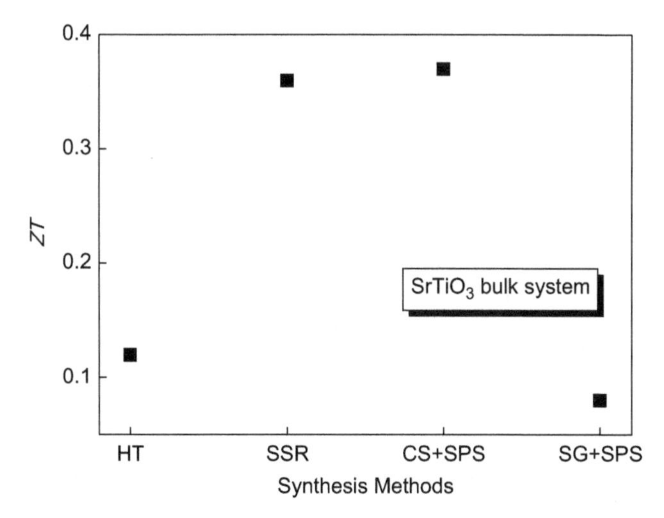

Figure 6.16 Figure of merit for the $SrTiO_3$ system produced by different methods. HT: hydrothermal; SSR: conventional solid-state reaction technology; CS + SPS: combustion synthesis and SPS; SG + SPS: sol–gel and SPS.[65–68]

of precursors as starting materials. It can offer homogeneous particle growth (at the nanoscale), a uniform size distribution and monodispersive particles. Therefore, this is one of the most favourable techniques for making single-phase ceramics. Na/La-doped $CaCo_{3.8}Cu_{0.2}O_9$ ceramics have reportedly been synthesized by sol–gel and SPS.[66] The sol–gel method started with precursors in stoichiometric proportions. The precursors were dissolved in aqueous solutions; heat was applied, and the pH was adjusted. The mixture solutions were stirred continuously. This yielded a xerogel, which was calcinated at high temperature in air. The obtained powders were then consolidated into bulk ceramics by SPS. From the above description, we can see that the sol–gel method is very cost effective and easy to handle and set up, and can yield predefined stoichiometric compounds. Conventional SSR and the sol–gel method are simple techniques in many types of synthesis methods for oxide thermoelectric materials. Many other methods, such as hydrothermal,[67] combustion,[68] sol–gel,[66] *etc*, are available, but we do not discuss them in detail. For the $SrTiO_3$-based materials, we list only the highest figure of merit obtained using each synthesis method in Figure 6.16.

6.5 Thermoelectric Devices

6.5.1 Power Generation

A typical structure of a thermoelectric (TE) device comprised of both n-type and p-type elements is shown in Figure 6.17. For simplicity, only a pair of n-type and p-type elements are presented in the figure,[1] but a real device for

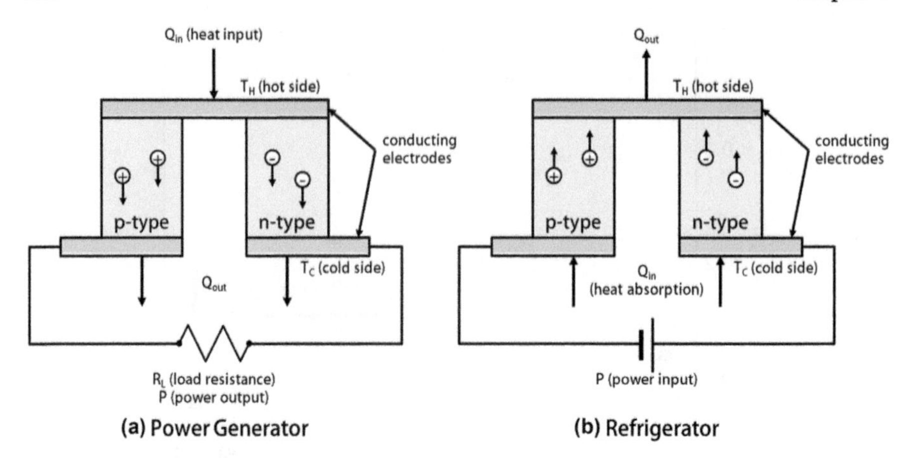

Figure 6.17 Two modes of thermoelectric device operation. (a) Power generation, (b) refrigeration.

practical use generally consists of multiple pairs of elements to meet the specific performance requirement. A thermoelectric device works as a power generator when a temperature gradient is applied across the device, *i.e.* vertically from the top to the bottom in Figure 6.17(a), and the resulting Seebeck voltage is applied to an external load resistance producing electric power. If the temperatures of the hot and cold sides are stabilized and reached at the steady-state, the net heat input and power output are expressed by eqn (6.12) and eqn (6.13), respectively, as

$$Q_{in} = K\Delta T + SIT_H - \frac{1}{2}I^2R \tag{6.12}$$

$$P = I^2R_L \tag{6.13}$$

where K is the total thermal conductance of the TE device, $\Delta T = T_H - T_C$ is the temperature difference across the device, I is the net current flowing in the electric circuit, and R and R_L are the internal resistance of the device and the load resistance, respectively. $S = S_p - S_n$ is the total Seebeck coefficient. Note that the Seebeck coefficient of the p-type element S_p has a positive sign, while that of the n-type element S_n has a negative sign, so that they are added up in magnitude for S. The net current I is obtained by the ohmic law in the electric circuit with the total Seebeck voltage $S\Delta T$ as

$$I = \frac{R + R_L}{S\Delta T} \tag{6.14}$$

The energy conversion efficiency is expressed as

$$\eta = \frac{P}{Q_{in}} = \frac{I^2R_L}{K\Delta T + SIT_H - \frac{1}{2}I^2R} \tag{6.15}$$

Assuming that the temperature of each side is fixed, the maximum total energy conversion efficiency can be obtained by adjusting the ratio of the load resistance to the internal resistance of the device, $\mu = R_L/R$. From $d\eta/d\mu = 0$, the optimal resistance ratio that maximizes the efficiency is evaluated as

$$\mu_{\max\eta} = \sqrt{1 + \frac{S^2 \sigma T_{avg}}{k}} = \sqrt{1 + ZT_{avg}} \tag{6.16}$$

where $T_{\text{avg}} = (T_H + T_L)/2$ is the average temperature of the hot and cold sides, and ZT_{avg} is the thermoelectric figure of merit of the material at the average temperature. The resulting maximum total energy conversion efficiency is obtained by

$$\eta_{\max} = \left(1 - \frac{T_C}{T_H}\right) \frac{\sqrt{1 + ZT_{avg}} - 1}{\sqrt{1 + ZT_{avg}} + \frac{T_C}{T_H}} = \eta_C \frac{\sqrt{1 + ZT_{avg}} - 1}{\sqrt{1 + ZT_{avg}} + \frac{T_C}{T_H}} \tag{6.17}$$

where η_C is the Carnot efficiency, which is the theoretical maximum efficiency of an energy convertor between a heat sink and a heat source. The thermoelectric figure of merit has a very important role in this equation: it is an intrinsic index representing the performance of a device made from a certain thermoelectric material. As for the maximum power output, the resistance ratio μ should satisfy $dP/d\mu = 0$, which results in $\mu = 1$. Then the maximum power output is obtained as

$$P_{\max} = \frac{S^2 \Delta T^2}{4R} \tag{6.18}$$

6.5.2 Refrigeration

Thermoelectric refrigeration is operated by an external input power that moves electrons and holes away from the target surface to cool down the surface. For the thermoelectric refrigerator shown in Figure 6.17(b), the rate of heat absorption (or the cooling power) at the cold side and input electric power are expressed by eqn (6.19) and eqn (6.20), respectively:

$$Q_{in} = -K\Delta T + SIT_C - \frac{1}{2} I^2 R \tag{6.19}$$

$$P = SI\Delta T + I^2 R \tag{6.20}$$

The coefficient of performance (COP) of the TE refrigerator is defined as the cooling power divided by the input electric power such that

$$COP = \frac{-K\Delta T + SIT_C - \frac{1}{2} I^2 R}{SI\Delta T + I^2 R}. \tag{6.21}$$

The COP can be optimized by adjusting the input current. The optimized current $I_{max\ COP}$ that maximizes COP, and the corresponding maximum COP are expressed in eqn (6.22) and eqn (6.23), respectively:

$$I_{\max COP} = \frac{S\Delta T}{R\sqrt{1+ZT_{avg}} - 1} \tag{6.22}$$

$$COP_{\max} = COP_C \frac{\left(\sqrt{1+ZT_{avg}} - \frac{T_H}{T_C}\right)}{\left(\sqrt{1+ZT_{avg}} + 1\right)} \tag{6.23}$$

where $COP_C = T_C/(T_H - T_C)$ is the Carnot coefficient of performance. The maximum COP depends on the thermoelectric figure of merit ZT. Note that the COP can be negative when the temperature of the hot side is much higher than that of the cold side, in which the heat energy dissipated to the cold side by thermal conduction is larger than the Peltier cooling, so that the net cooling power becomes negative. When one wants to maximize the cooling power Q_{in} instead of COP, the optimal current $I_{max\ Qin}$ is obtained when $dQ_{in}/dI = 0$, such that

$$I_{\max Q_{in}} = \frac{ST_C}{R} \tag{6.24}$$

and the corresponding maximum cooling power, and the COP are obtained to be

$$Q_{in,\max} = -K\Delta T + \frac{S^2 T_C^2}{2R} \tag{6.25}$$

$$COP_{\max Q_{in}} = \frac{\frac{1}{2} ZT_C^2 - \Delta T}{ZT_H T_C}. \tag{6.26}$$

6.5.3 Fabrication of Thermoelectric Devices

Conventional thermoelectric devices are fabricated using bulk thermoelectric materials. Most of the thermoelectric materials with high figures of merit are semiconducting alloys[51,58,69] or nanostructured bulk materials.[20,21] These materials need to be scalable for a large scale production, so bulk materials are ideal.[21] However, in applications that require a low-power refrigeration for a very small area, *e.g.* hotspot cooling for micro-processors, thin-film materials are ideal for thermoelectric device fabrication since thin-film based TE can be batch-fabricated with the chip and can produce large cooling power density.[74] Organic thermoelectric materials composed of conducting polymers such as poly(3,4-ethylenedioxythiophene)[70–72] are usually made in the form of thin films. These organic materials can be useful for thin-film thermoelectric devices,[73] but their very low figures of merit have hindered their widespread use until now. Thus, most thin-film thermoelectric devices are manufactured with high-figure-of-merit inorganic semiconducting materials.[16,31,49,74] Superlattices that have superior thermoelectric properties in the cross-plane direction have also been used for thin-film thermoelectric devices.[75] On the basis of a new approach to increase the figure of merit by lowering the dimensionality of thermoelectric materials,[3] many studies have investigated 1D thermoelectric nanowires of silicon[17,19] or bismuth.[76] Then thermoelectric devices with nanowires[77,78] or a hybrid of nanowires and bulk[79] have been suggested for new thermoelectric applications. Still, despite the low figures of merit of organic materials, organic thermoelectric applications are attracting attention because the material and manufacturing costs are much lower than those of inorganic high-figure-of-merit materials. After many research efforts on the conventional issues of organic devices, such as the low efficiency, they are now being intensively investigated in terms of low-cost manufacturing techniques, such as the roll-to-roll procedure.[80–82]

Several important aspects need to be considered in the device fabrication process. Thermal stress is an issue for thin-film devices; it has been studied by Huang *et al.*,[83] Hori *et al.*[84] and others. According to these studies, thermal stress induced by a large temperature difference across the device may damage the thin-film structure due to its short device length. Thus, when designing a thin-film thermoelectric device, the effects of normal and sheer stresses under the working temperature profile across the device should also be considered. One of the most important things to consider when designing thermoelectric devices is the choice of electrode material.[8] It is necessary to first consider the available thermoelectric materials and operating temperatures. The metallurgic bonding between the electrodes and the thermoelectric materials can be significantly degraded at high temperatures, which is related to the lifespan of the device. To guarantee reliability and mechanical stability of the electrical and/or thermal contacts of the device, the thermal expansion coefficients of those two coupled materials should be comparable to avoid any significant stress during the thermal cycles. The most common electrode material for PbTe, which is widely used in applications, is iron because its thermal expansion coefficient is very similar to that of PbTe. However, iron is easily diffused into n-type PbTe, degrading the thermoelectric properties, so a nickel barrier is installed between the iron electrode and the semiconductor to prevent Fe diffusion at high temperatures.[8] The most common electrode metal for Bi_2Te_3 is lead solder, which essentially makes mechanical contact.[8]

Another important issue for thermoelectric devices is the cost. Many studies[85,86] have considered the use of thermoelectric devices as low in cost as other popular energy conversion devices. This led to the optimisation of thermoelectric devices. The performance of thermoelectric devices depends not only on the material properties but also on the thermal environment in which the devices are running. That is why it is essential to optimize the thermal resistance of the device, including the electrodes and substrates of the hot and cold sides in the heat flow direction, in order to generate the maximum electric power output and convert thermal energy into electricity with a high efficiency or COP at a reasonably low cost. There has been much work on investigating the optimisation of thermoelectric devices. Most has been focused on mathematical modelling for the optimisation of the thermal resistance in power generation[87–89] and refrigeration.[90] Power generation can be improved compared with that of conventional thermoelectric devices by modifying the shape of the thermoelectric material.[91,92] Some research has been conducted on examining the cost of thermoelectric devices because the energy conversion efficiency and power generation of such devices are not yet comparable with those of principal modern generators such as Rankine cycle steam engines or Stirling engines. Yazawa and Shakouri[93] suggested the possibility of commercializing thermoelectric generators by investigating their cost-per-Watt optimisation. Significant work has also been done to improve the cooling power of refrigeration applications. Meng *et al.*[94] suggested a new configuration for a thermoelectric refrigerator assisted by a thermoelectric generator. The allocation of the thermoelements in this two-stage combined thermoelectric device was optimized to give the highest COP given the number of thermoelements. Astrain *et al.*[90] investigated the use of a thermal resistance network to optimize the performance of thermoelectric refrigerators. They optimized the performance of heat dissipation of the thermosyphon system assisted by the thermoelectric refrigerator by analytical calculations and experiments.

6.5.4 Applications to Waste Heat Recovery

The energy source of thermoelectric generation is heat itself. Thermoelectric devices can be installed in various exothermic systems such as heat engines or catalytic converters in vehicles to recover waste heat and convert it to electricity. Many people have recently become interested in waste heat recovery for automobiles using thermoelectric generators. The heat energy wasted by hot exhaust gas from a vehicle amounts to 27.7% of the total energy consumed,[95] and it is known that using exhaust energy from an automobile engine with a heat exchanger results in fuel savings of up to 34%.[96] BMW and Ford are well-known commercial passenger vehicle companies, and a thermoelectric generator built by Amerigon was attached to the exhaust gas pipe in their vehicles and produced around 500 W of electricity from the waste heat.[97] Another interesting application of waste heat recovery is the cooling of overheated water from a hot spring. Panasonic developed a thermoelectric generator for use in hot water pipes, which generated 2.5 W of electricity per 10 cm of pipe.[98]

6.6 Summary

In this chapter, we have covered various topics in thermoelectrics such as transport theory, schemes to enhance thermoelectric performance, material synthesis methods, device theory and possible applications. Thermoelectric performance has been improved by using nanostructures since the 2000s. Nanostructures alleviated the interdependency between three thermoelectric properties, *i.e.* the Seebeck coefficient, electrical conductivity and thermal conductivity, and they were particularly effective in reducing thermal conductivity. We also presented ways to enhance the power factor, such as the quantum confinement effect, electron filtering, band convergence and resonant levels. It is expected that combining these two technologies could lead to high performance thermoelectric materials. To realize these transport theories, various synthesis procedures have been scrutinized. Some of these efforts have been introduced in this chapter with an emphasis on the thermoelectric figure of merit dependency over crystal structures. Finally, device theory and fabrication were presented. We hope that this book chapter can significantly help scholars working in the field of thermoelectrics.

Acknowledgements

This work was supported by the Mid-career Researcher Program (No. 2011-0028729) through a National Research Foundation of Korea (NRF) grant funded by the Ministry of Education, Science and Technology (MEST) and Low Observable Technology Research Center Program of Defense Acquisition Program Administration and Agency for Defense Development.

References

1. D. M. Rowe, *CRC handbook of thermoelectrics*, CRC Press, Boca Raton, FL, 1995.
2. G. Chen and A. Shakouri, *J. Heat Transfer*, 2002, **124**, 242–252.
3. M. S. Dresselhaus, G. Chen, M. Y. Tang, R. G. Yang, H. Lee, D. Z. Wang, Z. F. Ren, J. P. Fleurial and P. Gogna, *Adv. Mater.*, 2007, **19**, 1043–1053.
4. H. J. Goldsmid, *Thermoelectric refrigeration*, Plenum Press, New York, 1964.
5. W. Kim, *Mater. Res. Innovations*, 2011, **15**, 375–385.
6. A. Majumdar, *Science*, 2004, **303**, 777.
7. D. M. Rowe, *CRC Handbook of Thermoelectrics: Macro to Nano*, CRC/Taylor and Francis, Florida, 2006.
8. C. Wood, *Rep. Prog. Phys.*, 1988, **51**, 459–539.
9. G. S. Nolas, J. Sharp and H. J. Goldsmid, *Thermoelectrics: Basic principles and new materials developments*, Springer, Berlin, Germany, New York, 2001.
10. F. J. DiSalvo, *Science*, 1999, **285**, 703.
11. C. B. Vining, *Nat. Mater.*, 2009, **8**, 83.
12. K. Yazawa, A. Ziabari, K. Yee Rui, A. Shakouri, V. Sahu, A. G. Fedorov and Y. Joshi, *2012 13th IEEE Intersociety Conference on Thermal and*

Thermomechanical Phenomena in Electronic Systems(ITherm), IEEE, New York, 2012.
13. C. J. Vineis, A. Shakouri, A. Majumdar and M. G. Kanatzidis, *Adv. Mater.*, 2010, **22**, 3970–3980.
14. L. D. Hicks and M. S. Dresselhaus, *Phys. Rev. B*, 1993, **47**, 16631.
15. L. D. Hicks and M. S. Dresselhaus, *Phys. Rev. B*, 1993, **47**, 12727.
16. R. Venkatasubramanian, E. Siivola, T. Colpitts and B. O'Quinn, *Nature*, 2001, **413**, 597–602.
17. A. I. Boukai, Y. Bunimovich, J. Tahir-Kheli, J. K. Yu, W. A. Goddard and J. R. Heath, *Nature*, 2008, **451**, 168–171.
18. T. C. Harman, P. J. Taylor, M. P. Walsh and B. E. LaForge, *Science*, 2002, **297**, 2229–2232.
19. A. I. Hochbaum, R. K. Chen, R. D. Delgado, W. J. Liang, E. C. Garnett, M. Najarian, A. Majumdar and P. D. Yang, *Nature*, 2008, **451**, 163–165.
20. K. F. Hsu, S. Loo, F. Guo, W. Chen, J. S. Dyck, C. Uher, T. Hogan, E. K. Polychroniadis and M. G. Kanatzidis, *Science*, 2004, **303**, 818–821.
21. B. Poudel, Q. Hao, Y. Ma, Y. C. Lan, A. Minnich, B. Yu, X. A. Yan, D. Z. Wang, A. Muto, D. Vashaee, X. Y. Chen, J. M. Liu, M. S. Dresselhaus, G. Chen and Z. F. Ren, *Science*, 2008, **320**, 634–638.
22. R. Y. S. Wang, PhD thesis, University of California, Berkeley, 2008.
23. G. J. Snyder and E. S. Toberer, *Nat. Mater.*, 2008, **7**, 105–114.
24. F. P. Incropera, *Fundamentals of heat and mass transfer*, John Wiley, Hoboken, NJ, 6th edn, 2007.
25. N. W. Ashcroft and N. D. Mermin, *Solid state physics*, Holt Rinehart and Winston, New York, 1976.
26. J. P. Heremans, V. Jovovic, E. S. Toberer, A. Saramat, K. Kurosaki, A. Charoenphakdee, S. Yamanaka and G. J. Snyder, *Science*, 2008, **321**, 554–557.
27. P. M. Wu, J. Gooth, X. Zianni, S. F. Svensson, J. G. Gluschke, K. A. Dick, C. Thelander, K. Nielsch and H. Linke, *Nano Lett.*, 2013, **13**, 4080–4086.
28. G. D. Mahan and L. M. Woods, *Phys. Rev. Lett.*, 1998, **80**, 4016–4019.
29. G. D. Mahan, J. O. Sofo and M. Bartkowiak, *J. Appl. Phys.*, 1998, **83**, 4683–4689.
30. A. Shakouri, *Annu. Rev. Mater. Res.*, 2011, **41**, 399–431.
31. D. Vashaee and A. Shakouri, *Phys. Rev. Lett.*, 2004, **92**, 106103.
32. M. D. Ulrich, P. A. Barnes and C. B. Vining, *J. Appl. Phys.*, 2001, **90**, 1625–1631.
33. D. Vashaee and A. Shakouri, *J. Appl. Phys.*, 2004, **95**, 1233–1245.
34. R. Kim, C. Jeong and M. S. Lundstrom, *J. Appl. Phys.*, 2010, **107**, 054502–054508.
35. J.-H. Bahk, Z. Bian and A. Shakouri, *Phys. Rev. B*, 2013, **87**, 075204.
36. Y. Z. Pei, X. Y. Shi, A. LaLonde, H. Wang, L. D. Chen and G. J. Snyder, *Nature*, 2011, **473**, 66–69.
37. W. Kim, R. Wang and A. Majumdar, *Nano Today*, 2007, **2**, 40–47.
38. P. D. Maycock, *Solid-State Electron.*, 1967, **10**, 161–168.

39. W. S. Capinski, H. J. Maris, T. Ruf, M. Cardona, K. Ploog and D. S. Katzer, *Phys. Rev. B*, 1999, **59**, 8105–8113.
40. D. Y. Li, Y. Y. Wu, P. Kim, L. Shi, P. D. Yang and A. Majumdar, *Appl. Phys. Lett.*, 2003, **83**, 2934–2936.
41. P. G. Klemens, in *Solid State Physics*, ed. S. Frederick and T. David, Academic Press, New York, 1958, vol. 7, pp. 1–98.
42. H. C. v. d. Hulst, *Light scattering by small particles*, Dover Publications, New York, 1981.
43. C. L. Tien, A. Majumdar and F. M. Gerner, *Microscale energy transport*, Taylor & Francis, Washington, D.C., 1998.
44. M. Law, J. Goldberger and P. Yang, *Annu. Rev. Mater. Res.*, 2004, **34**, 83–122.
45. W. Kim and A. Majumdar, *J. Appl. Phys.*, 2006, **99**, 084306.
46. W. Kim, S. L. Singer, A. Majumdar, J. M. O. Zide, D. Klenov, A. C. Gossard and S. Stemmer, *Nano Lett.*, 2008, **8**, 2097–2099.
47. P. Y. Yu and M. Cardona, *Fundamentals of semiconductors: physics and materials properties*, Springer, Berlin, New York, 3rd edn, 2001.
48. S. T. Huxtable, A. R. Abramson, C. L. Tien, A. Majumdar, C. LaBounty, X. Fan, G. H. Zeng, J. E. Bowers, A. Shakouri and E. T. Croke, *Appl. Phys. Lett.*, 2002, **80**, 1737–1739.
49. R. Venkatasubramanian, T. Colpitts, E. Watko, M. Lamvik and N. ElMasry, *J. Cryst. Growth*, 1997, **170**, 817–821.
50. W. Kim, J. Zide, A. Gossard, D. Klenov, S. Stemmer, A. Shakouri and A. Majumdar, *Phys. Rev. Lett.*, 2006, **96**, 045901.
51. K. Biswas, J. Q. He, I. D. Blum, C. I. Wu, T. P. Hogan, D. N. Seidman, V. P. Dravid and M. G. Kanatzidis, *Nature*, 2012, **489**, 414–418.
52. L. D. Ivanova and Y. V. Granatkina, *Inorg. Mater.*, 2000, **36**, 672–677.
53. C. B. Vining, W. Laskow, J. O. Hanson, R. R. Vanderbeck and P. D. Gorsuch, *J. Appl. Phys.*, 1991, **69**, 4333–4340.
54. S. Bathula, M. Jayasimhadri, N. Singh, A. K. Srivastava, J. Pulikkotil, A. Dhar and R. C. Budhani, *Appl. Phys. Lett.*, 2012, **101**, 213902–213905.
55. K. Biswas, J. Q. He, Q. C. Zhang, G. Y. Wang, C. Uher, V. P. Dravid and M. G. Kanatzidis, *Nat. Chem.*, 2011, **3**, 160–166.
56. M. Orihashi, Y. Noda, L. D. Chen, T. Goto and T. Hirai, *J. Phys. Chem. Solids*, 2000, **61**, 919–923.
57. B. Yu, Q. Zhang, H. Wang, X. Wang, H. Wang, D. Wang, H. Wang, G. J. Snyder, G. Chen and Z. F. Ren, *J. Appl. Phys.*, 2010, **108**, 016104–016103.
58. Y. Z. Pei, A. LaLonde, S. Iwanaga and G. J. Snyder, *Energy Environ. Sci.*, 2011, **4**, 2085–2089.
59. M. Yoshimura, M. Kaneko and S. Sōmiya, *J. Mater. Sci. Lett.*, 1985, **4**, 1082–1084.
60. Q. Zhang, F. Cao, W. S. Liu, K. Lukas, B. Yu, S. Chen, C. Opeil, D. Broido, G. Chen and Z. F. Ren, *J. Am. Chem. Soc.*, 2012, **134**, 10031–10038.

61. V. K. Zaitsev, M. I. Fedorov, E. A. Gurieva, I. S. Eremin, P. P. Konstantinov, A. Y. Samunin and M. V. Vedernikov, *Phys. Rev. B*, 2006, **74**, 045207.
62. W. Liu, X. J. Tan, K. Yin, H. J. Liu, X. F. Tang, J. Shi, Q. J. Zhang and C. Uher, *Phys. Rev. Lett.*, 2012, **108**, 166601.
63. S. K. Bux, M. T. Yeung, E. S. Toberer, G. J. Snyder, R. B. Kaner and J. P. Fleurial, *J. Mater. Chem.*, 2011, **21**, 12259–12266.
64. S. A. Rowlands, A. K. Hall, P. G. Mccormick, R. Street, R. J. Hart, G. F. Ebell and P. Donecker, *Nature*, 1994, **367**, 223–223.
65. H. C. Wang, C. L. Wang, W. B. Su, J. A. Liu, Y. Sun, H. Peng and L. A. M. Mei, *J. Am. Ceram. Soc.*, 2011, **94**, 838–842.
66. Y. Ou, J. Peng, F. Li, Z. X. Yu, F. Y. Ma, S. H. Xie, J. F. Li and J. Y. Li, *J. Alloys Compd.*, 2012, **526**, 139–144.
67. S. Katsuyama, Y. Takiguchi and M. Ito, *J. Mater. Sci.*, 2008, **43**, 3553–3559.
68. A. Kikuchi, N. Okinaka and T. Akiyama, *Scr. Mater.*, 2010, **63**, 407–410.
69. A. D. LaLonde, Y. Z. Pei and G. J. Snyder, *Energy Environ. Sci.*, 2011, **4**, 2090–2096.
70. D. Kim, Y. Kim, K. Choi, J. C. Grunlan and C. H. Yu, *ACS Nano*, 2010, **4**, 513–523.
71. O. Bubnova, Z. U. Khan, A. Malti, S. Braun, M. Fahlman, M. Berggren and X. Crispin, *Nat. Mater.*, 2011, **10**, 429–433.
72. F. X. Jiang, J. K. Xu, B. Y. Lu, Y. Xie, R. J. Huang and L. F. Li, *Chin. Phys. Lett.*, 2008, **25**, 2202–2205.
73. S. Logothetidis, *Mater. Sci. Eng., B*, 2008, **152**, 96–104.
74. I. Chowdhury, R. Prasher, K. Lofgreen, G. Chrysler, S. Narasimhan, R. Mahajan, D. Koester, R. Alley and R. Venkatasubramanian, *Nat. Nanotechnol.*, 2009, **4**, 235–238.
75. A. Shakouri, *Proc. IEEE*, 2006, **94**, 1613–1638.
76. J. Heremans and C. M. Thrush, *Phys. Rev. B*, 1999, **59**, 12579–12583.
77. A. R. Abramson, W. C. Kim, S. T. Huxtable, H. Q. Yan, Y. Y. Wu, A. Majumdar, C. L. Tien and P. D. Yang, *J. Microelectromech. Syst.*, 2004, **13**, 505–513.
78. W. Wang, F. L. Jia, Q. H. Huang and J. Z. Zhang, *Microelectron. Eng.*, 2005, 77, 223–229.
79. J. Keyani, A. M. Stacy and J. Sharp, *Appl. Phys. Lett.*, 2006, **89**, 233106.
80. M. Yanaka, B. M. Henry, A. P. Roberts, C. R. M. Grovenor, G. A. D. Briggs, A. P. Sutton, T. Miyamoto, Y. Tsukahara, N. Takeda and R. J. Chater, *Thin Solid Films*, 2001, **397**, 176–185.
81. A. Laskarakis and S. Logothetidis, *J. Appl. Phys.*, 2006, **99**, 066101.
82. S. Logothetidis, *Rev. Adv. Mater. Sci.*, 2005, **10**, 387–397.
83. M. J. Huang, P. K. Chou and M. C. Lin, *Sens. Actuators, A*, 2006, **126**, 122–128.
84. Y. Hori, D. Kusano, T. Ito and K. Izumi, *18th International Conference on Thermoelectrics, International Thermoelectric Society*, 1999, pp. 328–331.
85. M. Zebarjadi, K. Esfarjani, M. S. Dresselhaus, Z. F. Ren and G. Chen, *Energy Environ. Sci.*, 2012, **5**, 5147–5162.

86. S. B. Riffat and X. L. Ma, *Appl. Therm. Eng.*, 2003, **23**, 913–935.
87. A. Sisman and H. Yavuz, *Energy*, 1995, **20**, 573–576.
88. J. C. Chen, Z. J. Yan and L. Q. Wu, *J. Appl. Phys.*, 1996, **79**, 8823–8828.
89. J. C. Chen, Z. J. Yan and L. Q. Wu, *Energy*, 1997, **22**, 979–985.
90. D. Astrain, J. G. Vian and M. Dominguez, *Appl. Therm. Eng.*, 2003, **23**, 2183–2200.
91. A. Z. Sahin and B. S. Yilbas, *Energy Convers. Manage.*, 2013, **65**, 26–32.
92. K. Yazawa and A. Shakouri, *J. Appl. Phys.*, 2012, **111**, 024509.
93. K. Yazawa and A. Shakouri, *Environ. Sci. Technol.*, 2011, **45**, 7548–7553.
94. F. K. Meng, L. G. Chen and F. R. Sun, *Cryogenics*, 2009, **49**, 57–65.
95. J. C. Conklin and J. P. Szybist, *Energy*, 2010, **35**, 1658–1664.
96. R. Saidur, M. Rezaei, W. K. Muzammil, M. H. Hassan, S. Paria and M. Hasanuzzaman, *Renewable Sustainable Energy Rev.*, 2012, **16**, 5649–5659.
97. U. S. Department of Energy, *in Thermoelectric Waste Heat Recovery Program for Passenger Vehicles* 2012, vol. 18.
98. D. Kennedy and R. Osuga, DIGINFO TV, http://www.diginfo.tv/v/12-0224-r-en.php, 2012.
99. H. Wang, J. Bahk, C. Kang, J. Hwang, K. Kim, J. Kim, P. Burke, J. E. Bowers, A. C. Gossard, A. Shakouri and W. Kim, *Proc. Natl. Acad. Sci. U. S. A.*, 2014, **111**, 10949–10954.
100. K. Kishimoto, M. Tsukamoto and T. Koyanagi, *J. Appl. Phys.*, 2002, **92**, 5331.
101. Z. Shen, M. Johnsson, Z. Zhao and M. Nygren, *J. Am. Ceram. Soc.*, 2002, **85**, 1921.

CHAPTER 7

Piezoelectric Energy Harvesting Nanofibers

JIYOUNG CHANG[a,b] AND LIWEI LIN*[b]

[a] Department of Physics, University of California at Berkeley, Berkeley, CA 94720, USA; [b] Berkeley Sensor and Actuator Center, University of California at Berkeley, Berkeley, CA 94720, USA
*Email: lwlin@me.berkeley.edu

7.1 Introduction

After decades of developments in the miniaturization of portable and wireless devices, new power sources beyond rechargeable batteries have become important topics for current and future stand-alone devices and systems. Specifically, ideal power sources should be scalable for the power demands of various portable devices without the necessity of a recharging process or replacement. Recent works in the field of nanomaterials have shown good progress toward self-powered energy sources by scavenging energy from ambient environments (solar, thermal, mechanical vibration, *etc.*). In particular, the use of piezoelectric generators by nanomaterials as a robust and simple solution for mechanical energy harvesting has attracted a lot of attention. One of the earliest nanogenerators for possible energy scavenging applications from mechanical strain utilized piezoelectric zinc oxide (ZnO) nanowires.[1] By coupling their semiconducting and piezoelectric properties, mechanical strains can be converted into electricity. In recent years, numerous research groups have demonstrated results in the field of mechanical energy scavenging using nanomaterials with different architectures, including: film-based, nanowire-based and nanofiber-based

RSC Nanoscience & Nanotechnology No. 35
Hierarchical Nanostructures for Energy Devices
Edited by Seung Hwan Ko and Costas P Grigoropoulos

Published by the Royal Society of Chemistry, www.rsc.org

nanogenerators. Film-based nanogenerators are often made by the spin-on or thin-film deposition methods.[2,3] Mechanical strains due to bending, vibration or compression of the thin-film structure can be the sources of energy generation. Nanowire-based nanogenerators[4] are typically made of semiconducting materials such as ZnO,[1,5,6] ZnS,[7] GaN[8,9] or CdS.[10,11] These piezoelectric nanowires have been demonstrated to build up an electrical potential when mechanically strained by AFM tips,[1] zig-zag electrodes[12] or compliant substrates[13] to convert mechanical strains into electricity. The third group of nanogenerators is based on nanofibers often constructed by the electrospinning process to be discussed in detail in this chapter.

In terms of piezoelectric materials, lead zirconate titanate (PZT), which is a ceramic material that exhibits exceptionally good piezoelectric properties, has been studied as fiber-based energy harvester recently.[14–16] Even though PZT has very high piezoelectric property compared to other polymer based materials, it has been mainly studied as a film-based structure and typically integrated with microelectromechanical systems (MEMS) cantilevers with proof masses.[17–19] These energy harvesters utilize the mechanical resonances of cantilevers to induce larger mechanical strain for energy generation. However, PZT fibers generally require high temperature annealing (>600 °C) to enhance the piezoelectric property.[20] Furthermore, PZT fibers made by the electrospinning process require mixing of PZT with solvents that lower the density of the PZT and lead to lower overall power efficiency. On the other hand, organic nanofibers made of polymeric polyvinylidene fluoride (PVDF) have been studied as nanogenerators.[21–23] Compared with the aforementioned nanomaterials, PVDF nanofibers have a unique good combination of material properties: flexibility, biocompatibility and availability in ultra-long lengths, various thicknesses and shapes, and they are lightweight, making them an interesting candidate for energy harvesting applications in wearable and/or implantable devices.

This chapter will describe piezoelectric nanofiber nanogenerators while nanowire and nanofiber-based nanogenerators and their fabrication processes have been extensively discussed in several review papers.[4,24,25] Furthermore, electrospun piezoelectric PVDF nanofibers are the main materials to be reviewed as they are the most commonly studied piezoelectric nanofiber nanogenerators. The key fabrication methodology to make PVDF nanofibers is the electrospinning process such that a key section in this chapter discusses the details of electrospinning processes to make piezoelectric nanofibers for high yield, scalable, and cost effective nanofiber nanogenerators.

Ceramic PZT and polymeric PVDF are two key piezoelectric materials which have been successfully demonstrated in the making of nanofiber nanogenerators. In these demonstrations, either near-field electrospinning (NFES)[26] or the conventional far-field electrospinning (FFES) process has been the key manufacturing tool to produce nanofibers.[27] For the NFES process, a continuous single nanofiber can be deposited in a controllable manner for nanogenerators while the FFES process can make dense nanofiber networks on large areas for nanogenerator demonstrations. In general, a poling process including both electrical poling and mechanical stretching is required for piezoelectricity at moderate temperature. However, the high electrostatic field used in the electrospinning process could provide *in situ* electric poling and mechanical stretching to produce piezoelectric nanofibers. Here, key achievements in nanofiber nanogenerators made of PVDF and PZT are described and discussed.

7.2 Nanofiber Fabrication Methods for Energy Harvesting Applications

Electrospinning can make fibers from solutions or melts to provide the foundation for piezoelectric nanofiber nanogenerators as discussed in the previous section. The diameters of electrospun fibers range from tens of nanometres to micrometres and a large number of different materials have been produced by electrospinning, including synthetic/natural polymers, polymer alloys, polymer composites as well as metals and ceramics.[27] The versatile possibilities in the selections of materials and easy process setup of electrospinning could provide unique pathways to enhance the performance of nanogenerators. Furthermore, the electrospinning process has high throughputs for mass production in various applications such as filtration, wound dressing, bio-scaffolds and medical implants, to name a few.[28–30] In recent years, electrospinning has also been applied in the making of micro/nano devices such as field effect transistors,[31] sensors,[32] actuators[33] *etc.*[34]

7.2.1 Far-field Electrospinning

The typical setup for conventional electrospinning includes four major components,[35,36] a syringe pump to maintain the constant flow rate of the polymer solution, a dispense needle that is connected to a high voltage supply as a cathode, a high voltage power supply unit, and a collector electrode that collects electrospun nanofibers. When a high voltage is applied, a strong electrostatic field is established between the needle tip and the collector electrode. The electrostatic force attracts the polymer melt out of the needle and is balanced by the surface tension force of the polymer melt. When the electrostatic force surpasses the surface tension force, a thin liquid/melt jet is ejected from the droplet toward the collector electrode. For a short distance immediately below the droplet, the polymer jet is stable and can be utilized in the near-field electrospinning process.[26] If the collector electrode is placed far away (tens of centimetres) from the needle tip, the jet will undergo a whipping and chaotic process to deposit nanofibers randomly on the collector electrode, which is the conventional far-field electrospinning process. The solvent evaporates and the polymer stream solidifies to form thin solid polymer fibers on the collector electrode. For far-field electrospinning, the typical inner diameter of the dispense needle is in the order of a few hundred μm and the applied voltage is in the range of several tens of kilovolts and the needle-to-collector distance is tens of centimetres. The formation of thin fibers is based on the uniaxial stretching of the viscoelastic solution by the electrostatic force for materials with suitable properties, such as conductivity, which can be influenced by salt solutions, and surface tension, which can be changed by adding surfactants. One key characteristic of far-field electrospinning is the random and chaotic distribution of nanofibers on the collector electrode.

7.2.2 Modified Far-field Electrospinning

For applications such as nanofiber nanogenerators, good fiber alignment could be necessary to improve the energy generation efficiencies. Research groups have demonstrated methods for aligned depositions of nanofibers with modified conventional electrospinning processes. For example, Boland *et al.* used a rotating collector to control the alignment of deposited nanofibers.[29] In their experiments, poly(glycolic acid) and collagen were electrospun on a cylinder collector rotating at a speed of 1000 rpm and 4500 rpm, respectively. The use of fast-spinning collectors in the form of drums,[37] wheel like disks,[38] and wire drums[39] have all shown various levels of success in controlling the deposition positions of nanofibers. These methods could be potential manufacturing approaches to make continuous aligned nanofibers for nanogenerator applications.

Instead of using mechanical means to improve the conventional electrospinning process for better controlled deposition positions, researchers have also exploited electric fields to guide the deposition locations of nanofibers, including the usage of two parallel electrodes on the collector[40] and the designs of one or several charged rings as auxiliary electrodes.[41] In the process, nanofibers were attracted by the two collector electrodes and were deposited to the left and right-side electrodes back and forth repeatedly to give good alignment.

7.2.3 Near-field Electrospinning

Another method to control deposition of nanofibers is the so-called near-field electrospinning (NFES),[26] where the needle-to-collector distance is reduced to enhance the controllability of the fiber deposition positions as illustrated in Figure 7.1(a). The needle-to-collector distance is reduced to mm range and the applied voltage is reduced to the order of 1 kV. The reduction in distance and the increased electric field (as a result of a much shorter distance) make it possible to control nanofiber deposition on the collector by utilizing the stable liquid jet region. In the earlier stage of the near-field electrospinning process, a dip-pen type approach was adopted as shown in Figure 7.1(a); the experimental image photo is shown in Figure 7.1(b). In this case, repeated dipping into the polymer solution was necessary to obtain more polymer sources, which interrupted the deposition process. Continuous near-field electrospinning was later developed[42] by

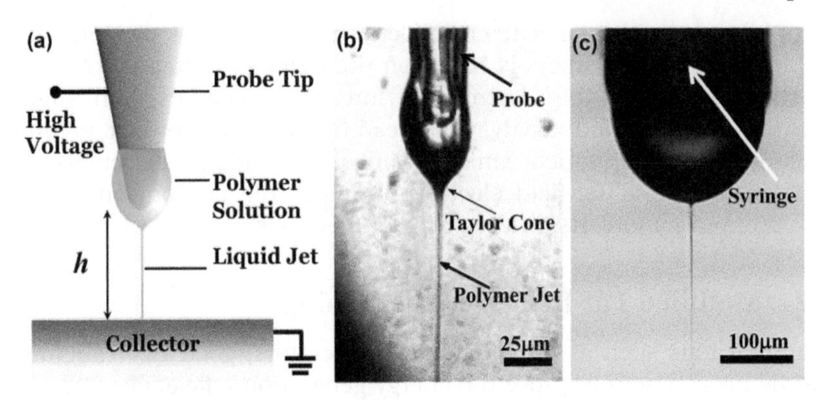

Figure 7.1 (a) Schematic diagram of the setup of the near-field electrospinning (NFES) process where the probe-to-collector distance, h, is reduced to mm range to utilize the stable polymer jet region for better controllability of the deposition locations. The polymer is attached to the top of the tungsten electrode in a manner analogous to that of a dip-pen. (b) An optical image photo showing the NFES with the dip-pen approach in operation. Reprinted with permission from ref. 26. Copyright (2006) American Chemical Society. (c) An optical image photo showing the continuous NFES in operation by using a syringe to supply the polymer solution continuously.
Reprinted with permission from ref. 42. Copyright 2008, American Institute of Physics.

using a syringe instead of a probe as shown in Figure 7.1(c) such that polymer solution can be supplied continuously to deposit continuous nanofibers. These innovative modifications make continuous NFES to maintain the continuous characteristics of conventional FFES with superior controllability of the deposition locations. These and other similar approaches on NFES[43–45] can be the foundations to construct nanofiber nanogenerators previously unachievable by FFES such as parallel arrays of nanofibers for nanogenerators.[23]

The piezoelectricity of PVDF comes from its β-phase crystalline structure. For PVDF thin films, a post process including both electric poling and mechanical stretching is required to promote the formation of the β-phase PVDF. For electrospun PVDF nanofibers, the electrospinning process can provide both high electrical field and mechanical stretching *in situ* without a post-poling process. As a result, most literature reports, as summarized in Table 1, have shown piezoelectricity or high contents of a β-phase structure in electrospun PVDF nanofibers. For example, the near-field electrospinning process has an average electrical field about one order of magnitude larger than the far-field electrospinning process and this could enhance piezoelectricity for nanogenerator applications.[22] Other nanofibers in Table 7.1 are from the far-field electrospinning process and only one has poor dipole alignments by using the second harmonic generation (SHG) method to detect the polarity of the nanofiber.[46] As can be seen, all other reports of the

Table 7.1 Fabrication parameters of PVDF-based nanofibers.

Material	Molecular weight	Solvent	Method	Typical bias	Tip-to-substrate	Mean field strength	Piezo-electricity?	Ref.
PVDF	172 000 (16% wt)	DMF	FFES	15 kV	15 cm	10^5 V m^{-1}	Yes	21
PVDF	534 000 (20% wt)	DMSO (50%) + acetone (50%)	NFES	1 kV	1 mm	10^6 V m^{-1}	Yes	22
PVDF	534 000 (20% wt)	DMF (60%) + acetone (40%)	FFES	12 kV	10 cm	1.2×10^5 V m^{-1}	No	46
PVDF	534 000 (12% wt)	DMF (40%) + acetone (60%)	FFES	12 kV	15 cm	0.8×10^5 V m^{-1}	Yes	47
PVDF	275 000 (various)	DMF + acetone various ratios	FFES	13 kV	15 cm	0.87×10^5 V m^{-1}	Yes	48
PVDF	Solef 1100 (20% wt)	DMF (60%) + acetone (40%)	FFES	15 kV	15 cm	0.75×10^5 V m^{-1}	Yes	49
P(VDF-TrFE)	(77 : 23 mol%)	Butan-2-one	FFES	20 kV	10 cm	2×10^5 V m^{-1}	Yes	50
PVDF	687 000 (10–20% wt)	DMF	FFES	20 kV	15 cm	1.3×10^5 V m^{-1}	Yes	51
PVDF + CNT	115 000 (20%)	DMF (60%) + acetone (40%)	FFES	15 kV	N/A	N/A	Yes	52
PVDF	Foraflon® 4000HD (various)	DMF + acetone various ratios	FFES	10 kV	3 cm	3.3×10^5 V m^{-1}	Yes	53
PVDF	268 000	DMF (60%) + acetone (40%)	FFES	15 kV	20 cm	0.75×10^5 V m^{-1}	Yes	54
PVDF + MWCNT	600 000	DMF + acetone	FFES	18 kV	15 cm	N/A	Yes	55
PVDF	534 000	DMSO + acetone + fluorosurfactant	NFES	1.2 kV	0.5–1 mm	N/A	Yes	56
P(VDF-TrFE) + $BaTiO_3$	Solef 1010, 70/30(Solvay)	DMF + methylethylketone (MEK)	FFES	~35 kV	~30 cm	N/A	Yes	57

far-field electrospinning process have either measurements on the strong contents of β-phase PVDF,[21,22,32–34,59–62] or electrical signals of piezoelectric effects in applications such as sensors[37] or energy harvesters.[21,22]

7.2.4 Mechanical Forcespinning

Another process, mechanical forcespinning, can also construct fibers.[58,59] In a recent study, the roles of mechanical stretching and electric poling for the conversion from α- to β-phase PVDF has been investigated for fibers fabricated by the electrospinning and mechanical forcespinning process using FTIR and XRD spectroscopic techniques. Figure 7.2(a) and (c) show the fabricated fibers from the 'baseline' polymer solution (16 wt% and $V_{NMP}/V_{acetone} = 5/5$) by electrospinning and forcespinning processes, respectively, while Figure 7.2(b) illustrates the force spinning process.[60] The electrospun fibers are constructed under an electrical bias of 7.5 kV with a polymer solution flow of 60 μL per hour and electrode-to-collector distance

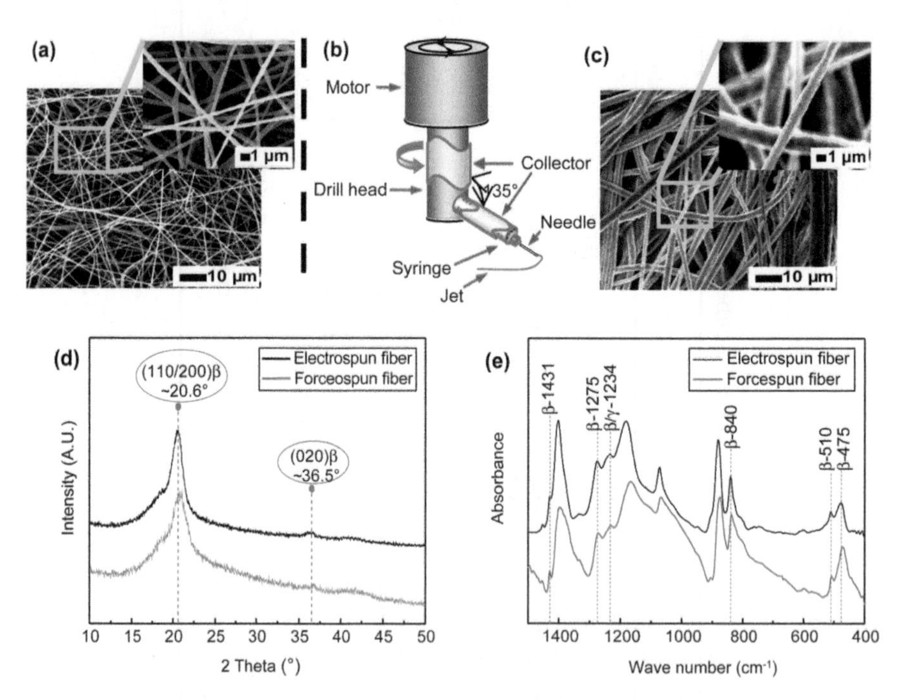

Figure 7.2 (a) SEM photos of PVDF fibers from prototype (7.5 kV, 60 μL per hour 0 cm) electrospinning; (b) schematic diagram of the forcespinning setup; (c) SEM photos of PVDF fibers from high-speed (11 000 rpm) forcespinning; (d) XRD patterns and (e) infrared spectra of the above electrospun fiber mat and forcespun PVDF fiber mat. The same polymer solution (16 wt%, $V_{NMP}/V_{acetone} = 5/5$) is used for both spinning experiments. Reproduced from ref. 60 with permission from the Royal Society of Chemistry.

of 10 cm and it is observed that they have a relatively uniform diameter in the range of 315 to 740 nm. On the other hand, the forcespinning process is conducted under a high speed of 11 000 rpm and the resulting fibers have diameters that varies greatly ranging from 0.95 to 3.35 μm. The XRD results in Figure 7.2(d) show a very strong diffraction peak at $2\theta = 20.6°$ and a weak peak at $2\theta = 36.5°$ corresponding to 110/200 and 020 reflections of the orthorhombic β-phase crystal, respectively, in both electrospun and forcespun fiber mats, indicating the presence of a ferroelectric β-phase.[58,59] Meanwhile, the characteristic absorption bands of the β-phase at 475, 510, 840, and 1275 cm^{-1} are also observed in Figure 7.2(e) under FTIR. These results suggest that pure mechanical stretching can result in a high fraction of all-trans β-phase at 95% while electrospun fibers from the same material system can also reach a high fraction of β-phase at approximately 99%.[60] In summary, the results imply the importance of mechanical stretching in the formation of the β-phase in PVDF fibers while further studies are needed to have a full understanding of the piezoelectricity of nanofibers.

7.3 Characterization Method for Piezoelectricity

The applications of piezoelectric nanofibers strongly rely on good piezoelectric properties. While the formation the β-phase is the key to enhancing piezoelectricity for PVDF, the PZT nanofibers need to be in the pervoskite phase. Therefore, piezoelectric analysis techniques and tools will be necessary to characterize crystal/molecular structure and inherent piezoelectric property of electrospun nanofibers for optimal parameters during the electrospinning processes. Instruments such as X-ray diffraction analysis (XRD), Fourier transform infrared spectroscopy (FTIR), piezoresponse force microscopy (PFM) and second harmonic generation microscopy (SHG) have all been utilized for this purpose and they are briefly discussed below.

7.3.1 X-Ray Diffraction (X-Ray)

X-Ray diffraction (XRD) has been commonly used for the analysis of the crystalline structure of materials. Baji *et al.*[65] have analyzed PVDF fibers fabricated by far field electrospinning using XRD. It was found that the β-phase is most abundant in electrospun PVDF nanofibers while other crystalline forms also existed. Electrospun nanofibers with smaller diameters were found to have higher contents of the β-phase probably due to the strong stretching effect. Other groups have also utilized XRD to analyze the structures of electrospun polar β-glycine nanofibers[61] and polycrystalline barium titanate nanofibers.[62] For PZT fibers, XRD is interesting because it can be used to confirm the presence of the pervoskite phase of the material. This phase is crucial for piezoelectric PZT. Using a metallo-organics decomposition (MOD) method for PZT fiber fabrication, the pervoskite phase is formed during the fiber annealing process between 600 to 850 °C.[63] Wang *et al.* used XRD analysis during the annealing process.[63] They observe that above 600 °C, the pervoskite phase starts to appear while coexisting with the pyrochlore phase. It is only above 850 °C that the pyrochlore phase is completely replaced by the pervoskite phase. Chen *et al.* conduct a similar experiment using a PZT sol–gel, a poly vinyl pyrrolidone and alcohol mixture.[64] The XRD results confirm the pure pervoskite phase of the fibers with an annealing temperature of 650 °C.

7.3.2 Fourier Transform Infrared Spectroscopy (FTIR)

Fourier transform infrared spectroscopy (FTIR) can be used to characterize both the dipole orientation and crystallographic structure of nanofibers based on the sensitivity of CF_2 orientation changes. For example, Mandal *et al.* have used FTIR to examine the dipole orientation of electrospun P(VDF-TrFE) nanofibers.[50] They have shown that the dipoles were aligned in the direction of the electrical field during the electrospinning process.[50] A comparison between the FTIR spectra of an as-spun fiber and a heat-treated electrospun fiber (heated above the Curie temperature in order to assure the random dipole alignments) have resulted in a difference in absorbance of perpendicular polarized light for CF_2 sensitive wavelengths. Baji *et al.* have also confirmed the presence of the crystal β-phase using FTIR in good correspondence to the results obtained using XRD.[65] Similarly Wang *et al.* use FTIR in order to examine the PZT fiber annealing process with which the decomposition temperature of organic groups as well as the appearance of the pervoskite phase can be determined.[66]

7.3.3 Piezoelectric Force Microscopy (PFM)

This method is the most direct process to measure and detect the piezoelectricity of the material. Piezoelectric force microscopy (PFM) measurements are based on the detection of a voltage induced deformation of the piezoelectric material. The data obtained allows quantification on the degree of the polarization as well as the polarization direction of the tested sample.[67] The PFM setup is comparable to atomic force microscopy (AFM) in that it uses a micro tip in contact with the surface while applying a sinusoidal bias voltage. In the case of a piezoelectric material with a poling direction in parallel to the direction of the PFM bias voltage, this leads to a detectable deformation of the material due to the piezoelectric response which is considered as the 'in phase' response. For an opposite poling direction at the contact area, the observed phase should shift by 180 degrees as the 'out of phase' response. Lateral deformations can also be sensed by the detection of the torsional responses of this same tip. PFM is particularly well suited for the analysis of individual nanofibers in contrast to other characterization tools because it allows local piezoelectric measurements. For example, in the work to study triglycine sulfate (TGS) nanocrystal embedded poly(ethylene) oxide (PEO) nanofibers, several images were taken.[68] The morphology of the nanofiber mat was first characterized by the AFM scan in

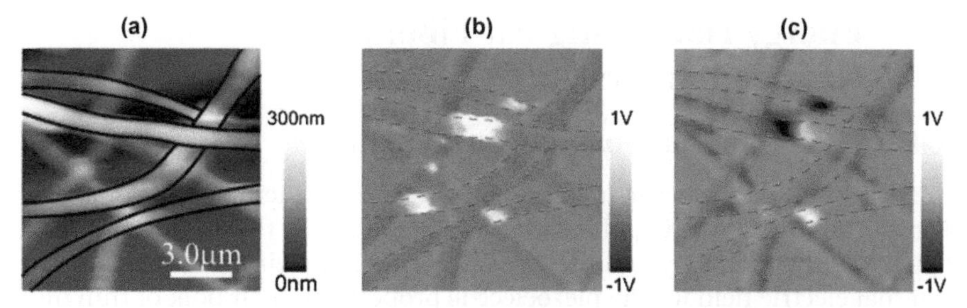

Figure 7.3 (a) AFM image of electrospun triglycine sulfate (TGS) nanocrystal embedded poly(ethylene) oxide (PEO) nanofibers showing their morphology by the far-field electorspinning process. (b) Vertical mode PFM image of the same sample area with results corresponding to the effective longitudial piezoelectric coefficient, $d_{33\text{eff}}$, of the nanofiber. (c) Lateral mode PFM image of the same sample corresponding to the effective shear piezoelectric coefficient, $d_{15\text{eff}}$, of the nanofiber.
Reprinted with permission from ref. 68. Copyright 2010, American Institute of Physics.

Figure 7.3(a). For the vertical, out-of-plane PFM measurements in Figure 7.3(b), the image contrast is roughly proportional to the effective longitudinal piezoelectric coefficient, $d_{33\text{eff}}$. The response in the lateral mode PFM is shown in Figure 7.3(c) that reflects the effective shear piezoelectric coefficient, $d_{15\text{eff}}$. The image shows a random distribution in both phase and amplitude of the response, which is attributed to the unsystematic orientation of crystals in the nanofiber. Similar experiments were conducted by Baji *et al.* for electrospun PVDF nanofibers.[65]

PFM is also used for characterizing PZT fibers. For example, Wang *et al.* determined the polarization domains of electrospun PZT fibers.[69] The domains of the original fibers were randomly oriented, up, down or parallel to the substrate on which the fiber was deposited. A post-poling is therefore necessary to implement these PZT fibers in a nanogenerator.

7.3.4 SHG and Raman Spectroscopy

The polarity of a piezoelectric material can also be characterized by SHG as demonstrated by Isakov *et al.* who analyzed a 532 nm second harmonic spectrum from a Nd : YAG laser on β-glycine and γ-glycine nanofibers.[61] The nonlinear optical responses were attributed to the embedded nanocrystals with their polar axis oriented in parallel with the fiber mat plane. The same technology has been used to measure the piezoelectric property of other nanofibers such as electrospun PVDF and α-helical poly(α-amino acid) fibers.[46] Raman spectroscopy can be used to detect various low frequency modes in a system. It is used to characterize material properties such as crystallographic orientation. In PZT fibers, the analysis tool has been applied to detect the formation of PZT during the annealing process.[63]

7.4 Energy Harvesting *via* Fiber-based Nanogenerators

7.4.1 PZT Nanofiber-based Nanogenerators

PZT is a good piezoelectric material; its crystalline structure is illustrated in Figure 7.4(a). An electric polarization of PZT can shift the Zr/Ti atom up or down and the atoms remain their positions after applying and removing an external electric field for the piezoelectric property. In their bulk or thin film formats, PZT can generate a higher voltage compared with other piezoelectric materials for sensing,[70] and actuation[71] and energy harvesting applications. However, as a ceramic material, bulk PZT is more fragile compared with organic PVDF while a recent study has shown that PZT nanowires have very good mechanical strength.[15] On the other hand, PVDF has superior piezoelectric properties compared with other types of polymeric materials due to its polar crystalline structure. In nature, the PVDF polymer consists of at least five different structural forms depending on the chain conformation of *trans* (T) and *gauche* (G) linkages. Figure 7.4(b) shows the crystalline structure of the α and β-phase, respectively. While the α-phase is known as the most abundant form in nature, the β-phase is responsible for most piezoelectric responses due to its polar structure with oriented hydrogen and fluoride (CH_2–CF_2) unit cells along with the carbon backbone. Typically, β-phase PVDF is acquired *via* electrical poling and mechanical stretching processes during the manufacturing process to align the dipoles in the crystalline PVDF structures as illustrated in Figure 7.4. However, the poling process is typically hard to implement within an electrospinning process due to the fibers' physical constraints.

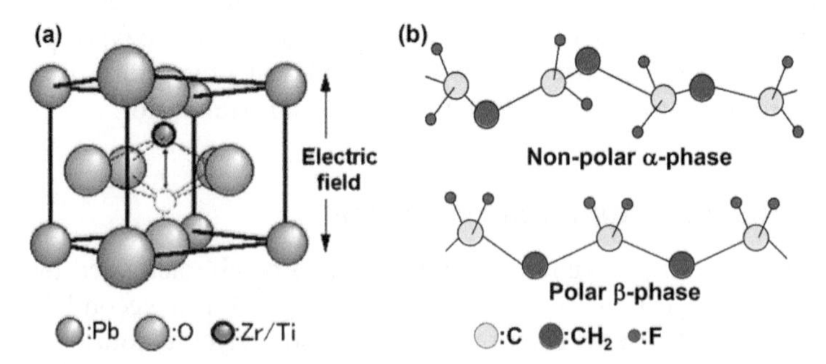

Figure 7.4 (a) A schematic diagram showing the crystalline structures of PZT. An electric polarization of PZT can shift the Zr/Ti atom up or down and the atoms remain in their positions after applying and removing an external electric field for the piezoelectric property. (b) Schematic diagrams showing crystalline structures of PVDF: (top) non-polar α-phase, and (bottom) polar β-phase. The dipoles in the non-polar, α-phase PVDF could be stretched and oriented by an electrical field to become the polar, β-phase structure under electrical poling and mechanical stretching. Reprinted with permission from ref. 72. Copyright 2012, Elsevier.

In the area of energy harvesting using PZT nanowires, Wang *et al.* have successfully demonstrated a series of nanogenerators using epitaxially grown PZT nanowires with outstanding performances.[1,73–75] These PZT nanowires have limitations in length as constrained by the fabrication method, similar to the ZnO nanowire-based nanogenerators. PZT microfibers have also been machined either by using a computer controlled dicing saw to cut fabricated PZT thin films by using mixed oxide powders with extrusion/suspension spinning, or by a wet-chemical sol–gel process *via* viscous plastic processing (VPP).[76] These fibers can be rectangular or circular with diameters from 50–500 μm as shown in Figure 7.5(a). A nanogenerator has been constructed in a similar fashion by using a lithography process to define PZT ribbons (5 μm in width and 500 nm in thickness) and a dry transfer process using a polydimethylsiloxane (PDMS) substrate as illustrated in Figure 7.5(b).[77] In two recent reports, on the other hand, by Chen *et al.*[14] and Zhang *et al.*,[16] the researchers have applied far-field

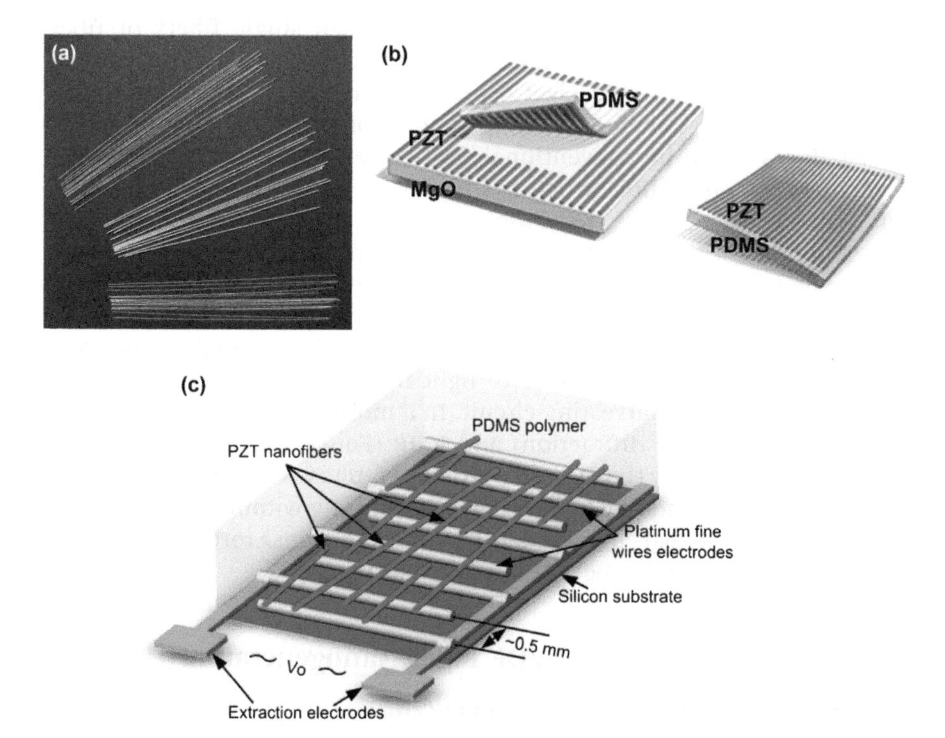

Figure 7.5 (a) Rectangular or circular PZT microfibers with different diameters.[78] Reprinted with permission from ref. 78. (b) A transfer process using a lithography process to define PZT ribbons and a PDMS substrate to transfer the PZT fibers.[77] Copyright (2010) American Chemical Society. (c) PZT nanofibers on two platinum comb-shaped electrodes.[14] The generator is completed with a PDMS cover.
Reprinted with permission from ref. 14. Copyright (2010) American Chemical Society.

electrospinning to produce long PZT nanofibers for nanogenerator applications. In the device demonstrated by Chen *et al.*, PDMS was used to cover electrospun PZT fibers on top of comb-shaped platinum electrodes as shown in Figure 7.5(c).[14] A post electric poling process was conducted at 140 °C for 24 hours at 4 V μm^{-1} between two adjacent platinum electrodes for enhanced piezoelectricity. When pressure was applied using a Teflon stack/human finger, the device was able to generate an output voltage of up to 1420 mV. Zhang *et al.* have also demonstrated a PZT nanowire-based nanogenerator using electrospun PZT nanofibers without using any post poling process.[16] These PZT nanofibers were contacted on either end using silver paste. A three-point bending test with an applied strain of 0.5% was used to create an output voltage of 170 mV, which can be attributed to the strain-induced charge of the PZT nanofibers.

7.4.2 Randomly Distributed PVDF Fiber Nanogenerators

In addition to the aforementioned well-organized single fibers or fiber arrays as nanogenerators, studies have shown that randomly distributed fibers made by the far-field electrospinning process and other processes can also be used for energy harvesting applications. The key demonstration is that fibers made by conventional far-field electrospinning could also form a β-phase crystalline structure without an additional poling process. For example, Fang *et al.* have demonstrated a PVDF membrane type nanogenerator with conventional electrospinning in which the PVDF membrane was fabricated by a network of electrospun PVDF nanofibers.[21] Without an additional poling process, a 140 μm-thick PVDF membrane was able to generate up to 7 V under compression strain at a high strain rate. The energy output was sufficient to light up an LED with the help of a capacitor type energy harvesting circuit. In a more fundamental study, Baji *et al.* used XRD (X-ray diffraction) and FTIR (Fourier transform infrared) measurements to show that a good portion of β-phase PVDF nanofibers did exist in the fibers made by far-field electrospinning without a post poling process.[65] The hysteresis loop under PFM (piezoresponse force microscopy) further confirmed the ferroelectricity of nanofibers.

7.4.3 Orderly Patterned PVDF Fiber Nanogenerators

7.4.3.1 Single PVDF Fiber Nanogenerator

During the typical commercial piezoelectric PVDF thin-film production process, a high electrical potential and mechanical stretching are applied at a raised temperature for enhanced piezoelectricity.[79] PVDF nanofibers fabricated by the conventional electrospinning process are under a high bias voltage (>10 kV) which could transform some non-polar α-phase structures to polar β-phase structures for piezoelectricity.[46–49] The near-field electrospinning (NFES) process as shown in Figure 7.6(a) also possesses an

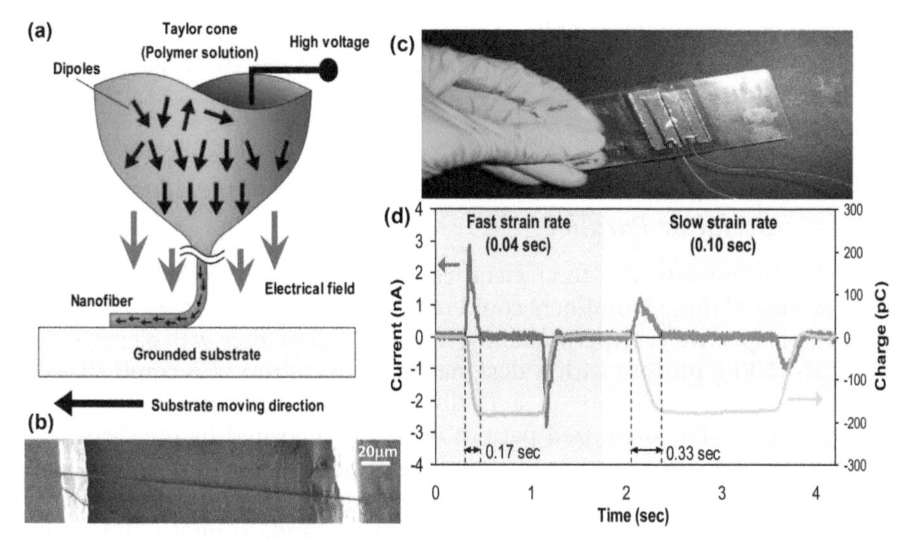

Figure 7.6 (a) A schematic diagram of the near-field electrospinning process showing possible dipole directions (black arrows) and electrical field direction (red arrows). The PVDF polymer solution experiences mechanical stretching and *in situ* electrical poling during the formation of nanofibers due to the high electrostatic field toward the substrate. (b) SEM microphoto showing an electrospun PVDF nanofiber on two contact pads. (c) An optical image of a single nanofiber nanogenerator on top of a plastic substrate with two electrical wires. (d) Testing results of a PVDF nanofiber nanogenerator under a same mechanical strain but with varying strain rate. As the strain rate increases, a higher output current is produced while the total charges generated in both cases remain approximately the same.

inherent high electric field with *in situ* mechanical stretching for the possible alignment of dipoles along the longitudinal direction of the nanofiber.[22] In the experimental demonstration, a single PVDF nanofiber was electrospun across two adjacent electrodes placed 100–600 μm apart and fixed to the contact electrode on either side, using conductive silver paste as shown in Figure 7.6(b). It is noted that if the dipoles are aligned along the longitudinal axis of the fiber, the d_{33} mode could be responsible for the main piezoelectric responses when the fiber is compressed or stretched in the axial direction. Figure 7.6(c) shows an optical photo of the electrospun PVDF nanofiber on top of a flexible plastic substrate with two contact electrodes and electrical wires. Experimentally, a single PVDF nanofiber-based nanogenerator was able to generate 0.5–3 nA of current and 5–30 mV of voltage under repeated long term reliability tests without a noticeable performance degradation. The study also validated that both voltage and current generation are related to the strain rate. When keeping the total strain at the same level, higher electrical responses were monitored under higher strain rate while the total accumulated charges remained about the same under the same magnitude of applied strain as shown in Figure 7.6(d).

7.4.3.2 *Multiple PVDF Fiber Nanogenerators Connected in Serial or Parallel*

In order to increase the total electrical outputs, either serial or parallel connections of these nanofibers could result in multiplied voltage or current outputs, respectively. Experimentally, a single PVDF fiber was electrospun using the NFES process with a designed pattern on top of a comb-shaped metal electrode fabricated on a flexible polymer substrate as shown in Figure 7.7(a).[23] The organized pattern was well deposited by the near-field electrospinning process using a computer-controlled x-y stage on the collector in Figure 7.7(b). Figure 7.7(c) briefly explains the fabrication process on top of a polymer substrate. It started with a standard photolithography process to define the comb-shaped electrode. A room-temperature silicon dioxide coating was conducted and followed by the deposition of the gold electrode layer (a thin chromium layer was used as the adhesion layer). The near-field electrospinning process was conducted to deposit PVDF

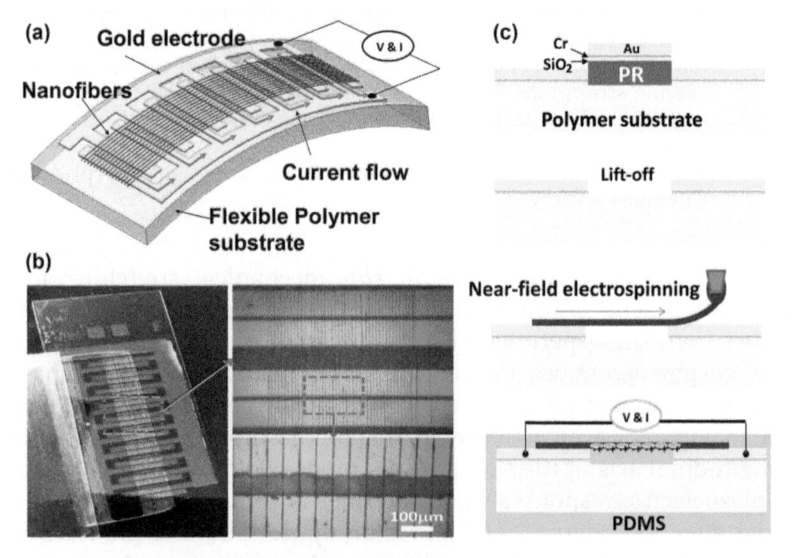

Figure 7.7 (a) A schematic diagram showing an arrayed PVDF fiber structure and a comb-shaped electrode on top of a flexible substrate to increase the electrical current outputs by parallel connection. A continuous near-field electrospinning process is used to control the depositions of these fibers. (b) Fabrication results showing the energy harvester unit. A total of 500 energy harvesting sections formed by 50 fibers and 10 electrodes are found to produce 30 nA of current. (c) Fabrication process.
Reprinted with permission from ref. 72. Copyright (2012) Elsevier.

nanofibers and PDMS was used to encapsulate the whole device. The lift-off process was performed to remove the photoresist (PR). The device was designed to amplify the current output by using the parallel connection to collect possible current generations from each individual nanofiber. A single energy harvester was composed of a total of 500 sections formed by 50 parallel fibers and 10 pairs of electrodes and the monitored peak current was about 30 nA. Key accomplishments in this work include: capability to place aligned nanofibers in an orderly manner, possibility to achieve higher electrical outputs in scavenging energy by using nanofibers connected in parallel, and feasibility of electric poling after the nanofiber deposition process using the comb-shaped electrodes.[23]

7.4.3.3 PVDF Fiber Nanogenerators by the Hollow Cylindrical Near-field Electrospinning Process

A hollow cylindrical near-field electrospinning (HCNFES) process has been proposed to address production and performance issues encountered previously in either far-field electrospinning (FFES) or near-field electrospinning (NFES) processes.[80,81] By the introduction of a rotating glass tube collector, PVDF fibers with small diameters, smooth surface morphology, high density, and good piezoelectricity have been accomplished using the HCNFES process as a potential building block to construct energy harvesters. In the HCNFES process, a rotating glass tube collector (diameter: 20 mm; thickness: 1 mm; length: 200 mm) is used to collect the electrospun fibers. A copper foil is placed in the internal wall of the glass tube and an electrical brush is attached as a grounding electrode. The glass tube collector significantly reduces the occurrence of short circuit. By employing a DC motor to turn the tube collector, and controlling the uniaxial movement, this method is able to rapidly and continuously collect electrospun PVDF fibers. In the prototype experiments, a high voltage of 10–16 kV is used and the tip-to-tube distance is about 0.5 mm. The tube collector rotates at velocities of 900–1900 rpm and the corresponding tangential speed on the surface is 942.3–1989.3 mm s^{-1}. The X-Y control platform has a motion speed of 2 mm s^{-1} and a travel distance of 50 mm. The fiber arrays fabricated by HCNFES can have controllable structural thickness based on layer-by-layer assembly as show in Figure 7.8(a) and (b). After a long period of collection time, there could be slight randomness as illustrated in Figure 7.8(b). The main reason could be the presence of residual charges on the electrospun fibers. One can easily remove the tube collector and extract the nonwoven fiber fabric (NFF). Furthermore, it is observed that under higher mechanical stretching speed, ultra-thin PVDF fibers (less than 1 μm) can be constructed and the extra mechanical stretching seems to help alleviate the surface defects such as holes and voids. When the flexible textile-fiber-based PVDF harvester is subjected to different stretch–release cycling frequencies of 2–7 Hz with 0.05% strain, electricity can be generated and recorded.

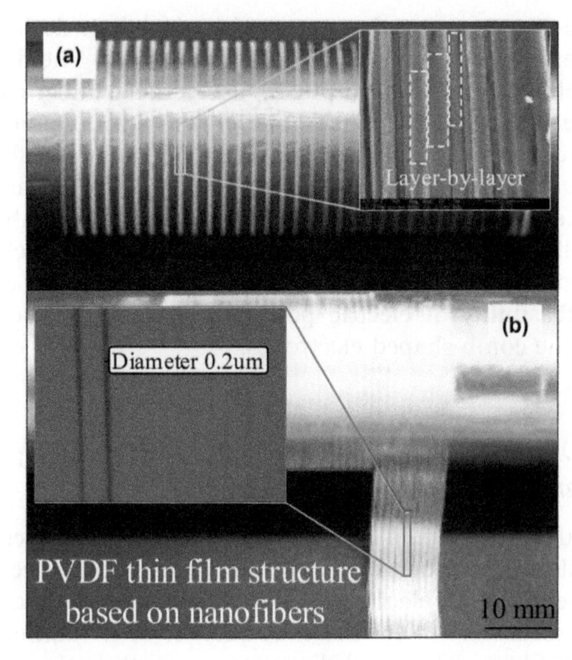

Figure 7.8 (a) Optical photos of the electrospun PVDF fiber arrays with well-controlled patterns on a glass tube collector using the HCNFES process with different diameters. The SEM image shows densely packed and well-aligned PVDF fiber arrays. (b) Ultra-long electrospun PVDF fibers are rolled around the glass tubes.

7.4.4 Towards System Level Demonstrations

In the area of utilizing conventional far-field electrospinning processes for PVDF nanogenerators, Hansen *et al.* have demonstrated an energy harvesting system combining a PVDF nanogenerator and a biofuel cell.[82] The PVDF nanofiber network was fabricated using a modified far-field electrospinning process as shown in Figure 7.9(a) to align individual nanofibers on to a Kapton film. The integrated system with both a piezoelectric PVDF nanofiber nanogenerator and a biofuel cell went through an in-plane post poling process at 0.2 MV cm^{-1} for 15 minutes to enhance the piezoelectric response. With a fixed strain rate of 1.67% per second, voltage and current outputs as high as 20 mV and 0.3 nA were recorded, respectively, as shown in Figure 7.9(b). This work expands on the aforementioned PVDF nanofiber nanogenerators using near-field electrospinning with a new pathway to utilize conventional electrospinning of PVDF nanofibers for nanogenerator applications. Furthermore, this is also the first demonstration of the integration of nanofiber nanogenerators with another energy system—a biofuel cell. More recent efforts in integrated energy harvesting systems have been applied to wearable applications with interesting demonstration platforms. Zeng *et al.* developed all fiber based systems in which PVDF was mixed with

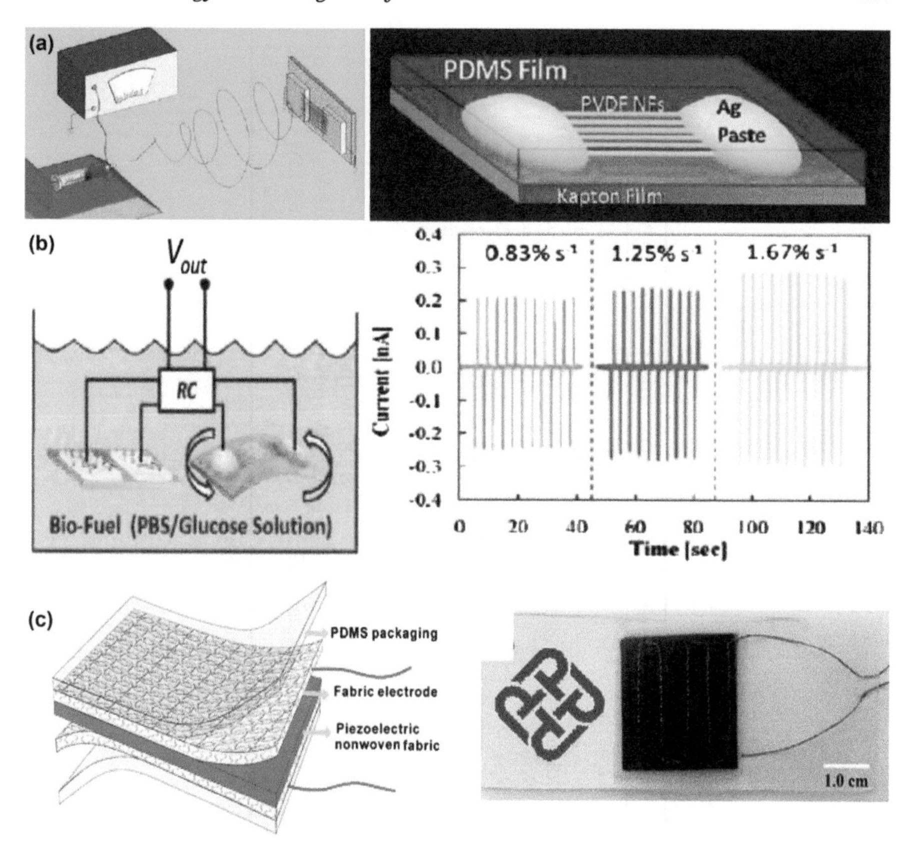

Figure 7.9 (a) A PVDF nanogenerator system is fabricated using far-field electrospinning and an individual nanofiber is placed manually on a Kapton film by fixing both ends with silver paste. (b) With the PMDS cover and a post electric poling process, the nanogenerator generates 20 mV of voltage and 0.3 nA of current under 1.67% per second strain rate. Furthermore, this nanogenerator is integrated with a biofuel cell to boost the total energy output.[82] Reprinted with permission from ref. 82. Copyright (2010) American Chemical Society. (c) The randomly distributed nanofibers (PVDF + $NaNbO_3$) are sandwiched between electrodes and PDMS to generate electrical power *via* the piezoelectric response. Reproduced from ref. 83 with permission from the Royal Society of Chemistry.

$NaNbO_3$ and showed reproducible energy harvesting (3.4 V and 4.4 μA) even after 1 000 000 compression testing as shown in Figure 7.9(c).[83]

7.4.5 State-of-the-art Piezoelectric Fiber-based Nanogenerators

Table 7.2 summarizes the key characteristics of nanofiber nanogenerators from aforementioned work. As can be observed in the table, electrospinning

Table 7.2 Summary of nanofiber nanogenerators.[a]

Material	Synthesis method	Diameter (nm)	Strain/ strain rate	Peak current	Peak voltage	Ref.
PZT	FFES (M)	60	12%	N/A	1.63 V	14
	FFES (S)	100	N/A	N/A	0.4 mV	15
	FFES (M)	50–150	0.5%	N/A	0.17 V	16
	FFES (M)	370	N/A	45 nA	6 V	84
	FFES (M)	N/A	N/A	53 μA/23.5 μA cm^{-2}	209 V	85
PVDF	FFES (M)	187	34 mm s^{-1} (at 5 Hz)	4 μA cm^{-2}	2.21 V	21
	NFES (S)	500–6500	0.085%	0.5–3 nA	5–30 mV	22
	NFES (M)	1000–2000	N/A	30 nA	0.2 mV	23
	FFES (M)	600	0.05%, 1.67% per second	0.3 nA	20 mV	82
P(VDF-TrFE)	FFES (M)	60–120	N/A	N/A	400 mV	50
PVDF + MWCNT	FFES (M)	300–1000	10 mm min^{-1}	81.8 nW (power)	6 V	55
P(VDF-TrFE) + $BaTiO_3$	FFES (M)	400–1000	N/A	25 μW (power)	5.02 V	57
ZnO + P(VDF-TrFE)	Hydrothermal	500–1000	0.1%/2.3% per second	2 nA cm^{-2}	~55 mV	86
$NaNbO_3$ + PVDF	FFES (M)	132–714	N/A	4.2 μA	3.2 V	83

[a]FFES: far-field electrospinning, NFES: near-field electrospinning, (S): single fiber, (M): multiple fibers.

has been used in all demonstrations to make continuous, long nanofibers either in an orderly fashion (NFES) or as randomly distributed networks (FFES). Nanogenerators using single nanofiber nanogenerators are marked as 'S', those using multiple nanofibers are marked 'M'. The diameters of these fibers are from as small as 60 nm to a few micrometres. While some reports didn't provide information on the applied strain or strain rate and some didn't record the peak current values, all have measured the peak voltage values from a wide range of less than 1 mV to 2.21 V (large number of nanofibers). In order to provide better prospects for future directions stemming from the current results, manufacturing methodologies, material properties and experimental procedures/characterizations are to be discussed in the following sections.

It is noted in a recent report that a PZT-based nanofiber nanogenerator has demonstrated up to 209 V peak voltage as shown in Figure 7.10.[76]

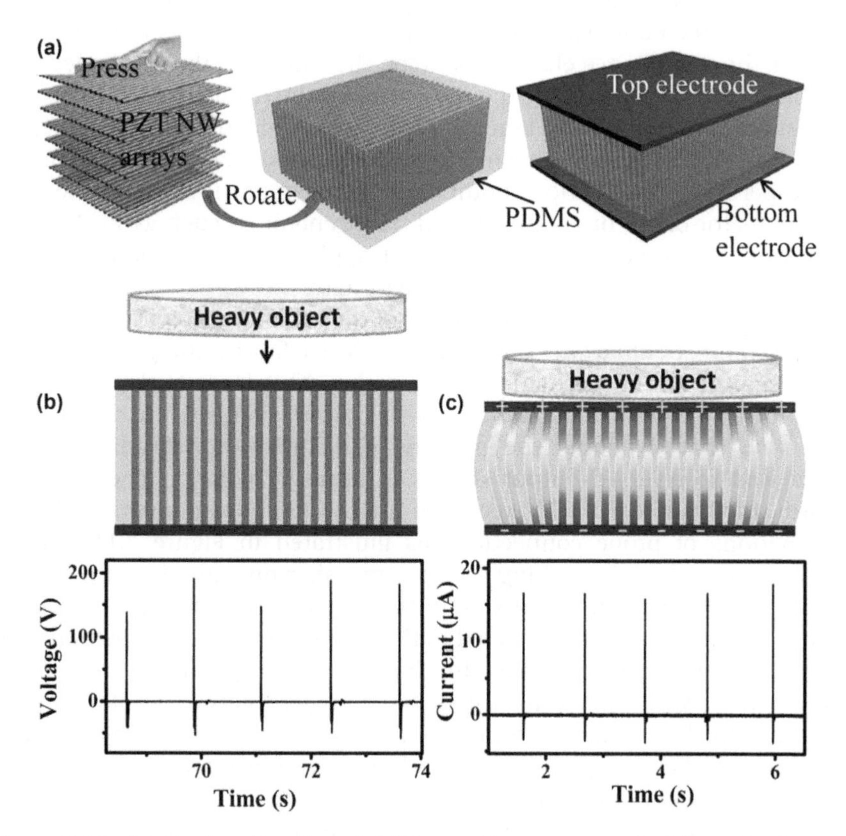

Figure 7.10 (a) Regionally aligned PZT nanofibers are stacked layer by layer to enhance the total electrical output. The stacked layers are then rotated 90 degrees for electrode connections. (b) Impact pressure ($\sim$0.53 MPa) by a heavy object generate strains in the stacked layers that can produce a local potential difference of up to 209 V. The serial connection, however, does not increase the total current output at about 20 μA. Reprinted with permission from ref. 85. Copyright (2013) American Chemical Society.

The multi-stacking structure can multiply output voltage to a high value. Impact pressure (~0.53 MPa) by a heavy object generated strains in the stacked layers. The serial connection, however, did not increase the total current output, which was measured to be about 20 μA, while the electrical power generated by the system was high enough to power LEDs.

7.5 Discussion

7.5.1 Measurement Artifacts

In order to test the performance of the nanogenerators, electrical outputs generated by mechanical inputs are often conducted for these nanofibers made of different materials. These signals are often very small and are difficult to measure due to the nature of the small size of the nanofibers. The background noise or artifacts can easily overshadow the true signals. It is especially problematic when dealing with single nanofiber structures. For example, the capacitance changes between the wires and the electrodes and possible electrical coupling of the measurement instruments could surpass the real nanofiber signals. Therefore, it is important to filter out or reject noises generated from the surrounding experimental environment. Some of these experimental validation conditions have been previously proposed for semiconductor-based or ceramic-based nanogenerators such as ZnO nanowires, while PVDF nanogenerators do not apply to some of the additional requirements such as the Schottky behavior test.

Switching polarity criterion: the polarity of the generated potential from the nanogenerator should be the same after the device fabrication process under the same direction of mechanical deformation. This leads to the electrons' flow direction to be independent from the connection wires to the instrument such that switching the polarity of the measurement instrument should lead to reversed output signals in terms of polarity. Therefore, to confirm the validity of the recorded piezoelectric responses, different combinations of probe connections as illustrated in Figure 7.11(a) with experiments on both stretching and compressing operations of the nanogenerators should be characterized. In the forward connection, the positive and negative probes were connected to the positive and negative potential of the nanogenerator, respectively. In this case, a single PVDF nanofiber nanogenerator with known polarities as shown is used as the illustration example.[22] In the backward connection, this connection is reversed. Since the polarity of the nanogenerator is fixed, this switching polarity test should generate electrical outputs with reversed responses.

Linear superposition criterion: when two devices are connected in serial, the voltage response has to be sum of each response as illustrated in Figure 7.11 (b), while the current response should be sum of each response when connected in parallel as illustrated in Figure 7.11 (c). Single PVDF nanofiber nanogenerators are used as the examples. All data are measured using the single PVDF nanofiber nanogenerator operated under the same strain, strain

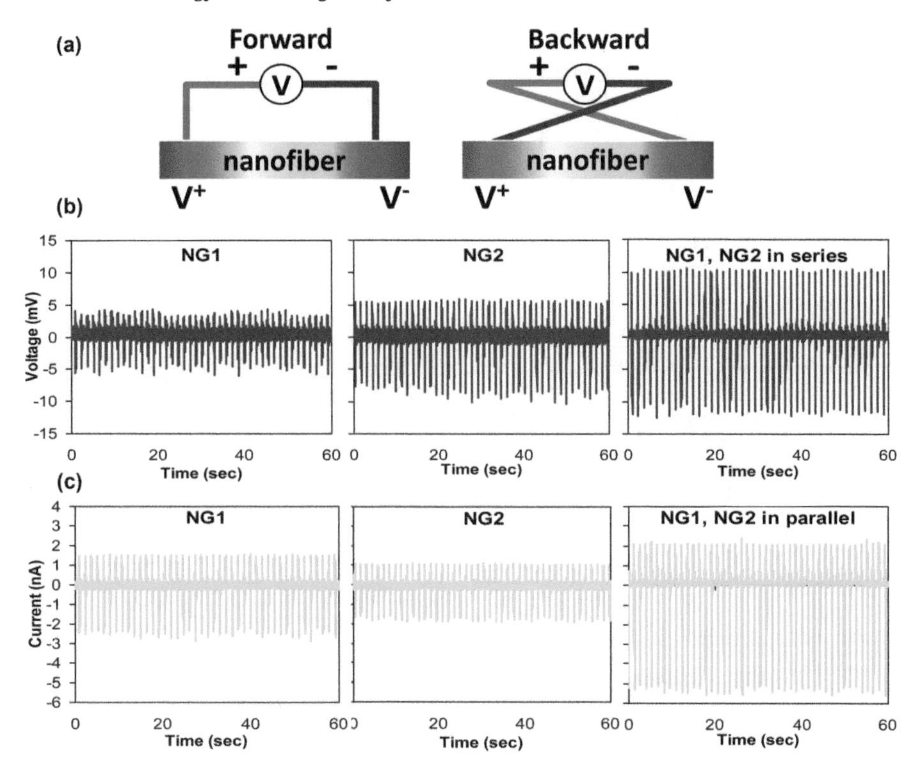

Figure 7.11 (a) Schematic illustrations of a forward connection and backward connection for the switching polarity tests. Here, a single nanofiber with known polarity is used as an example. (b) Output voltages of nanogenerator #1 and nanogenerator #2 subject to continuous mechanical stretch and release. When the two nanogenerators are connected in serial, their output voltages constructively add up. (c) Output currents of nanogenerator #1 and nanogenerator #2 subject to continuous mechanical stretch and release. When the two nanogenerators are connected in parallel, their output currents constructively add up. All data are measured using the single PVDF nanofiber nanogenerator operated under the same strain, strain rate, and frequency.
Reprinted with permission from ref. 22. Copyright (2010) American Chemical Society.

rate, and frequency. Similarly, connecting two identical devices in serial with opposite polarity should result in nearly zero response in voltage measurement.

Artifacts due to contact electrodes: in order to measure the electrical outputs from the nanogenerators, it is necessary to have at least two electrical contact pads (or multiple electrodes) typically placed on top of a substrate with nanofibers deposited on top of them as illustrated in Figure 7.12(a). The mechanical actuation of the substrate nanogenerator could result in a capacitance change due to the varying distance of adjacent contact electrodes and add to the true output signals as artifacts. Specifically, the total

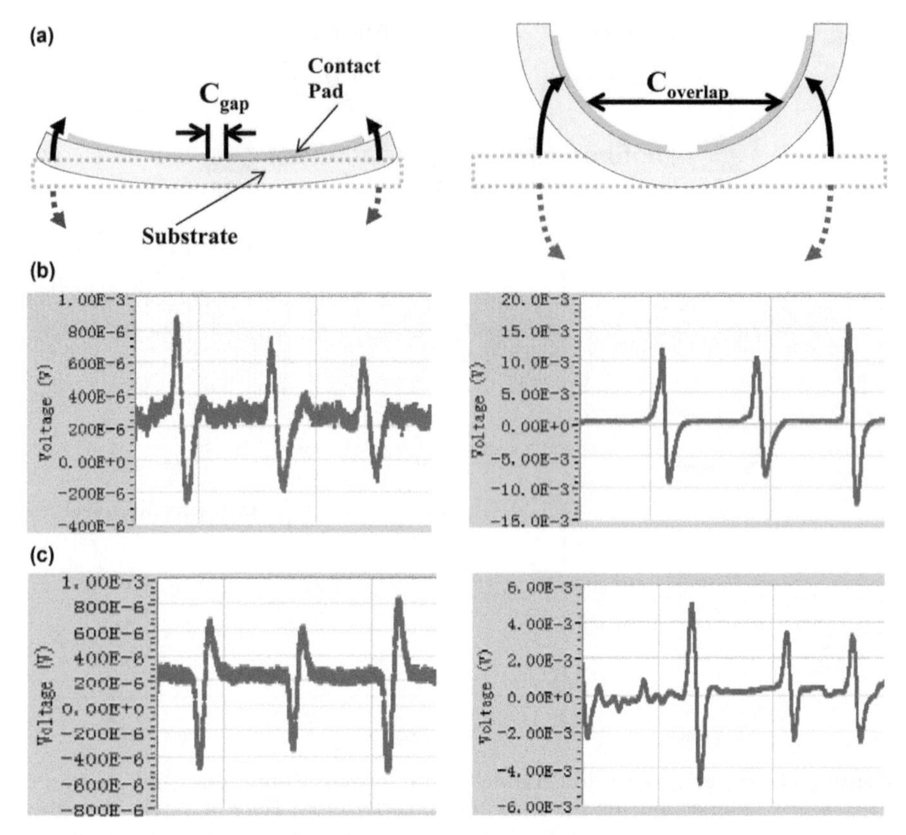

Figure 7.12 (a) Schematic diagram showing the possible artifacts of nanogenerator outputs due to the contact electrodes from the sources of C_{gap} and $C_{overlap}$ under small (left) and large (right) substrate actuation tests. (b) and (c) Experimental voltage outputs from upward and downward actuation tests, respectively, with small (left figures) and large (right figures) substrate actuation. Under small actuation, C_{gap} could be the dominating factor to change the polarity of the output signals of upward and downward actuation as shown. Under large actuation, $C_{overlap}$ could be the dominating factor and no polarity change is found due to upward or downward actuation as $C_{overlap}$ increases in both actuation directions.

capacitance of the device due to contact pads is the sum of the capacitance of the gap between adjacent electrodes and that of overlapping area as illustrated in Figure 7.12(a).

$$C_{\text{contact pad}} = C_{\text{gap}} + C_{\text{overlap}} \tag{7.1}$$

Since the magnitude of capacitance is proportional to the area and inversely proportional to the gap distance, a small contact electrode area and large gap distance are preferable to reduce the impact of artifacts due to the contact pads. Under a small actuation as shown on the left side of Figure 7.12(a), $C_{overlap}$ could be insignificant while under a large actuation as shown on the right side of Figure 7.12(a), $C_{overlap}$ could dominate the total responses. By switching the actuation direction upwards or downwards, the output responses due to the electrode gap will have opposite effects as C_{gap} between two electrodes decreases or increases, respectively. On the other hand, $C_{overlap}$ should always increase due to the increased overlapping area under either the upward or downward actuation. Figure 7.12(b) shows experimental results based on upward actuation and releasing to the original flat position under small actuation (left) and large actuation (right), respectively. Figure 7.12(c) shows results from the same tests based on downward actuation and releasing to the original flat position. The specimen used in these tests has an electrode area of 1000 μm×1000 μm and a cross-sectional area in a gap of 100 nm×1000 μm without depositing any nanofiber. The output signals from small actuation (left side of figures(b) and (c)) are smaller than 1 mV, which can be considered as noise when compared with the typical voltage outputs of a single nanofiber at tens of mV[22] while the clear change in the polarities of the output signals suggests the outputs come from the changes in C_{gap}. On the other hand, under large actuation, the contribution of C_{gap} is small and the changes of $C_{overlap}$ dominate the output signals while either upward or downward actuation should increase $C_{overlap}$. As expected, on the right side of Figure 7.12(b) and (c), larger outputs up to 10 mV were observed while the polarity changes are the same in both figures under either upward or downward actuation. Nevertheless, these effects could be easily filtered out if the 'switching polarity criterion' is checked.

7.5.2 Energy Conversion Efficiency

The energy conversion efficiency of nanofiber nanogenerators is an important measurement for energy scavenging applications. It can be calculated by comparing the output electrical energy with the input mechanical energy from the applied strain. The electric energy W_e, can be estimated by integrating the product of output voltage and current of the nanogenerator. The elastic strain energy can be estimated by using (in the case of a single nanofiber):

$$W_s = \frac{1}{2} E A \varepsilon^2 L_0 \tag{7.2}$$

where E is Young's modulus of the material, A is the cross-sectional area, ε is the strain applied on the material, and L_0 is the length of the material. Several reports have found that nanostructures could have high energy generation efficiency in nanogenerator applications, including PVDF nanofibers,[22] ZnO nanowires[1] and PZT nanofibers (the piezoelectric constant for nanofibers was 0.079 mV N^{-1} as compared with PZT bulk material at

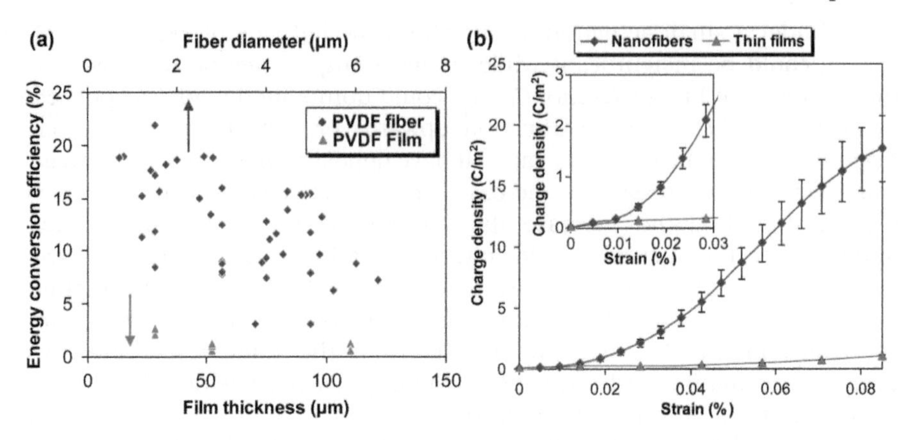

Figure 7.13 (a) Plots of measured energy conversion efficiency of PVDF nanogenerators and thin films with different feature sizes. (b) Experimental results of PVDF thin film and nanofiber charge density (generated charges divided by electrode area) with respect to applied strain. The charge density of the PVDF nanofiber increases nonlinearly when the applied strain is larger than *ca.* 0.01%. The inset shows the details under small strains.[22]
Reprinted with permission from ref. 22. Copyright (2010) American Chemical Society.

0.025 mV N^{-1}).[14] In the study of single PVDF nanofibers, 45 PVDF nanofiber nanogenerators have been tested that were 600 nm–6.5 μm in diameter and 100–600 μm in length.[22] The average conversion efficiency was calculated as 12.5%, a value that is much greater than typical energy harvesters made from reported piezoelectric PVDF thin films (0.5–4%)[87,88] and commercial PVDF thin films (0.5–2.6%) tested under the same conditions as PVDF nanofibers as illustrated in Figure 7.13(a). The efficiency of the PVDF thin film was estimated using the same method/experiment as the PVDF nanogenerators. However, the d_{31} mode was utilized in the PVDF thin film instead of the d_{33} mode to accommodate experimental setup. The general trend in Figure 7.13(a) indicates that nanogenerators with smaller diameters exhibit higher energy conversion efficiencies, even with variable piezoelectric properties resulting from slight differences in processing conditions. Several possible reasons have been proposed for the enhanced electromechanical response, including a possible higher degree of crystallinity and chain orientation,[89] a difference in elastic boundary conditions (thin film/bulk samples have metal electrodes on both ends as constraints), the physical size of the nanogenerators, which could promote size-dependent piezoelectricity such as 'flexoelectricity',[90] and possible nonlinear piezoelectric responses coming from extrinsic contributions known as 'domain wall motion'.[91] For example, a much smaller domain wall motion barrier in PVDF nanofibers (*ca.* 0.01%) as shown in Figure 7.13(b) was observed and it resulted in large piezoelectric responses for strain higher than 0.01% for higher energy conversion efficiency.

7.5.3 Enhancing Strong Piezoelectricity in Nanofibers

As the most important measure of all piezoelectric materials, piezoelectricity in nanofibers should be investigated thoroughly as it relates strongly to their performance, including the energy conversion efficiency. However, various processing parameters and conditions easily affect the piezoelectricity not only in nanofibers but also in large scale thin films. For example, many of the commercially available PZT or PVDF thin films/products have been processed with unknown ingredients and/or process parameters. Although there are a wide variety of studies in the literature on the piezoelectric properties of laboratory-made piezoelectric nanofibers, many of them lack details in processing data. Furthermore, most of the published reports only characterize piezoelectricity of nanofibers without specific characterizations on the specific piezoelectric mode information. Differences in material compositions and possible experimental variations/inaccuracies could be the fundamental reason that there are a wide range of nanogenerator results on nanofiber nanogenerators in Table 7.1 as well as piezoelectric properties as listed in Table 7.2. For example, molecular weight and concentration, solvent types and mixing percentage, electrospinning methodologies, applied bias and electrode-to-collector distance could all affect the piezoelectric properties of electrospun nanofibers and there is a wide range of these parameters used by different groups as shown in Table 7.1. In the work of Yee *et al.*,[92] it is concluded that electrospinning of PVDF from DMF–acetone solution would promote the formation of a β-phase and aligned electrospinning using a rotation disk would result in the *c*-axis of the β-phase crystallites in orientation along the fiber axis. They also suggest that the formation of the β-phase is likely to be by the electric field instead of the mechanical and shear force. In another study using a modified rotating disk collector, they further suggest that effective stretching by the rotation disk could help the formation of a highly oriented β-phase, which is quite different from their previous report.[93] A study by Zhong *et al.* on PVDF/polyacrylonitrile and PVDF/polysulfone nanofibers also has a different conclusion.[94] They found that mechanical stretching is more effective than electric poling to induce a ferroelectric phase. Clearly, these recent reports indicate a lack of uniform and optimal processing parameters and lack of fundamental understanding of the piezoelectric effects of nanofibers.

In several different approaches, researchers have worked on enhancing the piezoelectric effects of nanofibers by adjusting solvents or adding other nanomaterials. For example, Andrew and Clarke have shown that β-phase PVDF fibers can be electrospun directly from dimethyl formamide (DMF) solution with a maximum β-phase fraction of 0.75[51] and the addition of well-dispersed ferrite ($Ni_{0.5}Zn_{0.5}Fe_2O_4$) nanoparticles can form nanofibers with an overall crystalline fraction made up solely of ferroelectric β- and γ-phases.[95] Huang *et al.* have found that adding a low concentration of SWCNTs in the electrospinning process can induce highly oriented β-form crystallites.[52] Studies by Costa *et al.* concluded that a low evaporation rate is in favor of β-phase PVDF while a high evaporation rate is in favor of α-phase PVDF.[53] These are some examples from papers published recently on the studies and possible enhancement of β-phase PVDF nanofibers. It is noted that many of the aforementioned reports were conducted by different experimental setups such as different solvents, concentrations, electrical bias and collectors. Therefore, it is possible that contradictory conclusions could have been drawn on specific processing parameters or materials. It would be desirable to have more uniform and optimized processing parameters with good characterizations on the piezoelectric effects of electrospun nanofibers.

7.6 Summary

Obviously, current and future stand-alone devices could all take advantage of self-powering energy harvesters and piezoelectric nanofibers as discussed in this chapter. In order to produce cost-effective piezoelectric nanofibers, the electrospinning process is a good candidate as various materials and nanoparticles have been successfully fabricated as nanofibers and/or nanofibers with embedded nanoparticles. The conventional electrospinning process has been modified for good alignment and position controllability as well as the feasibility to embed nanoparticles. The nanofibers obtained, such as PVDF and PZT for good piezoelectricity, are highly flexible and easy to fabricate for possible integration in implantable and/or flexible devices, as well as textile applications such as electric clothing. This provides tremendous opportunities in various fields for fiber-based nanogenerators. Several groups have demonstrated prototypes of nanofiber nanogenerators as reviewed in this work. Beyond the current feasibility studies mainly from academic institutes, further development of fiber based nanogenerators will be necessary for practical applications. Here, several future prospects/directions are discussed.

High power nanogenerators: most of the current nanogenerators (including nanofiber nanogenerators) are limited to low power generation in laboratory environments. Some recent studies have advanced the records to higher power outputs[73] as well as using real mechanical actuation sources such as the human heartbeat[87] and mice[96] to drive active devices such as LEDs.[21] Nevertheless, the electrical power generated by current nanogenerators made of nanostructures is often too small to have practical usage for commercial hand-held systems such as an electrical watch, which typically consumes electrical power in the range of a few μW. Therefore, if and when the power generated by the current single nanofiber nanogenerator in the recorded range of 10^{-11} Watt can be boosted up to the range of 10^{-6} Watt, probably by multiple-nanofibers, many practical applications could attract strong commercial interest in nanofiber nanogenerators.

Energy storage/regulation systems: similar to all electrical power generation devices, suitable energy storage/regulation systems will be required for nanofiber nanogenerators to store the generated energy and to release it at the right time. For example, repetitive deformation of a piezoelectric nanostructure is essential to generate energy and the generated energy could be accumulated in an energy storage system such as a rechargeable battery or supercapacitor for later release. For example, a ZnO-nanowire nanogenerator has been connected to a capacitor and macro scale electrical circuitry for a system demonstration with discrete wires.[75] An integrated system that combines an integrated nanowire nanogenerator, integrated circuitry and an integrated battery/supercapacitor on a single chip could be an important technology challenge for future developments.

Fundamentals on piezoelectricity: as stated and discussed in the earlier section, the fundamentals of the piezoelectricity of nanofibers have to be investigated systematically in all aspects to produce nanofibers with high piezoelectricity. Specifically, there is still no persuasive evidence to clarify the mechanism of the $\alpha \rightarrow \beta$ phase transformation during the electrospinning process and published literature often provides conflicting information. Some groups have attributed this transformation to the whole stretching process of the jet due to the comparatively high fraction of β-phase observed in smaller-diameter fibers from lower-concentration solutions,[43,49] while other groups relate the cause of β-phase formation to the solvent evaporation rate[97] or the ion current,[98] rather than the jet elongation experienced during the electrospinning process, because of a predominant β-phase obtained in films consisting of small (charged) droplets. In addition, some other works remain problematic with regard to identifying the formation of crystalline phases during the electrospinning process, especially the β- and γ-phases. Hence, further systematic studies and characterizations for the electrospun fibers are required. These include optimizations in process parameters such as materials, solvents, concentrations, electrical bias, needle-to-electrode distance as well as electrospinning methodologies. Furthermore, detailed analyses on the electrospun nanofibers have to be fully characterized with tools such as XRD, FTIR, PFM and SHG and Raman spectroscopy. As such, nanofiber nanogenerators with high energy conversion efficiency could be consistently built in the future.

Wearable electronics: energy harvesters powered by human motion could be interesting and important as devices for wearable electronics, such as smart watches and displays. Directly charging and powering the wearable electronics *via* human motion could be advantageous over battery-powered systems, given that the form-factor and weight of wearable electronics are limited. As such, piezoelectric fiber-based nanogenerators could provide a potential breakthrough with the possibility of being integrated with fabrics for these applications.

References

1. Z. L. Wang and J. Song, *Science*, 2006, **312**, 242–246.
2. C. Sun, J. Shi, D. J. Bayerl and X. Wang, *Energy Environ. Sci.*, 2011, **4**, 4508–4512.

3. K.-I. Park, S. Xu, Y. Liu, G.-T. Hwang, S.-J. L. Kang, Z. L. Wang and K. J. Lee, *Nano Lett.*, 2010, **10**, 4939–4943.
4. B. Kumar and S.-W. Kim, *J. Mater. Chem.*, 2011, **21**, 18946–18958.
5. Y. Qin, X. Wang and Z. L. Wang, *Nature*, 2008, **451**, 809–813.
6. S. N. Cha, J.-S. Seo, S. M. Kim, H. J. Kim, Y. J. Park, S.-W. Kim and J.-M. Kim, *Adv. Mater.*, 2010, **22**, 4726–4730.
7. M.-Y. Lu, J. Song, M.-P. Lu, C.-Y. Lee, L.-J. Chen and Z. L. Wang, *ACS Nano*, 2009, **3**, 357–362.
8. C.-T. Huang, J. Song, W.-F. Lee, Y. Ding, Z. Gao, Y. Hao, L. Chen and Z. L. Wang, *J. Am. Chem. Soc.*, 2010, **132**, 4766–4771.
9. X. Wang, J. Song, F. Zhang, C. He, Z. Hu and Z. Wang, *Adv. Mater.*, 2010, **22**, 2155–2158.
10. Y.-F. Lin, J. Song, Y. Ding, S.-Y. Lu and Z. L. Wang, *Appl. Phys. Lett.*, 2008, **92**, 022105.
11. Y.-F. Lin, J. Song, Y. Ding, S.-Y. Lu and Z. L. Wang, *Adv. Mater.*, 2008, **20**, 3127–3130.
12. X. Wang, J. Song, J. Liu and Z. L. Wang, *Science*, 2007, **316**, 102–105.
13. H.-K. Park, K. Y. Lee, J.-S. Seo, J.-A. Jeong, H.-K. Kim, D. Choi and S.-W. Kim, *Adv. Funct. Mater.*, 2011, **21**, 1187–1193.
14. X. Chen, S. Xu, N. Yao and Y. Shi, *Nano Lett.*, 2010, **10**, 2133–2137.
15. X. Chen, S. Xu, N. Yao, W. Xu and Y. Shi, *Appl. Phys. Lett.*, 2009, **94**, 253113.
16. G. Zhang, S. Xu and Y. Shi, *Micro Nano Lett.*, 2011, **6**, 59–61.
17. D. Shen, J.-H. Park, J. Ajitsaria, S.-Y. Choe, H. C. Wikle and D.-J. Kim, *J. Micromechan. Microeng.*, 2008, **18**, 055017.
18. Z. Wang and Y. Xu, *Appl. Phys. Lett.*, 2007, **90**, 263512.
19. W. J. Choi, Y. Jeon, J.-H. Jeong, R. Sood and S. G. Kim, *J. Electroceram.*, 2006, **17**, 543–548.
20. X. Chen, S. Xu, N. Yao, W. Xu and Y. Shi, *Appl. Phys. Lett.*, 2009, **94**, 253113.
21. J. Fang, X. Wang and T. Lin, *J. Mater. Chem.*, 2011, **21**, 11088–11091.
22. C. Chang, V. H. Tran, J. Wang, Y.-K. Fuh and L. Lin, *Nano Lett.*, 2010, **10**, 726–731.
23. J. Chang and L. Lin, Large array electrospun PVDF nanogenerators on a flexible substrate, In *2011 16th International Solid-State Sensors, Actuators Microsystems Conference,* IEEE, Beijing, 2011, pp. 747–750.
24. Y. Qi and M. C. McAlpine, *Energy Environ. Sci.*, 2010, **3**, 1275–1285.
25. X. Wang, *Nano Energy*, 2012, **1**, 13–24.
26. D. Sun, C. Chang, S. Li and L. Lin, *Nano Lett.*, 2006, **6**, 839–842.
27. Z. Huang, *Compos. Sci. Technol.*, 2003, **63**, 2223–2253.
28. F. Dotti, A. Varesano, A. Montarsolo, A. Aluigi, C. Tonin and G. Mazzuchetti, *J. Ind. Text.*, 2007, **37**, 151–162.
29. E. Boland, G. Wnek, D. Simpson, K. Pawlowski and G. Bowlin, *J. Macromol. Sci., Part A: Pure Appl. Chem.*, 2001, **38**, 1231–1243.
30. J. Chen and G. Chang, *Colloids Surf. A*, 2008, **313–314**, 183–188.
31. R. Gonzalez and N. Pinto, *Synth. Met.*, 2005, **151**, 275–278.

32. B. Ding, M. Wang, X. Wang, J. Yu and G. Sun, *Mater. Today*, 2010, **13**, 16–27.
33. J. Pu, X. Yan, Y. Jiang, C. Chang and L. Lin, *Sens. Actuators, A*, 2010, **164**, 131–136.
34. J. M. Corres, Y. R. Garcia, F. J. Arregui, I. R. Matias and S. Member, *Sensors*, 2011, **11**, 2383–2387.
35. A. L. Yarin, S. Koombhongse and D. H. Reneker, *J. Appl. Phys.*, 2001, **90**, 4836.
36. D. H. Reneker, A. L. Yarin, H. Fong and S. Koombhongse, *J. Appl. Phys.*, 2000, **87**, 4531.
37. J. A. Matthews, G. E. Wnek, D. G. Simpson and G. L. Bowlin, *Biomacromolecules*, 2001, **3**, 232–238.
38. A. Theron, E. Zussman and A. L. Yarin, *Nanotechnology*, 2001, **12**, 384–390.
39. P. Katta, M. Alessandro, R. D. Ramsier and G. G. Chase, *Nano Lett.*, 2004, **4**, 2215–2218.
40. D. Li, Y. Wang and Y. Xia, *Nano Lett.*, 2003, **3**, 1167–1171.
41. J. Deitzel, *Polymer*, 2001, **42**, 8163–8170.
42. C. Chang, K. Limkrailassiri and L. Lin, *Appl. Phys. Lett.*, 2008, **93**, 123111.
43. C. Chang, K. Limkrailassiri and L. Lin, *Appl. Phys. Lett.*, 2008, **93**, 123111.
44. R. Yang, Y. Qin, C. Li, L. Dai and Z. L. Wang, *Appl. Phys. Lett.*, 2009, **94**, 022905.
45. G. Biagi, T. Holmgaard and E. Skovsen, *Opt. Express*, 2013, **21**, 4355–4360.
46. D. Farrar, K. Ren, D. Cheng, S. Kim, W. Moon, W. L. Wilson, J. E. West and S. M. Yu, *Adv. Mater.*, 2011, **23**, 3954–3958.
47. Y. R. Wang, J. M. Zheng, G. Y. Ren, P. H. Zhang and C. Xu, *Smart Mater. Struct.*, 2011, **20**, 045009.
48. J. Zheng, A. He, J. Li and C. C. Han, *Macromol. Rapid Commun.*, 2007, **28**, 2159–2162.
49. C. Ribeiro, V. Sencadas, J. L. G. Ribelles and S. Lanceros-Méndez, *Soft Matter*, 2010, **8**, 274–287.
50. D. Mandal, S. Yoon and K. J. Kim, *Macromol. Rapid Commun.*, 2011, **32**, 831–837.
51. J. S. Andrew and D. R. Clarke, *Langmuir*, 2008, **24**, 670–672.
52. S. Huang, W. A. Yee, W. C. Tjiu, Y. Liu, M. Kotaki, Y. C. F. Boey, J. Ma, T. Liu and X. Lu, *Langmuir*, 2008, **24**, 13621–13626.
53. L. M. M. Costa, *Mater. Sci. Appl.*, 2010, **1**, 246–251.
54. D. V. Isakov, E. de Matos Gomes, L. G. Vieira, T. Dekola, M. S. Belsley and B. G. Almeida, *ACS Nano*, 2011, **5**, 73–78.
55. H. Yu, T. Huang, M. Lu, M. Mao, Q. Zhang and H. Wang, *Nanotechnology*, 2013, **24**, 405401.
56. Z. H. Liu, C. T. Pan, L. W. Lin and H. W. Lai, *Sens. Actuators, A*, 2013, **193**, 13–24.
57. J. Nunes-Pereira, V. Sencadas, V. Correia, J. G. Rocha and S. Lanceros-Méndez, *Sens. Actuators, A*, 2013, **196**, 55–62.

58. S. Padron, R. Patlan, J. Gutierrez, N. Santos, T. Eubanks and K. Lozano, *J. Appl. Polym. Sci.*, 2012, **125**, 3610–3616.
59. B. Vazquez, H. Vasquez and K. Lozano, *Polym. Eng. Sci.*, 2012, **52**, 2260–2265.
60. T. Lei, X. Cai, X. Wang, L. Yu, X. Hu, G. Zheng, W. Lv, L. Wang, D. Wu, D. Sun and L. Lin, *RSC Adv.*, 2013, **3**, 24952.
61. D. Isakov, E. de, M. Gomes, I. Bdikin, B. Almeida, M. Belsley, M. Costa, V. Rodrigues and A. Heredia, *Cryst. Growth Des.*, 2011, **11**, 4288–4291.
62. J. Mccann, J. Chen, D. Li, Z.-G. Ye and Y. Xia, *Chem. Phys. Lett.*, 2006, **424**, 162–166.
63. Y. Wang and J. J. Santiago-Avilés, *Integr. Ferroelectr.*, 2011, **126**, 60–76.
64. X. Chen, N. Yao, Y. Shi and L. Zang, *Energy Efficiency and Renewable Energy Through Nanotechnology*, Springer London, London, 2011.
65. A. Baji, Y.-W. Mai, Q. Li and Y. Liu, *Nanoscale*, 2011, **3**, 3068–3071.
66. Y. Wang and J. J. Santiago-Avilés, *Nanotechnology*, 2004, **15**, 32–36.
67. S. Kalinin and D. Bonnell, *Phys. Rev. B.*, 2002, **65**, 1–11.
68. D. V. Isakov, E. D. M. Gomes, B. G. Almeida, I. K. Bdikin, A. M. Martins and A. L. Kholkin, *J. Appl. Phys.*, 2010, **108**, 042011.
69. J. J. Santiago-Aviles, Y. Wang, R. Furlan and I. Ramos, *Appl. Phys. A: Mater. Sci. Process.*, 2004, **78**, 1043–1047.
70. F. P. Sun, Z. Chaudhry, C. Liang and C. A. Rogers, *J. Intell. Mater. Syst. Struct.*, 1995, **6**, 134–139.
71. S. Jung and S. Kim, *Precis. Eng.*, 1994, **16**, 49–55.
72. J. Chang, M. Dommer, C. Chang and L. Lin, *Nano Energy*, 2012, 356–371.
73. Y. Hu, L. Lin, Y. Zhang and Z. L. Wang, *Adv. Mater.*, 2011, **24**, 110–114.
74. S. Xu, B. J. Hansen and Z. L. Wang, *Nat. Commun.*, 2010, **1**, 93.
75. D. Choi, K. Y. Lee, M.-J. Jin, S.-G. Ihn, S. Yun, X. Bulliard, W. Choi, S.-Y. Lee, S.-W. Kim, J.-Y. Choi, J.-M. Kim and Z. L. Wang, *Energy Environ. Sci.*, 2011, **4**, 4607–4613.
76. C. R. Bowen, R. Stevens, L. J. Nelson, A. C. Dent, G. Dolman, B. Su, T. Button, M. Cain and M. Stewart, *Smart Mater. Struct.*, 2006, **15**, 295–301.
77. Y. Qi, N. T. Jafferis, K. Lyons, C. M. Lee, H. Ahmad and M. C. McAlpine, *Nano Lett.*, 2010, **10**, 524–528.
78. Piezoceramic fibers – Fraunhofer Institute for Ceramic Technologies and Systems IKTS, (n.d.).
79. G. T. Davis, J. E. McKinney, M. G. Broadhurst and S. C. Roth, *J. Appl. Phys.*, 1978, **49**, 324446.
80. Z. H. Liu, C. T. Pan, L. W. Lin, J. C. Huang and Z. Y. Ou, *Smart Mater. Struct.*, 2014, **23**, 025003.
81. Z. H. Liu, C. T. Pan, L. W. Lin, H. W. Li, C. A. Ke, J. C. Huang, *et al.*, Mechanical properties of piezoelectric PVDF/MWCNT fibers prepared by flat/hollow cylindrical near-field electrospinning process, in *the 8th Annual IEEE International Conference on Nano/Micro Eng. Mol. Syst.*, IEEE, Suzhou, 2013, pp. 707–710.
82. B. J. Hansen, Y. Liu, R. Yang and Z. L. Wang, *ACS Nano*, 2010, **4**, 3647–3652.

83. W. Zeng, X.-M. Tao, S. Chen, S. Shang, H. L. W. Chan and S. H. Choy, *Energy Environ. Sci.*, 2013, **6**, 2631.
84. W. Wu, S. Bai, M. Yuan, Y. Qin, Z. L. Wang and T. Jing, *ACS Nano*, 2012, **6**, 6231–6235.
85. L. Gu, N. Cui, L. Cheng, Q. Xu, S. Bai, M. Yuan, W. Wu, J. Liu, Y. Zhao, F. Ma and Y. Qin, *Nano Lett.*, 2013, **13**, 91–94.
86. M. Lee, C.-Y. Chen, S. Wang, S. N. Cha, Y. J. Park, Y. J. Park, *et al.*, *Adv. Mater.*, 2012, **24**, 1759–1764.
87. E. Häsler, L. Stein and G. Harbauer, *Ferroelectrics*, 1984, **60**, 277–282.
88. T. Starner, *IBM Syst. J.*, 1996, **35**, 618–629.
89. S.-Y. Gu, Q.-L. Wu, J. Ren and G. J. Vancso, *Macromol. Rapid Commun.*, 2005, **26**, 716–720.
90. M. Majdoub, P. Sharma and T. Cagin, *Phys. Rev. B*, 2008, 77, 125424.
91. N. Bassiri-Gharb, I. Fujii, E. Hong, S. Trolier-McKinstry, D. V. Taylor and D. Damjanovic, *J. Electroceram.*, 2007, **19**, 49–67.
92. W. A. Yee, M. Kotaki, Y. Liu and X. Lu, *Polymer*, 2007, **48**, 512–521.
93. W. A. Yee, A. C. Nguyen, P. S. Lee, M. Kotaki, Y. Liu, B. T. Tan, S. Mhaisalkar and X. Lu, *Polymer*, 2008, **49**, 4196–4203.
94. G. Zhong, L. Zhang, R. Su, K. Wang, H. Fong and L. Zhu, *Polymer*, 2011, **52**, 2228–2237.
95. J. S. Andrew and D. R. Clarke, *Langmuir*, 2008, **24**, 8435–8438.
96. R. Yang, Y. Qin, C. Li, G. Zhu and Z. L. Wang, *Nano Lett.*, 2009, **9**, 1201–1205.
97. H. Horibe, Y. Sasaki, H. Oshiro, Y. Hosokawa, A. Kono, S. Takahashi, *et al.*, *Polym. J.*, 2013, **46**, 104–110.
98. M. Nasir, H. Matsumoto, M. Minagawa, A. Tanioka, T. Danno and H. Horibe, *Polym. J.*, 2007, **39**, 670–674.

CHAPTER 8

Hierarchical Nanostructures for Photo-Electro-Chemical Cells

DAEHO LEE*[a,b] AND COSTAS P. GRIGOROPOULOS[a]

[a] Department of Mechanical Engineering, University of California, Berkeley, CA 94720-1740, USA; [b] Department of Mechanical Engineering, Gachon University, Seongnam-si, Gyeonggi-do 461-701, South Korea
*Email: dhl@gachon.ac.kr

8.1 Meaning of "Hierarchical Structure"

The term "hierarchy", which was originated from a Greek word (hierarkhia [ίεραρχία]: leader of sacred rites) has been widely used in a variety of fields including social science, education, zoology, science of religion, computer science, math, *etc*. The definition and meaning of "hierarchy" or "hierarchical" vary depending on the referenced field. Words such as *rank*, *class*, *food chain*, *pyramid*, *caste system*, *many branches* may be associated with the term "hierarchy".

In general, the "hierarchical" structure in nanoscience and engineering can be defined as an integrated structure that is composed of many, low-dimensional nano-building blocks.[1] In other words, a hierarchical nanostructure has a configuration of higher dimension assembled from lower dimensional nanomaterials including nanoparticles (0-D), nanowires/rods/tubes (1-D) and nanosheets (2-D). Therefore a "hierarchical structure" invokes a 3-D image resembling trees or flowers having very dense branches with extremely high surface areas. If both backbones and branches are

RSC Nanoscience & Nanotechnology No. 35
Hierarchical Nanostructures for Energy Devices
Edited by Seung Hwan Ko and Costas P Grigoropoulos

Published by the Royal Society of Chemistry, www.rsc.org

Figure 8.1 Example of the hierarchical structure in nature that is composed of a trunk with 1st generation branches (a) and multiple generation branches (b) attached to it.

wire-type structures, tree-like hierarchical structures[2,3] are formed while spherical particles comprise urchin-like or chestnut-like hierarchical structures[4] when combined with wire-type branches. In this chapter, the words "hierarchical structure" rather signify structures with a distinct backbone (also called a trunk or core) to which a plethora of branches are attached as seen in Figure 8.1. Therefore, tetrapod,[5,6] cauliflower[7] and chrysanthemum[8] structures are not examined in this chapter, though they are also often referred to as hierarchical structures in a broader sense.

Among various materials and applications, this chapter focuses on hierarchical nanostructures of materials for photo-electro-chemical (PEC) cell applications, which are largely classified as photovoltaics and water splitting for fuel (hydrogen) generation.

8.2 Importance of the "Hierarchical" Structure in PEC Cell Applications

Tweaking and improving the properties of existing materials could enhance cell efficiency by a relatively modest factor. In contrast, structural reconfiguration such as introducing nanowires instead of nanoparticles, for example, has shown remarkable results in efficiency improvement.[9] It has been verified through numerous research studies that 1-D crystalline nanostructures such as nanowires and nanotubes significantly improve the electron diffusion length by providing a direct conduction pathway to the rapid collection of photogenerated electrons, thereby diminishing the possibility of charge recombination during interparticle percolation.

In the same context, utilizing hierarchical structures suggests a promising path towards achieving even higher cell performance. It is not surprising that the cell efficiency has been enhanced by a factor of 4 or 5 through introducing photoelectrodes with hierarchical nanowire structures in lieu of straight nanowires.[2,3,10] Hierarchical structures provide a substantially increased surface area, hence enabling higher light harvesting while allowing a longer effective path for the photons to be absorbed in the photoelectrodes by stronger scattering and trapping enhancement. Moreover, the resulting larger surface area boosts the chemical reactions and enhances diffusion of chemical species (ions or gases) into the interfaces and surfaces in PEC cells.[11]

Hierarchical structures constructed by hybridizing two or more different materials provide another opportunity. For example, integrating materials of different band structures enables absorption of photon energy over a broader range. In addition, each semiconductor conducts the individual half reactions with effective charge separation in water splitting PEC cells.

8.3 Basic Mechanism of PEC Cells

A PEC cell is a device that directly generates electrical energy and converts solar energy into chemical fuels such as hydrogen by the water splitting process.[12] For instance, PEC cells may be referred to as "PEC cells for water splitting" in order to distinguish from "photovoltaics" (PVs).[13,14] However, for the purpose of this chapter, PEC cells include both cases since both photovoltaics and water splitting cells are devices driven by fundamental PEC processes.

Both PEC cells for PVs and fuel (hydrogen) production are based on two key characteristics of semiconductors: (1) electron–hole pairs are generated when photons are absorbed, (2) space charge, or a built-in electric field near the surface or interface of a semiconductor efficiently separates those photogenerated carriers, thereby preventing recombination. However, the working mechanisms in PVs and water splitting PEC cells are different in several aspects. The main difference is that PVs are regenerative cells since they induce no net chemical change upon converting light directly into electricity, while water splitting PEC cells involve reduction and oxidation at each photoelectrode by splitting water into hydrogen and oxygen. More details are discussed in the next sections.

8.3.1 PEC Cell for Electrical Energy Generation

One of the most well-known types of this cell is the dye-sensitized solar cell (DSSC), also called the dye-sensitized photovoltaic cell or "Grätzel cell" named after Michael Grätzel, the pioneer researcher of this type of solar cell.

The working principle of the electrical energy-generating PEC cell is well documented[12,15,16] and can be summarized as follows based on a cell with an n-type semiconductor with which most of this type of PEC cells are developed. As illustrated in Figure 8.2,[12] dyes or photosensitizers (S) with broad absorption spectra that are adsorbed at the surface of a semiconductor such as TiO_2 and ZnO absorb photons and generate excited electrons ($S^0 + h\nu \rightarrow S^*$, $S^* \rightarrow S^+ + e^-$). The excited electrons travel through the conduction band of the semiconductor to the conducting back plate, and then flow to the external load and re-enter the cell on a metal electrode (counter electrode) to reduce the redox relay molecules in the electrolyte (*e.g.* $I_3^- + 2e^- \rightarrow 3I^-$ in I_3^-/I^- redox couples). For the efficient injection of electrons from the dye to the semiconductor, the conduction band edge of the semiconductor should be placed lower than excited-state energy level of the dye. Finally, the oxidized dye molecules are regenerated, accepting electrons and oxidizing the redox relay molecule (*e.g.* $S^+ + e^- \rightarrow S^0$, $3I^- \rightarrow I_3^- + 2e^-$ in I_3^-/I^- redox couples). Therefore, there is no net chemical change behind this cycle. Many kinds of redox mediators such as vanadium(II)/vanadium(III),[17,18] sulfide/polysulfide[19,20] and cobalt(II)/cobalt(III)[21,22] have been reported and analogous considerations applied to the chemical reactions. For high power conversion (*i.e.* photoconversion or light conversion) efficiency, maximizing photoabsorption while minimizing energy loss due to recombination of electrons with either the oxidized dye molecules or other chemical species in the electrolyte is required. To meet these conditions, the three most critical requirements for the semiconductor photoanode are (1) a large surface area for higher dye loading, (2) high electronic mobility for rapid charge transport and (3) high stability in acidic dye solution for sufficient dye adsorption over an extended period of time. The photoconversion efficiency, one of the key

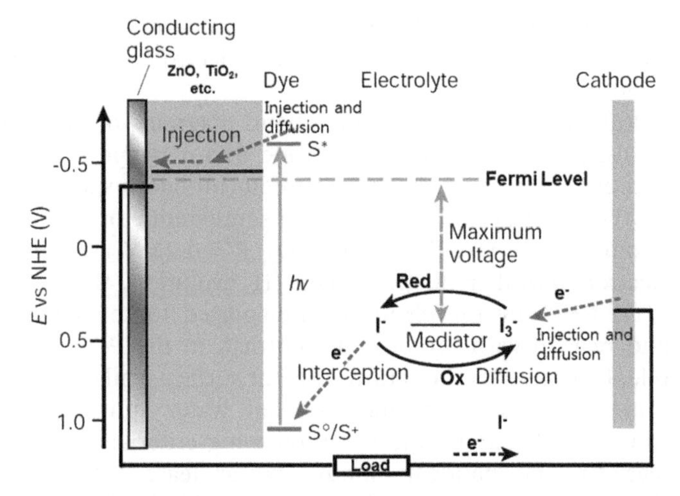

Figure 8.2 Illustration of the working mechanism of the dye-sensitized solar cell in the electrolyte containing I_3^-/I^- redox couples.
Reproduced with permission from ref. 12. (Copyright 2001, NPG.)

performance indicators, is measured by recording the current–voltage behavior of the photovoltaic cell when the cell is irradiated under AM 1.5 G (air mass 1.5 global) illumination and calculated according to the following equations:[23]

$$\text{Photoconversion efficiency } (\eta) = \frac{FF \times V_{OC} \times J_{SC}}{P_{in}} \tag{8.1}$$

$$FF = \frac{V_{max} \times I_{max}}{V_{OC} \times I_{SC}} \tag{8.2}$$

where FF is the fill factor, V_{OC} (V) is the open-circuit voltage, J_{SC} (mA cm^{-2}) is the short-circuit current density, P_{in} (mW cm^{-2}) is the intensity of the illumination, V_{max} (V) and I_{max} (mA) are the voltage and current at the maximum power output respectively.

8.3.2 PEC Cell for Fuel (Hydrogen) Generation

At least one of the electrodes of a PEC cell for hydrogen generation by water splitting is a semiconductor. A conventional PEC cell is composed of an n-type semiconductor as a photoanode and a metal (usually platinum) electrode as a cathode immersed in an aqueous solution of a salt (liquid electrolyte). Figure 8.3 shows the energy band structure of the PEC cells demonstrating how the band level of the electrodes changes upon galvanic contact, irradiation and applied bias.[24] Band bending occurs in electron transfer between the semiconductor and the electrolyte so that equilibrium is established between the Fermi levels. Upon irradiation with photon energy equal to or greater than the band gap of the material, photogenerated carriers are separated by the applied external potential and/or the depletion (space-charge) layer formed between the electrode and liquid electrolyte. Depending on the band structure of the photoelectrode, external potential may or may not be necessary.

Figures 8.4(a) and (b) show the energy band diagrams of PEC cells composed of an n-type semiconductor electrode and a p-type semiconductor electrode that acts as a photoanode and photocathode respectively. In the photoanode, the holes accumulated on the semiconductor surface oxidize water (O_2 evolution, $2H_2O + 4h^+ \rightarrow 4H^+ + O_2$, $E^0 = 1.23$ V *vs.* NHE [normal hydrogen electrode]) while reduction occurs (H_2 evolution, $4H^+ + 4e^- \rightarrow 2H_2$, $E^0 = 0.00$ V *vs.* NHE) by the electrons transported to the metal counter electrode and reacted with protons. In contrast, in the PEC cell with the photocathode, H_2 evolves at the semiconductor surface while O_2 is produced at the counter electrode. For efficient solar water splitting, the semiconductor photoanode should meet the following criteria:[14,25] (1) for practical applications, the bandgap should be at least 1.9 eV, which is determined by the minimum potential for water oxidation (1.23 eV) plus the thermodynamic losses (0.3–0.4 eV) and the overpotential required for sufficiently fast reaction kinetics (0.4–0.6 eV), (2) the conduction band and the

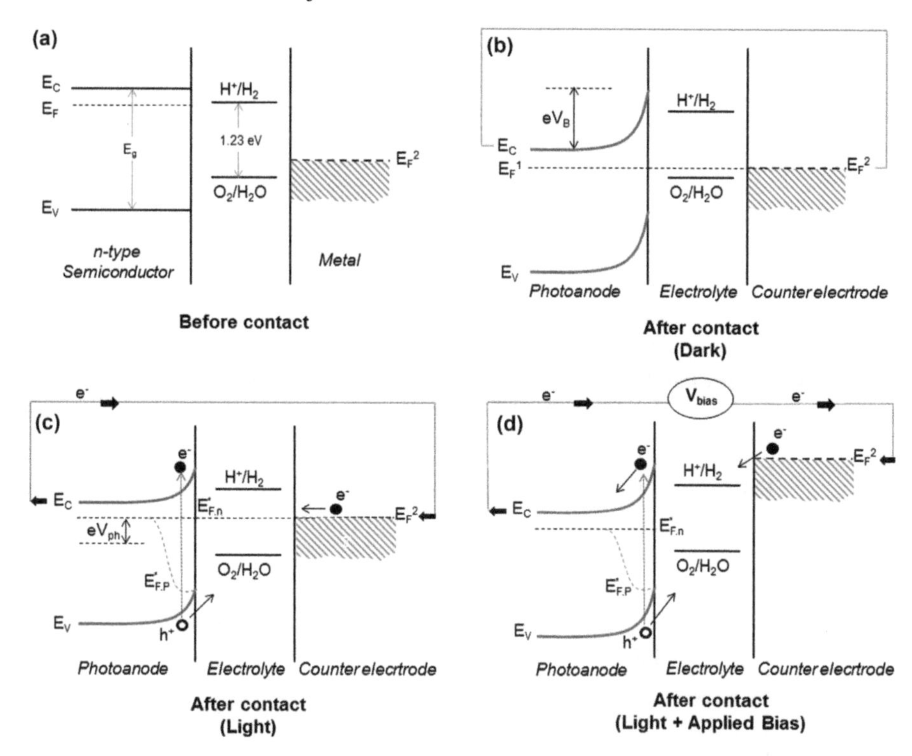

Figure 8.3 Energy band diagram of a PEC cell for water splitting composed of an n-type semiconductor and a metal counter electrode (a) before galvanic contact, (b) after galvanic contact without illumination, (c) after galvanic contact with illumination, and (d) after galvanic contact with illumination and an applied bias. E_C: energy of conduction band minimum, E_V: energy of valence band maximum, E_F: Fermi level, vacuum energy; E_C, energy of conduction band minimum; $E^*_{F,n}$: quasi Fermi level for electrons, $E^*_{F,p}$: quasi Fermi level for holes, V_B: surface potential, V_{ph}: internal photovoltage.

valence band edges of the semiconductor should straddle the electrochemical potentials of water reduction and oxidation, *i.e.* the conduction band edge should be more negative than the H_2 evolution potential, E^0 (H^+/H_2), and the valence band edge should be more positive than the O_2 evolution potential, E^0 (O_2/H_2O), otherwise external potential should be applied, (3) strong light absorption, (4) the electrode should be stable in the aqueous electrolyte in the dark and under illumination, (5) efficient charge transport in the photoelectrodes, (6) low cost. Though no material can cover all these requirements, semiconductor metal oxides such as TiO_2, WO_3 and Fe_2O_3 are good candidates since they are relatively inexpensive and are highly stable. However, since the conduction band of most metal oxides is less negative than the H_2 evolution potential energy as shown in Figure 8.5, external bias needs to be applied. Please note that the vertical axis of the diagram is indicated using the NHE (normal hydrogen electrode) as a reference and

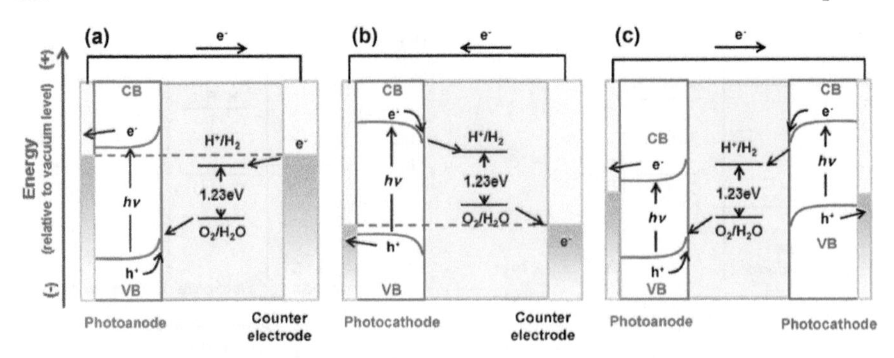

Figure 8.4 Energy band diagram of a PEC cell for water splitting composed of (a) an n-type semiconductor (photoanode) (b) a p-type semiconductor (photocathode) (c) n-type and p-type semiconductors (photoanode and photocathode).
Reproduced with permission from ref. 30.

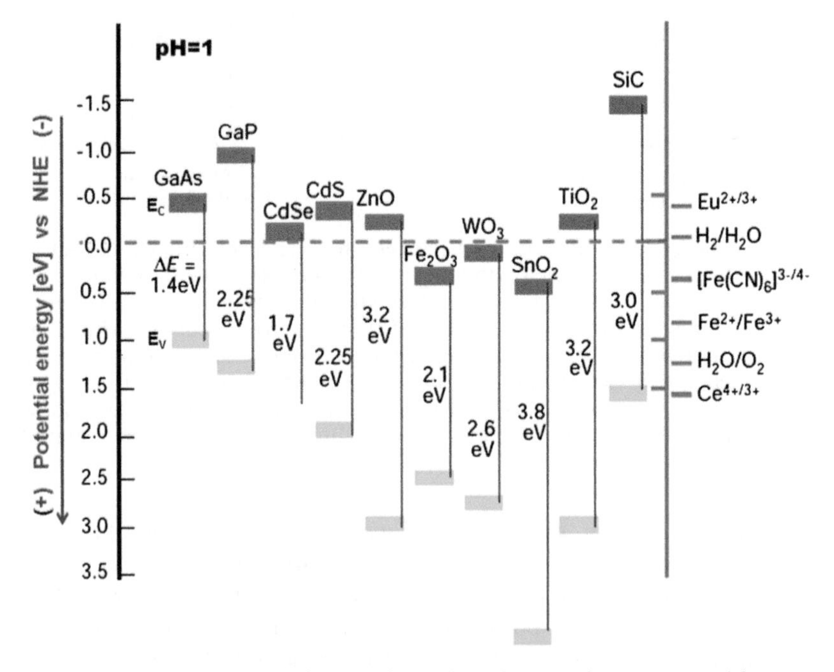

Figure 8.5 Band positions of several semiconductors in contact with aqueous electrolyte at pH 1. The energy level is indicated using the NHE (normal hydrogen electrode) as a reference.
Reproduced with permission from ref. 12. (Copyright 2001, NPG.)

negative value is located in the upper position. To exclude the external bias, PEC/PV tandem cells[26,27] or PEC cells consisting of a semiconductor photoanode and a semiconductor photocathode have been explored.[28] As for the latter, each semiconductor electrode absorbs photons simultaneously and minority carriers (holes at the anode and electrons at the cathode) are used for the water splitting by carrying out the individual half reactions while the majority carriers (electrons at the anode and holes at the cathode) are recombined at the ohmic contact between the materials.[28,29] Figure 8.4(c) illustrates the energy band diagram of a PEC cell composed of two semiconductor photoelectrodes.[30] This two-step excitation is similar to the mechanism of the "Z-scheme" reaction of a natural photosynthesis system in green plants and other organisms by which NADPH (nicotinamide adenine dinucleotide phosphate) and O_2 are generated at each photosystem. Since the bandgap of each electrode needs to cover only either O_2 or H_2 evolution potential, less constraint in the bandgap exists in the selection of the semiconductors.

The key performance parameter of PEC cells for hydrogen generation is the solar-to-hydrogen (STH) efficiency given by the following equation under unbiased conditions:[31–33]

$$\eta_{\mathrm{STH}} = \frac{N_{\mathrm{H_2}} \times \Delta G}{P_{\mathrm{in}} \times A} \tag{8.3}$$

where $N_{\mathrm{H_2}}$ (mol s^{-1}) is the hydrogen generation rate, ΔG (J mol^{-1}) (usually taken as 237 000 J mol^{-1} at room temperature[31]) is the Gibbs free energy per mol of hydrogen, P_{in} (W cm^{-2}) is the intensity of the illumination and A (cm^2) is the area of the illuminated electrode. In many studies, the applied bias photon-to-current efficiency (ABPE), defined per the following equation, is more widely used as an indicator of the PEC cell efficiency measured with an applied external bias.

$$\eta_{\mathrm{ABPE}} = \frac{P_{\mathrm{electrical}}^{\mathrm{out}} - P_{\mathrm{electrical}}^{\mathrm{in}}}{P_{\mathrm{in}}} = \frac{J_{\mathrm{ph}}(V_{\mathrm{redox}} - |V_{\mathrm{bias}}|)}{P_{\mathrm{in}}} \tag{8.4}$$

Here, J_{ph} (mA cm^{-2}) is the photocurrent density obtained under an applied bias V_{bias} (V) between the photoanode and the photocathode, V_{redox} (V) (usually taken to be 1.23 V, *i.e.* the value at pH 0 at room temperature) is the theoretical minimum potential difference required to split water and P_{in} (mW cm^{-2}) is the illumination intensity. Whether considering STH or ABPE efficiencies, one should be mindful that the respective values sometimes can be over- or underestimated, depending on measurement procedures such as using different light sources (xenon lamp *vs.* AM 1.5 G solar simulator) or different electrode systems (two electrode system *vs.* three electrode system), *etc.* Details of several factors that affect the efficiency calculation can be found elsewhere,[31,34] and further discussion is beyond the scope of this chapter.

8.4 Several Fabrication Strategies for Hierarchical Nanostructures

In most cases, hierarchical structures can be produced by combining or modifying conventional processes for nanostructure synthesis. Here, several representative fabrication strategies for hierarchical structures are introduced.

8.4.1 Combination of Hydrothermal Growth and Seed Engineering (Solution-phase Chemical Route)

The hydrothermal growth method is a low-cost, scalable and facile way to produce metal oxide nanostructures such as ZnO,[35] TiO_2,[36] In_2O_3[37] and WO_3.[38,39] Sometimes, the word "hydrothermal" is used only when the precursor solution is aqueous while the expression "solvothermal" is adopted when a non-aqueous precursor solution is utilized. However, in many instances, "hydrothermal method" has a broader meaning encompassing both cases.

For hierarchical structures synthesized *via* the hydrothermal method, nano-size branches are grown from either secondary seed nanoparticles[2] or nuclei generated on the surface of the backbone structure during an intermediate treatment.[10] ZnO is the most actively researched material as a hydrothermally grown metal oxide due to the low synthesis temperature and water-based non-toxic precursors. Therefore, many types of hierarchical structures have been reported for PEC applications, especially for DSSCs.

Ko *et al.*[2] and Herman *et al.*[3] reported ZnO "nanoforest" structures of high density, long-branched "tree-like" multi generation hierarchical photoanodes that can significantly increase power conversion efficiency due to substantially enhanced surface area, enabling higher dye loading and light harvesting as well as due to reduced charge recombination by direct conduction along the crystalline branches as shown in Figure 8.6. They have shown that complex and hierarchical ZnO nanowire photoanodes could be fabricated *via* a low-cost, all-solution processed hydrothermal method incorporating seed particle deposition steps and utilization of capping polymers. Lee *et al.*[10] reported high efficiency DSSCs with densely-packed, omni-directionally branched TiO_2 nanostructures grown by the hydrothermal method. Through the so-called "2 + 1 method", which comprises two hydrothermal growth steps with an intermediate high-concentration $TiCl_4$ treatment on upright backbone NWs (Figure 8.7(a)), the branches grew in all directions from densely distributed needle-like seeds on jagged cylindrical surfaces of the backbone NWs, resulting in a dramatically increased surface area as shown in Figure 8.7(b). There are also many variations of hierarchical TiO_2 nanowire structures through solution-phase routes. Details of fabrication methods have been reported.[32,38,40,41]

Zhuge *et al.*[42] demonstrated TiO_2 nanotubes coated with an outer nanocrystalline and mesoporous TiO_2 shell. They used hydrothermally-grown ZnO nanowires as a template for the layer-by-layer coated TiO_2 nanotube. The ZnO core is dissolved by mild etching in $TiCl_4$ solution to form TiO_2 nanotubes which transform to the hierarchical structure through additional layer-by-layer coating followed by hydrothermal crystallization (Figures 8.7(c) and (d)). The DSSC efficiency with the hierarchical TiO_2 nanotube structure is 30% higher than with the original nanotube arrays.

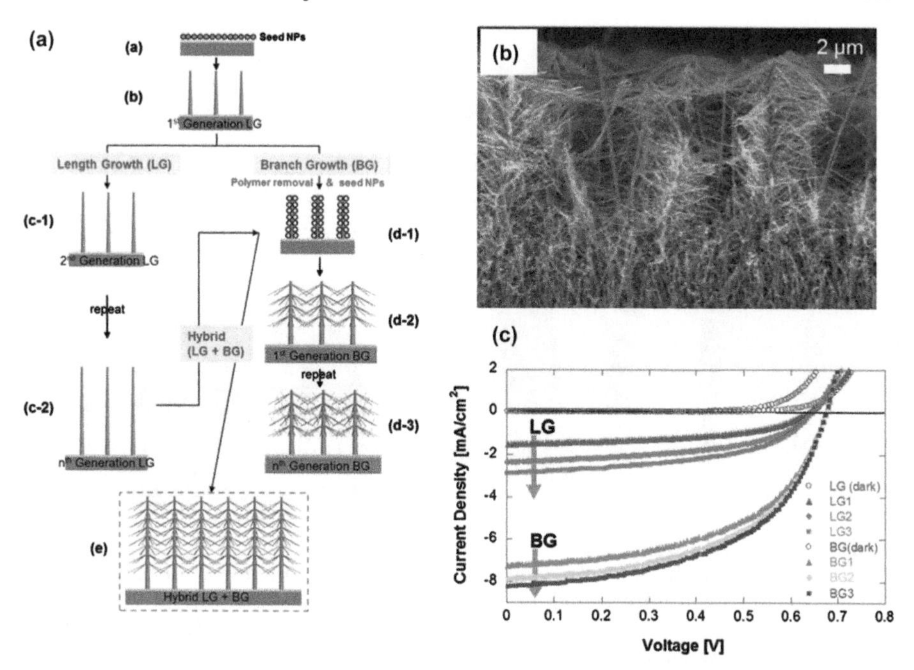

Figure 8.6 (a) Schematic illustration of the synthesis of a nanoforest photoanode with multigeneration branches *via* hydrothermal growth. Adapted with permission from ref. 2. (Copyright 2011, ACS.) (b) SEM image of a ZnO nanoforest. Adapted with permission from ref. 2. (Copyright 2011, ACS.) (c) J–V curve of DSSC with the ZnO nanoforest photoanode showing enhanced light conversion efficiency. Adapted with permission from ref. 2. (Copyright 2011, ACS.)

8.4.2 Repetition of Catalyst Deposition and Growth by Chemical Vapor Deposition

Another route for constructing hierarchical structures is sequential or repeating catalyst deposition during chemical vapor deposition processes. The vapor-liquid-solid (VLS)[43] method is one of the most well-known growth mechanisms for 1-D structures from chemical vapor deposition (CVD). For the VLS process, a catalytic liquid alloy phase is deposited or patterned on a substrate followed by introducing vapor phase source material in a vacuum chamber. The vapor source is rapidly adsorbed on the catalyst surface and diffuses into the catalyst to be supersaturated and nucleated at the liquid/solid interface, resulting in axial growth of a highly anisotropic 1-D crystal structure. By repeating the "catalyst deposition and growth" steps, new generations of branches are grown on the predefined backbones. The diameter, length and density of branches are controlled by differing catalyst size, reaction time and catalyst concentration. In this way, Wang *et al.*[44] fabricated branched nanowire structures of Si and GaN with gold and nickel catalysts respectively. These nanowire structures revealed that the

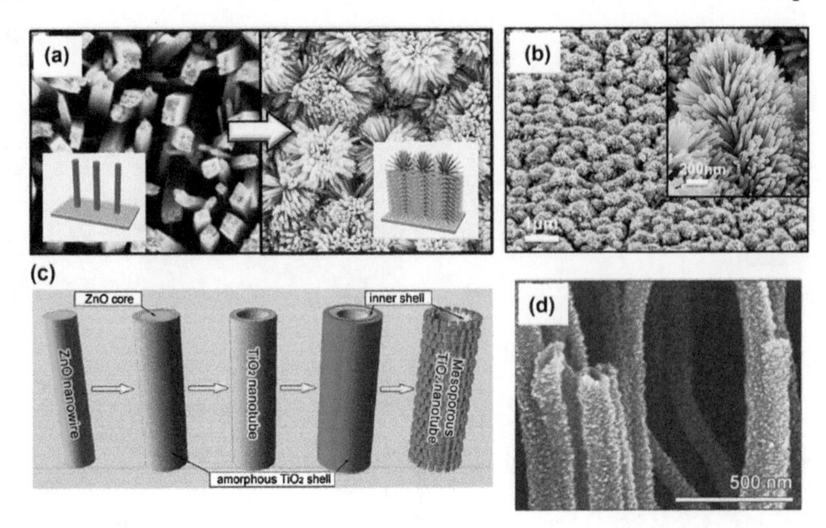

Figure 8.7 (a) Hierarchical evolution of upright TiO_2 nanowires. Adapted from ref. 10. (b) SEM image of the TiO_2 nanostructures with densely-packed and omnidirectional branches. Adapted from ref. 10. (c) Schematic illustration of the structural derivation of the hierarchical TiO_2 nanotube. Adapted with permission from ref. 42. (Copyright 2011, WILEY-VCH Verlag GmbH & Co. KGaA, Weinheim.) (d) Magnified SEM images of the TiO_2 hierarchical nanotubes. Adapted with permission from ref. 42. (Copyright 2011, WILEY-VCH Verlag GmbH & Co. KGaA, Weinheim.)

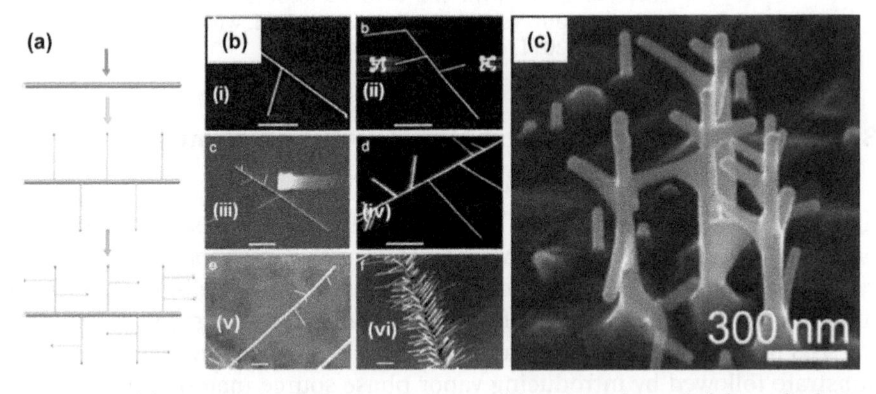

Figure 8.8 (a) Schematic diagram of the multistep syntheses of branched NW structures. Adapted with permission from ref. 44. (Copyright 2004, ACS.) (b) SEM images of branched Si nanowire (i–iv) and GaN nanowire structures (v–vi). Scale bars are all 1 μm. Adapted with permission from ref. 44. (Copyright 2004, ACS.) (c) SEM image of InAs nanotrees at a tilted angle. Adapted with permission from ref. 45. (Copyright 2006, ACS.)

single-crystal branches were epitaxially grown from the backbone nanowires suggesting a potential application for electronically active devices (Figures 8.8(a) and (b)). Dick *et al.*[45] also grew 3-D InAs networks, by a metal organic vapor phase epitaxy (MOVPE) method, decorated with branches grown from pre-alloyed binary Au-In seed particles on the trunk nanowires from e-beam patterned Au seed particles as shown in Figure 8.8(c).

Although this process is straightforward, the chamber needs to be opened between each catalyst deposition step. During this opening step, oxidation of pre-grown nanomaterials sometimes happens, which is deleterious to forming an epitaxial junction between the backbone and branch nanostructure.

8.4.3 Branch Formation by *in situ* Catalyst Generation

This method is distinguishable from the previous method in that the catalyst for the branches is provided *in situ* during nanowire growth. May *et al.*[46] synthesized highly branched InAs nanowires on a GaAs(111) substrate through MOVPE composed of backbone nanowires by VLS growth with a gold catalyst, and Mn-mediated branches. Self-assembled Mn clusters on growing nanowires initiated epitaxial branch growth, which results in highly branched dendritic structures as displayed in Figures 8.9(a) and (b).

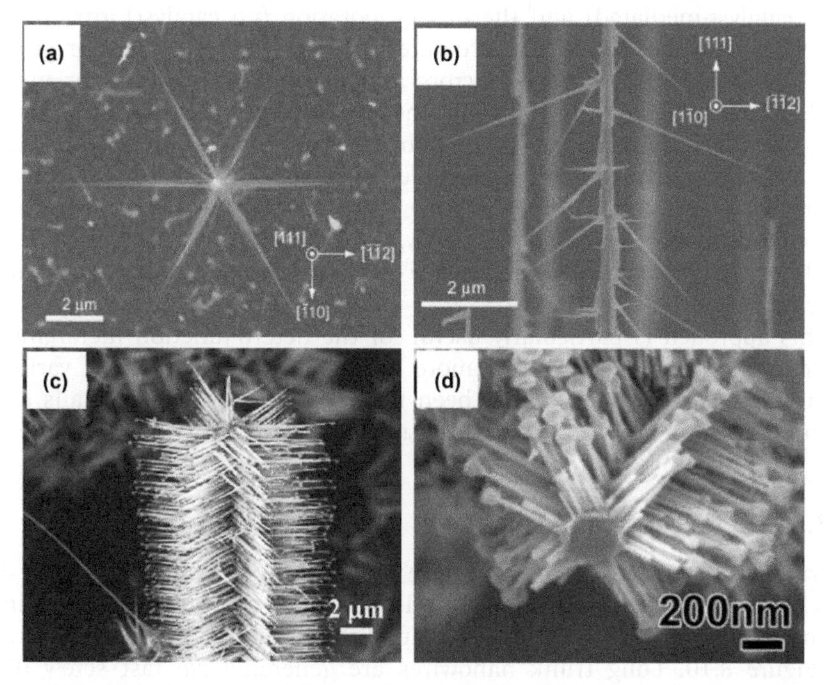

Figure 8.9 SEM image of (a) top view and (b) side view of InAs nanowire branches grown from a InAs backbone. Adapted with permission from ref. 46. (Copyright 2005, WILEY-VCH Verlag GmbH & Co. KGaA, Weinheim.) (c) A single column of ZnO nanopropeller arrays. Adapted with permission from ref. 47. (Copyright 2004, AIP Publishing LLC.) (d) 6-Fold hierarchical ZnO nanonail structures. Adapted from ref. 48.

Gao *et al.*[47] reported "nanopropeller" arrays of ZnO synthesized on a Al_2O_3 substrate by a two-step high temperature solid-vapor deposition process where ZnO and SnO_2 powders were reduced by graphite in a heated tube furnace ($ZnO(s) + C(s) \rightarrow Zn(v) + CO(v)$). Initially, ZnO backbone nanowires were grown through the VLS process aided by Sn as a catalyst at a lower temperature (600–700 °C) and then Sn vapors condensed on the surface of backbone nanowires at a higher temperature (700–800 °C) from which branch nanowires were grown with a six-fold symmetry (Figure 8.9(c)). Lao *et al.*[48] also synthesized various types of ZnO hierarchical structures such as ZnO "nanonails" on nanowires/nanobelts in a similar process to Gao (Figure 8.9(d)).

This method has an advantage in that the oven or vacuum chamber need not be opened for the sequential catalyst deposition, hence preventing oxidation of previously grown nanostructures or deposited catalysts.

8.4.4 Combination of VLS/Thermal Evaporation Method and Hydrothermal Growth

VLS (catalyst-mediated) and thermal evaporation (no catalyst) growth are widely used methods for 1-D structure material synthesis, VLS for group IV[49–51] and III–V[52,53] semiconductors and thermal evaporation for metal oxides such as WO_3,[54] Fe_2O_3[55] and ZnO.[56] Hydrothermal growth, as mentioned earlier, is a facile route for nanowire synthesis over a large area at a low cost and utilizing relatively simple equipment. To create hierarchical nanostructures, typically the backbone stems are built by VLS or a thermal evaporation process, and subsequently, branch nanowires are grown from seed materials attached on the surface of the backbone *via* a hydrothermal process.

This approach is particularly useful for the hetero-junction hierarchical structures composed of two or more different materials. Hence, various combinations of materials have been reported recently. More details and examples of this process will be discussed in section 8.4.7.

8.4.5 Defect-driven Growth

Bierman *et al.*[57] and Zhu *et al.*[58] have reported a new nanowire synthesis mechanism that results in a PbS and PbSe pine tree structure featuring branches that are spirally arranged around the trunk nanowires as depicted in Figure 8.10. Long trunk nanowires are generated *via* fast screw dislocation-driven growth where the axial screw dislocations provide self-perpetuating steps enabling catalyst-free 1-D crystal growth,[57,59] while branches are formed by slower VLS epitaxial growth. The helically rotating branches are a consequence of the strain of the axial dislocation called the Eshelby twist[60] developed in the trunk nanowire. To achieve synergy between the dislocation-driven and VLS mechanisms for hierarchical nanostructure

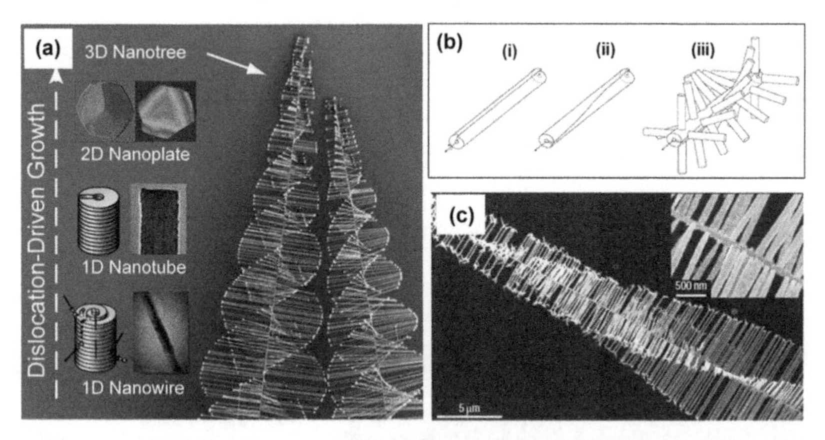

Figure 8.10 (a) Illustrations and micrographs showing formation of hierarchical nanostructures driven by screw dislocation. Adapted with permission from ref. 65. (Copyright 2013, ACS.) (b) Schematic illustration showing the formation of helically arranged nanowire branches: (i) a nanowire with an axial screw dislocation in the center. Both ends of the NW are clamped. (ii) One end of the NWs is free to rotate, resulting in an Eshelby twist. (iii) The branches are grown epitaxially onto the backbone nanowire by VLS growth. Adapted with permission from ref. 58. (Copyright 2008, NPG.) (c) SEM image of a single helically branched PbSe nanowire (inset, high-resolution SEM). Adapted with permission from ref. 58. (Copyright 2008, NPG.)

growth, the supersaturation condition should be in a range where both layer-by-layer growth and dislocation-driven growth coexist.[13]

The dislocation-driven nanowire growth mechanism was also expanded to other semiconducting materials such as ZnO,[61] In_2O_3,[62,63] CdS/CdSe[64] *etc.*[65]

8.4.6 Laser-induced Hydrothermal Growth (LIHG)

Conventional hydrothermal growth involves heating in a furnace or convection oven. However, furnace or oven heating is less energy-efficient and a rather slow process requiring quite long heating/cooling time. In contrast, laser provides a rapid heat source with ease of parametric control at a localized heating zone and furthermore enables selective heating by enabling choice of desired wavelength. Exploiting these laser characteristics, Yeo *et al.*[66] and In *et al.*[67] demonstrated laser-induced hydrothermal growth (LIHG) of ZnO nanowires. Compared with conventional hydrothermal growth, the nanowire growth rate dramatically increased and nanowire structures could be deposited and even patterned on designated spots very effectively and precisely by using a focused laser to induce a controlled temperature rise as displayed in Figures 8.11(a) and (b). More interestingly, very long ZnO nanowires, which have a high possibility of a significant increase of the surface area at a given projected area, have been produced in a short time with limited quantities of precursors (Figures 8.11(c) and (d)).[67]

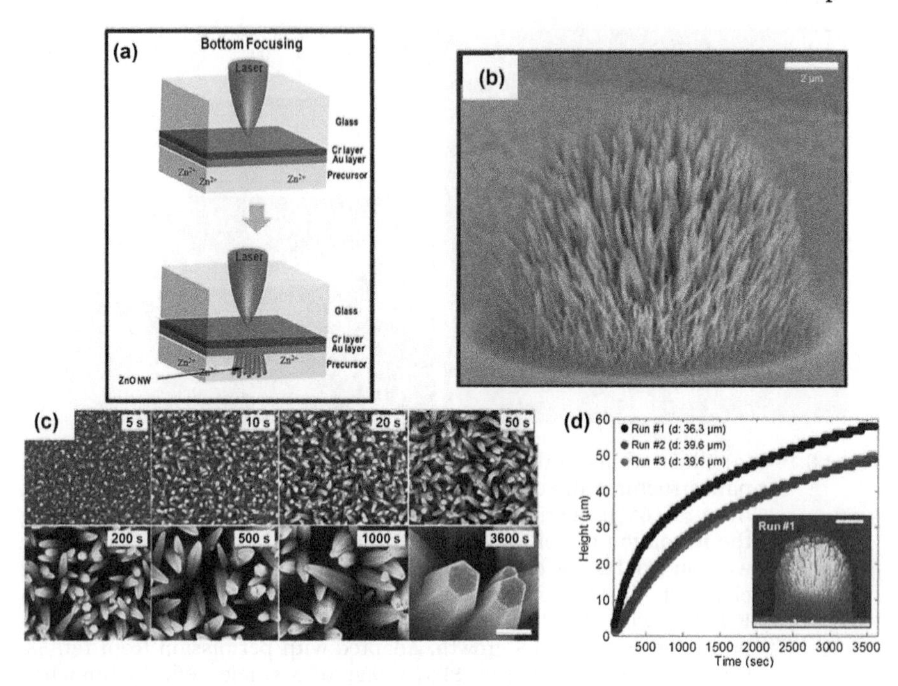

Figure 8.11 (a) Schematic diagram of the LIHG process by bottom laser focusing. Adapted with permission from ref. 66. (Copyright 2013, WILEY-VCH Verlag GmbH & Co. KGaA, Weinheim.) (b) Tilted SEM image of the ZnO nanowire array synthesized by LIHG. Adapted with permission from ref. 66. (Copyright 2013, WILEY-VCH Verlag GmbH & Co. KGaA, Weinheim.) (c) A series of SEM images of nanowires grown with varying growth time (from 5 to 3600 s). Adapted with permission from ref. 67. (Copyright 2013, WILEY-VCH Verlag GmbH & Co. KGaA, Weinheim.) (d) *In situ* kinetic curves that were obtained with different beam diameters under the same laser irradiation power. The inset shows the SEM image of Run #1 (scale bar: 20 μm). The maximum height is about 58.8 μm. Adapted with permission from ref. 67. (Copyright 2013, WILEY-VCH Verlag GmbH & Co. KGaA, Weinheim.)

Although hierarchical structure synthesis by the LIHG process has not been reported yet, the LIHG process is capable of generating hierarchical structures by just supplementing intermediate seeding procedures between each growth. Laser-induced hydrothermal growth might be more effective for fabricating compact devices. However, utilizing high power lasers and expanding the irradiated spot suggests a promising route towards a high throughput process.

8.4.7 Heterostructures Composed of Two or More Materials

Regarding photocatalytic activities, heterostructures composed of two or more materials with different bandgaps provide broadband absorption.

Moreover, branched nanowires from the backbone structure usually have a much smaller size and higher density,[2,10] resulting in a larger surface area as well as more efficient charge collection and separation. Therefore, backbone-branch NW type hierarchical heterostructures combine these advantages.

Several studies on hierarchical heterostructures have been reported, including ZnO/In_2O_3 (backbone/branch) by an evaporation method,[68] $BaCrO_4$ by a catanionic reserver micelle system,[69] and Si/SiO_2 by VLS method.[70] Among various materials, we will focus on heterostructures of the PEC cells for water splitting in this section.

Xi and coworkers synthesized Fe_3O_4/WO_3 (core–branch) heterostructures showing enhanced photoconversion capability with respect to pure WO_3 or Fe_3O_4 nanostructures as depicted in Figure 8.12(a).[4] The branches, WO_3 nanoplates, have a relatively narrower bandgap (2.6–2.8 eV) enabling photocatalytic activity in the visible range, while the Fe_3O_4 microsphere cores provide a fast transport route for the photogenerated charges due to the high

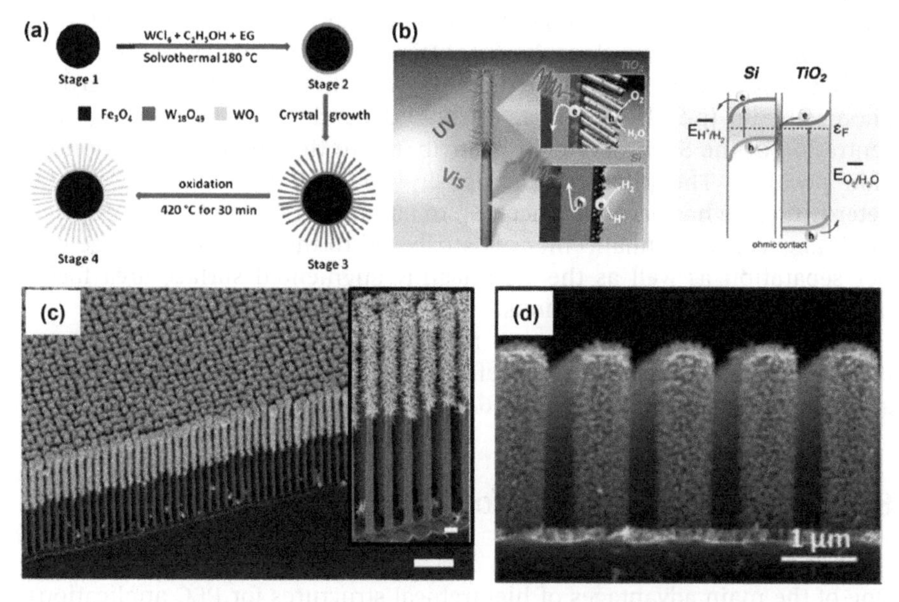

Figure 8.12 (a) Schematic illustration showing formation of the Fe_3O_4/WO_3 core–shell structures. Adapted with permission from ref. 4. (Copyright 2011, WILEY-VCH Verlag GmbH & Co. KGaA, Weinheim.) (b) Schematic illustration of the Si/TiO_2 nanotree heterostructure and working mechanism of water splitting. The electron–hole pairs separated at the semiconductor–electrolyte interface conduct the individual half reactions of water splitting with the help of co-catalysts, without any applied bias. Adapted with permission from ref. 28. (Copyright 2013, ACS.) (c) SEM image of the Si/TiO_2 nanotree heterostructures. Adapted with permission from ref. 28. (Copyright 2013, ACS.) (d) SEM image of Si/ZnO heterostructures.[1] Adapted with permission from ref. 71. (Copyright 2013, ACS.)

conductivity. TiO_2/Si heterostructures have also been demonstrated as a fully integrated system for PEC solar hydrogen production (Figure 8.12(b)).[28] Rutile TiO_2 nanowire branches serving as oxygen-generating photoanodes were hydrothermally grown from the p-Si backbone nanowires that comprised the hydrogen-generating photocathodes (Figure 8.12(c)). The Si nanowires were fabricated by a reactive-ion etching (RIE) process on patterned single crystalline Si wafers. Iridium nanoparticles and platinum nanoparticles were loaded on the TiO_2 and Si nanowires to diminish the reaction overpotentials. These results confirmed solar water splitting without applied bias on the Si/TiO_2 heterostructure, apparently due to the optimized and integrated design of the two materials since water splitting could not be observed on the constituent materials or their composites without integration. Kargar *et al.*[71] reported the synthesis of 3-D Si/ZnO branched nanowire photoelectrodes consisting of Si cores and ZnO branches as shown in Figure 8.12(d). The core p-Si nanowires were fabricated by nanoimprinting followed by dry etching and ZnO nanowire branches were grown subsequently by the hydrothermal growth method. The lack of stability of the Si/ZnO nanowire structures could be enhanced by corrosion-resistant thin ($\sim$20 nm) atomic layer deposition (ALD)-coated TiO_2 and a platinum co-catalyst. The Si/ZnO heterostructures exhibit photoanodic (p^+-Si/TiO_2) or photocathodic (p-Si/TiO_2) characteristics, depending on the doping concentration of the Si cores, which is attributed to the different depletion region position. The photocurrent density obtained from the Si/ZnO-NW heterostructure has been enhanced up to 80 times compared to Si/ZnO-seed (no branches). The enhancement is attributed to the promoted charge carrier separation as well as the significantly augmented surface area facilitating efficient chemical reactions.

Recently, combinations of other materials to build heterostructures are increasingly being reported. Some of them with their respective applications are listed in Table 8.1 in section 8.6.

8.5 Useful Parameters for Hierarchical Nanostructures

One of the main advantages of hierarchical structures for PEC applications is that they can realize electrodes with extremely large surface areas within a limited cell size. Therefore, it is convenient to define parameters expressing the density of the hierarchical structure and thereby the anticipated benefit in surface area that can be achieved from the given configuration.

The most widely used parameter is the "roughness factor (RF)", which is a dimensionless parameter defined as the ratio of the total surface area of the nanostructure to the projected substrate area[10,32] as illustrated in Figure 8.13. If N backbone nanowires (diameter: D, length: H) are grown on the substrate (area $= a \times b$) and n cylindrical branches (diameter: d, length: h)

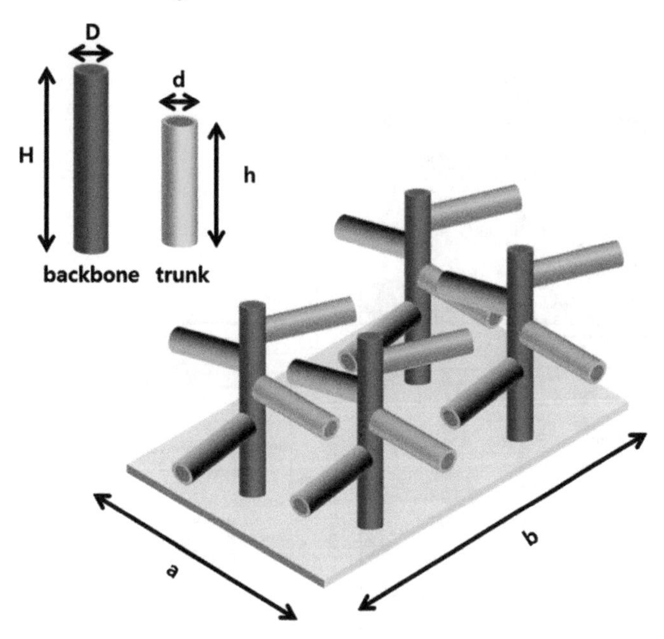

Figure 8.13 Simplified model to calculate the roughness factor of a hierarchical structure.

are attached to the surface of each backbone, the RF is approximately calculated as the following equation:

$$\text{roughness factor (RF)} \approx \frac{N(\pi DH + 0.25\pi D^2) + nN(\pi dh + 0.25\pi d^2)}{ab} \tag{8.5}$$

For the high aspect ratio backbone and branch nanowires ($D \ll H$, $d \ll h$), the RF can be further simplified as follows:

$$\text{roughness factor (RF)} \approx \frac{N\pi(DH + ndh)}{ab} \tag{8.6}$$

Usually, nanowires do not have circular cross sections and are often tapered rather than having straight edges. However, the above expression can be employed to provide an estimate without incurring significant errors. Also, when the hierarchical structure has multi-generation branches, the above equation is readily adjustable.

Lee *et al.*[10] synthesized TiO_2 photoanodes of different RFs for DSSCs and verified that the DSSC efficiency ratio is in quite good agreement with the RF ratio as displayed in Figure 8.14. Many other studies also adopted the RF as an important parameter for PEC cell performance.[9,72–74] It is worth noting that a higher RF does not necessarily mean higher cell performance. For example, a 2 μm thick film consisting of 30 nm size nanoparticles has a roughness factor of about 210, which is much higher than that of a film of the same thickness comprising hierarchical nanowire structures (around

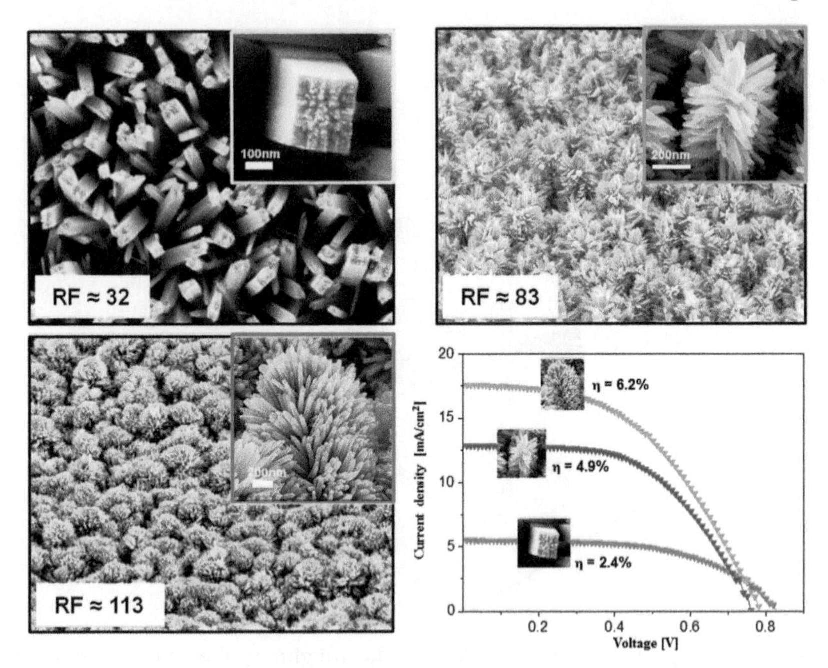

Figure 8.14 TiO_2 nanostructures of different RFs and their DSSCs performance. The DSSC efficiency ratio is in quite good agreement with the RF ratio. Reproduced from ref. 10.

130). However, the former shows worse PEC properties, which is attributed to factors other than the surface area.[32] Also, thick electrode layers have higher RF values; however, charge recombination tends to increase in a thick layer and is deleterious to the cell efficiency. Therefore, it makes more sense to use the RF value as a parameter of PEC cells comprising nanostructure layers of the same or similar thickness. For this reason, some literature have adopted the modified RF, which is defined as the total surface area over projected one per unit length of the film thickness with its dimension of [length^{-1}].[75–77]

8.6 Examples of Hierarchical Nanostructures for PEC Cell Applications

Representative examples of hierarchical structure photoelectrodes for PEC cell applications are provided in Table 8.1. Overall, the solution-phase route (hydrothermal) and vapor-phase route (VLS and evaporation) are two major processes used to build backbone or trunk nanostructures from various kinds of materials, while anisotropic etching prevails for Si backbone generation. The hydrothermal method is the most widely implemented route to generate branch nanostructures attached to the backbone. PEC cells adopting hierarchical structures show considerably enhanced performance compared with those with unbranched or non-hybridized electrodes.

Table 8.1 List of PEC Cell Applications with Hierarchical Nanostructure Photoelectrodes.

Material	Mechanism	Homo/ Hetero Structure	Representative Image	Applications	Ref.	Notes and Significance
TiO_2 (rutile)	Hydrothermal + seeding	Homo	Adapted from ref.10	Photovoltaics (photoanode)	10, 38, 40, 41	Rutile-phase TiO_2 photoanodes with densely-packed branches were fabricated on FTO (fluorine-doped tin oxide) glass by the hydrothermal method. Vertical backbone TiO_2 nanowires were synthesized, and then TiO_2 branches were evolved from the seed layer attached to the surface of the backbones. Up to 6.2% photoconversion efficiencies have been reported.[10]
TiO_2 (rutile)	Hydrothermal + seeding	Homo	Adapted with permission from ref.32 (copyright 2012, ACS)	Water splitting (photoanode)	32	0.49% ABPE efficiency was reported with a photoanode comprising branched rutile-phase TiO_2 nanorods that were fabricated *via* a two-step hydrothermal growth mediated by dip coating for seed nanoparticles. The efficiency is almost 50 and 3 times higher than that of PEC cell composed of nanoparticles and unbranched nanorods respectively.

Table 8.1 (*Continued*)

Material	Mechanism	Homo/ Hetero Structure	Representative Image	Applications	Ref.	Notes and Significance
TiO_2 (anatase)	Hydrothermal + seeding	Homo	Adapted with permission from ref.78 (copyright 2103, NPG)	Photovoltaics (photoanode)	78	Anatase-phase hierarchical TiO_2 arrays have been synthesized directly on FTO glass solely through a one-step hydrothermal process. Anatase-phase TiO_2 is known to show superior photovoltaic performance compared to rutile-phase TiO_2.[79]
ZnO	Hydrothermal + seeding	Homo	Adapted with permission from ref.2 (copyright 2011, ACS)	Photovoltaics (photoanode)	2, 3, 80	The hydrothermal method for ZnO nanostructures is much simpler than that for TiO_2 due to a low temperature and aqueous precursors. Significant improvement on the total ZnO nanowire surface area has been achieved by a facile hydrothermal method and seeding procedures. The seed layer was deposited either by drop casting of nanoparticles or electrospinning of ZnO nanofibers. Up to 5.2% power conversion efficiency has been reported for DSSC applications.[80]

ZnO	VLS + hydrothermal	Homo	 Adapted with permission from ref.81 (copyright 2011, ACS)	Water splitting (photoanode)	81	ZnO nanotetrapods were synthesized *via* a VLS process and etched followed by thermal decomposition of $Zn(AC)_2$ to generate tiny irregular spikes on the surface. Successive hydrothermal growth generated dense branches to form a high surface area hierarchical structure. Furthermore, enhanced ABPE efficiencies up to 0.31% have been demonstrated by nitrogen doping that narrows the bandgap of ZnO, thus increases visible light absorption.
ZnO	Electrodeposition[82] + hydrothermal	Homo	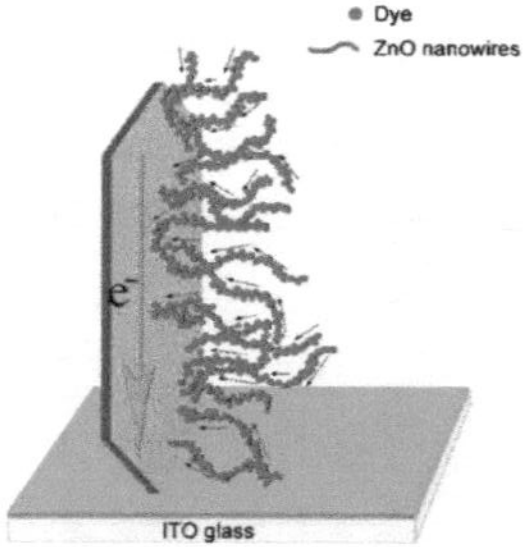 Dye; ZnO nanowires; e⁻; ITO glass; Adapted with permission from ref.83 (copyright 2010, ACS)	Photovoltaics (photoanode)	83	A new type of combination comprising a ZnO nanosheet backbone and ZnO nanowire branches has been demonstrated as a hierarchical structure for DSSC applications. A 4.8% power conversion efficiency has been reported.

Table 8.1 (*Continued*)

Material	Mechanism	Homo/ Hetero Structure	Representative Image	Applications	Ref.	Notes and Significance
Si/ZnO	Etching (Si) + hydrothermal (ZnO)	Hetero	(i) 500 nm Adapted with permission from ref.71 (copyright 2013, ACS)	Water splitting (photocathode)	71, 84	A 3-D Si(p-type)/ZnO heterostructure is used as a photocathode for H_2 production, which promotes the separation of photoinduced charge carriers due to the different band structure of the two materials: the conduction band of ZnO is lower than that of the p-type Si. Pt wire was used as a counter electrode and an applied bias is required.
Si/TiO_2	Etching (Si) + hydrothermal (TiO_2)	Hetero	Adapted with permission from ref.28 (copyright 2013, ACS)	Water splitting (photoanode + photocathode)	28	The most significant aspect of this study is that O_2 is generated at the doped Si nanowire (backbone) photoanode and H_2 at the rutile TiO_2 nanowires (branches) without any applied bias. Though the STH efficiency (0.12%) is still low, the results show the possibility of integrating individual nanocomponents into a functional system similar to the photosynthetic system in a chloroplast.

Si/InGaN	VLS(Si) + halide CVD[85,86] (InGaN)	Hetero	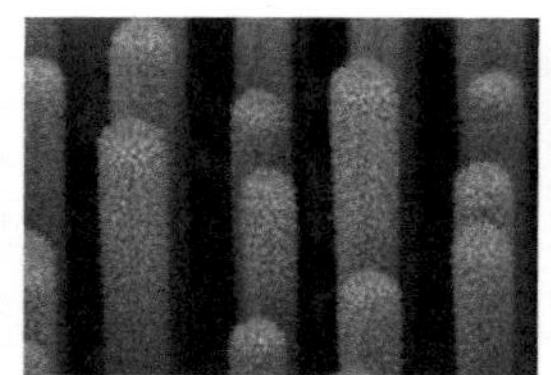 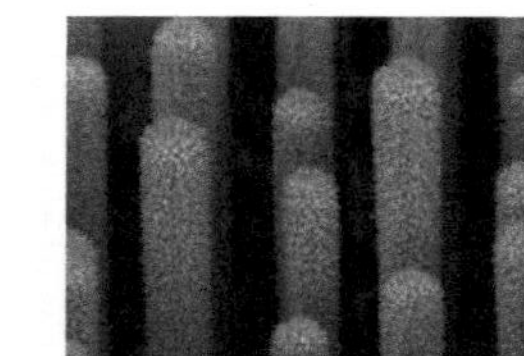 Adapted with permission from ref.87 (copyright 2012, ACS)	Water splitting (photoanode)	87	Si(n-type)/InGaN hierarchical arrays produce an n-type/n-type hetero-junction that can be used as a stable photoanode with a high surface area for water splitting PEC cells. The photocurrent density with hierarchical Si/InGaN structures increased by 5 times compared to the photocurrent density with InGaN nanowire arrays on planar Si.
WO_x/ZnO/CdS/CdSe	Thermal evaporation (WO_x) + hydrothermal (ZnO)	Hetero	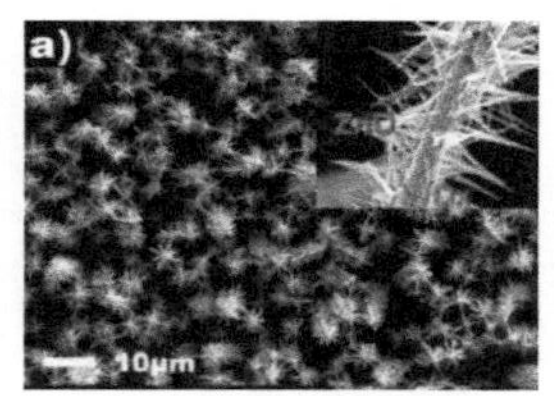 Adapted with permission from ref.88 (copyright 2013, ACS)	Water splitting (photoanode)	88	The WO_x/ZnO/CdSe/CdS composite structure is used as a photoanode. ZnO branches absorb UV irradiation while narrow bandgap CdSe/CdS photosensitizes the visible light region. The WO_x core with a low electrical resistivity acts as a charge transport material. The multicomponent composite forms a cascade band structure[89] that enables efficient charge transport with a large surface area.

Table 8.1 (*Continued*)

Material	Mechanism	Homo/ Hetero Structure	Representative Image	Applications	Ref.	Notes and Significance
Sb:SnO_2/TiO_2 and Sb:SnO_2/TiO_2/CdS	VLS (Sb:SnO_2) + chemical bath deposition (TiO_2) + electrodepositon (CdS)	Hetero	Adapted with permission from ref.91 (copyright 2014, Springer)	Water splitting (photoanode)	90, 91	A combination of Sb-doped SnO_2 (ATO) nanowire backbones and TiO_2 branches has been demonstrated as a photoanode for water splitting applications. Controlling the pH level of the $TiCl_4$ solution during the reaction process is a key factor for the morphology and the TiO_2 phase of the hybrid nanostructures.[90] Also, the hierarchical ATO/rutile-TiO_2 structures decorated with CdS quantum dot sensitizers showed enhanced water splitting PEC performances.[91]
CuO/ZnO	Thermal oxidation (CuO) + hydrothermal (ZnO)	Hetero	Adapted with permission from ref.92 (copyright 2013, ACS)	Water splitting (photocathode)	92	3-D ZnO/CuO heterojunction branched nanowires were employed as a photocathode for water splitting PEC cells. The hetero-junction photoelectrodes exhibit broadband photoresponses from the UV to near-IR region due to

						the different band gaps of two materials (CuO: ~1.4 eV, ZnO: ~3.4 eV). The backbone/branch structure shows a better performance than the core–shell structure due to its increased surface area.
$ZnO/TiO_2/CuO$	Multiple hydrothermal	Hetero	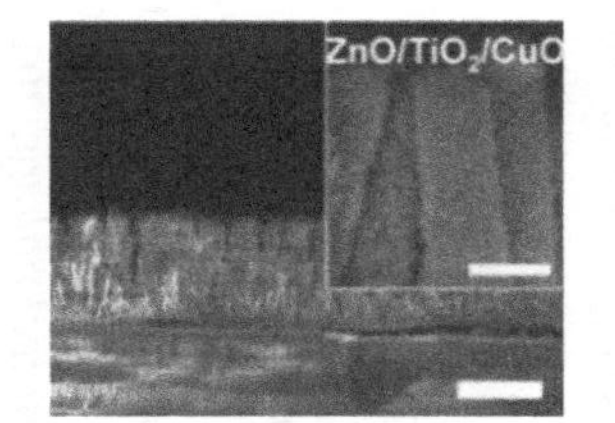 Adapted with permission from ref.93 (copyright 2012,WILEY-VCH Verlag GmbH & Co. KGaA, Weinheim)	Water splitting (photoanode)	93	A tree-like (trunk: Zn, bark: TiO_2, branches: CuO) hierarchical structure developed with a full solution-processable strategy and its application as a photoanode for water splitting has been demonstrated. The CuO branches absorb visible light, and the TiO_2 shell protects the ZnO trunk from erosion and gathers UV light simultaneously to generate photoelectrons, while the ZnO nanowire works as an efficient pathway for charge transport.

8.7 Summary

The photoelectrode is the main component of the PEC cell that absorbs incident photons to generate electron–hole pairs and generates either electricity directly or chemical fuel by redox reactions. Adopting hierarchical structure photoelectrodes by structural reconfiguration of the same material and/or integrating two or more materials provides a promising route to achieve significant enhancement of PEC cell efficiency. Several approaches to fabricate the hierarchical structure *via* either developing a new method or combining conventional processes have been reported.

A high density of hierarchical branches with a much smaller diameter increases the roughness factor dramatically and also provides longer effective pathways for photon absorption in the photoelectrode by enhanced light scattering and trapping, thereby resulting in better photoconversion efficiency. Moreover, the large contact area of the interface between the electrode and electrolyte boosts chemical reactions and diffusion of chemical species. Particularly promising is the hybridization of two or more different materials. Integrating materials having distinct band structures broadens the absorption spectrum and leads to effective charge separation and collection, enabling a wider selection of materials for the water splitting PEC cell configuration without external bias.

References

1. J.-H. Lee, *Sens. Actuators, B*, 2009, **140**, 319–336.
2. S. H. Ko, D. Lee, H. W. Kang, K. H. Nam, J. Y. Yeo, S. J. Hong, C. P. Grigoropoulos and H. J. Sung, *Nano Lett.*, 2011, **11**, 666–671.
3. I. Herman, J. Yeo, S. Hong, D. Lee, K. H. Nam, J. Choi, W. Hong, C. P. Grigoropoulos and S. H. Ko, *Nanotechnology*, 2012, **23**, 194005.
4. G. Xi, B. Yue, J. Cao and J. Ye, *Chem.–Eur. J.*, 2011, **17**, 5145–5154.
5. L. Manna, D. J. Milliron, A. Meisel, E. C. Scher and A. P. Alivisatos, *Nat. Mater.*, 2003, **2**, 382–385.
6. Y. Dai, Y. Zhang, Q. Li and C. Nan, *Chem. Phys. Lett.*, 2002, **358**, 83–86.
7. I. Cesar, K. Sivula, A. Kay, R. Zboril and M. Grätzel, *J. Phys. Chem. C*, 2008, **113**, 772–782.
8. M. Ye, H.-Y. Liu, C. Lin and Z. Lin, *Small*, 2013, **9**, 312–321.
9. M. Law, L. E. Greene, J. C. Johnson, R. Saykally and P. Yang, *Nat. Mater.*, 2005, **4**, 455–459.
10. D. Lee, Y. Rho, F. I. Allen, A. M. Minor, S. H. Ko and C. P. Grigoropoulos, *Nanoscale*, 2013, **5**, 11147–11152.
11. T. P. Chou, Q. Zhang, G. E. Fryxell and G. Z. Cao, *Adv. Mater.*, 2007, **19**, 2588–2592.
12. M. Grätzel, *Nature*, 2001, **414**, 338–344.
13. M. J. Bierman and S. Jin, *Energy Environ. Sci.*, 2009, **2**, 1050–1059.
14. R. van de Krol, in *Photoelectrochemical Hydrogen Production*, Springer, New York, 2012, pp. 13–67.

15. M. Grätzel, *J. Photochem. Photobiol. C: Photochem. Rev.*, 2003, **4**, 145–153.
16. K. Kalyanasundaram, *Dye-sensitized solar cells*, EPFL press, École polytechnique fédérale de Lausanne, 2010.
17. S. Licht, *J. Phys. Chem. B*, 2001, **105**, 6281–6294.
18. A. Heller, B. Miller and F. Thiel, *Appl. Phys. Lett.*, 1981, **38**, 282–284.
19. V. Jovanovski, V. Gonzãlez-Pedro, S Giménez, E. Azaceta, G. n. Cabañero, H. Grande, R. Tena-Zaera, I. n. Mora-Seró and J. Bisquert, *J. Am. Chem. Soc*, 2011, **133**, 20156–20159.
20. L. Li, X. Yang, J. Zhao, J. Gao, A. Hagfeldt and L. Sun, *J. Mater. Chem.*, 2011, **21**, 5573–5575.
21. D. Zhou, Q. Yu, N. Cai, Y. Bai, Y. Wang and P. Wang, *Energy Environ. Sci.*, 2011, **4**, 2030–2034.
22. S. M. Feldt, E. A. Gibson, E. Gabrielsson, L. Sun, G. Boschloo and A. Hagfeldt, *J. Am. Chem. Soc.*, 2010, **132**, 16714–16724.
23. Q. Zhang, C. S. Dandeneau, X. Zhou and G. Cao, *Adv. Mater.*, 2009, **21**, 4087–4108.
24. T. Bak, J. Nowotny, M. Rekas and C. Sorrell, *Int. J. Hydrogen Energy*, 2002, **27**, 991–1022.
25. R. van de Krol, Y. Liang and J. Schoonman, *J. Mater. Chem.*, 2008, **18**, 2311–2320.
26. J. Brillet, M. Cornuz, F. L. Formal, J.-H. Yum, M. Grätzel and K. Sivula, *J. Mater. Res.*, 2010, **25**, 17–24.
27. O. Khaselev and J. A. Turner, *Science*, 1998, **280**, 425–427.
28. C. Liu, J. Tang, H. M. Chen, B. Liu and P. Yang, *Nano Lett.*, 2013, **13**, 2989–2992.
29. Y. Li and J. Z. Zhang, *Laser Photonics Rev.*, 2010, **4**, 517–528.
30. T. Hisatomi, J. Kubota and K. Domen, *Chem. Soc. Rev.*, 2014, DOI: 10.1039/C3CS60378D.
31. R. Van de Krol and M. Grèatzel, *Photoelectrochemical hydrogen production*, Springer, New York, 2012.
32. I. S. Cho, Z. Chen, A. J. Forman, D. R. Kim, P. M. Rao, T. F. Jaramillo and X. Zheng, *Nano Lett.*, 2011, **11**, 4978–4984.
33. S. Choudhary, S. Upadhyay, P. Kumar, N. Singh, V. R. Satsangi, R. Shrivastav and S. Dass, *Int. J. Hydrogen Energy*, 2012, **37**, 18713–18730.
34. A. Murphy, P. Barnes, L. Randeniya, I. Plumb, I. Grey, M. Horne and J. Glasscock, *Int. J. Hydrogen Energy*, 2006, **31**, 1999–2017.
35. L. E. Greene, B. D. Yuhas, M. Law, D. Zitoun and P. Yang, *Inorg. Chem.*, 2006, **45**, 7535–7543.
36. X. Feng, K. Shankar, O. K. Varghese, M. Paulose, T. J. Latempa and C. A. Grimes, *Nano Lett.*, 2008, **8**, 3781–3786.
37. J. Yang, C. Lin, Z. Wang and J. Lin, *Inorg. Chem.*, 2006, **45**, 8973–8979.
38. H. Wang, Y. Bai, Q. Wu, W. Zhou, H. Zhang, J. Li and L. Guo, *Phys. Chem. Chem. Phys.*, 2011, **13**, 7008–7013.
39. Q. Zhang, L. Zhang, Y. Li and H. Wang, *CrystEngComm*, 2013, **15**, 5986–5993.
40. W.-P. Liao and J.-J. Wu, *J. Mater. Chem.*, 2011, **21**, 9255–9262.

41. F. Shao, J. Sun, L. Gao, S. Yang and J. Luo, *J. Mater. Chem.*, 2012, **22**, 6824–6830.
42. F. Zhuge, J. Qiu, X. Li, X. Gao, X. Gan and W. Yu, *Adv. Mater.*, 2011, **23**, 1330–1334.
43. C. Jiang, X. Sun, G. Lo, D. Kwong and J. Wang, *Appl. Phys. Lett.*, 2007, **90**, 263501–263503.
44. D. Wang, F. Qian, C. Yang, Z. Zhong and C. M. Lieber, *Nano Lett.*, 2004, **4**, 871–874.
45. K. A. Dick, K. Deppert, L. S. Karlsson, W. Seifert, L. R. Wallenberg and L. Samuelson, *Nano Lett.*, 2006, **6**, 2842–2847.
46. S. J. May, J. G. Zheng, B. W. Wessels and L. J. Lauhon, *Adv. Mater.*, 2005, **17**, 598–602.
47. P. X. Gao and Z. L. Wang, *Appl. Phys. Lett.*, 2004, **84**, 2883–2885.
48. J. Y. Lao, J. Y. Huang, D. Z. Wang and Z. F. Ren, *J. Mater. Chem.*, 2004, **14**, 770–773.
49. T. Stelzner, M. Pietsch, G. Andrä, F. Falk, E. Ose and S. Christiansen, *Nanotechnology*, 2008, **19**, 295203.
50. S. Kodambaka, J. Tersoff, M. Reuter and F. Ross, *Science*, 2007, **316**, 729–732.
51. Y. Wu and P. Yang, *J. Am. Chem. Soc.*, 2001, **123**, 3165–3166.
52. J. Bauer, V. Gottschalch, H. Paetzelt and G. Wagner, *J. Cryst. Growth*, 2008, **310**, 5106–5110.
53. T. Mårtensson, J. B. Wagner, E. Hilner, A. Mikkelsen, C. Thelander, J. Stangl, B. J. Ohlsson, A. Gustafsson, E. Lundgren and L. Samuelson, *Adv. Mater.*, 2007, **19**, 1801–1806.
54. Y. Baek and K. Yong, *J. Phys. Chem. C*, 2007, **111**, 1213–1218.
55. Y. Fu, J. Chen and H. Zhang, *Chem. Phys. Lett.*, 2001, **350**, 491–494.
56. B. Yao, Y. Chan and N. Wang, *Appl. Phys. Lett.*, 2002, **81**, 757–759.
57. M. J. Bierman, Y. K. A. Lau, A. V. Kvit, A. L. Schmitt and S. Jin, *Science*, 2008, **320**, 1060–1063.
58. J. Zhu, H. Peng, A. Marshall, D. Barnett, W. Nix and Y. Cui, *Nat. Nanotechnol.*, 2008, **3**, 477–481.
59. S. Jin, M. J. Bierman and S. A. Morin, *J. Phys. Chem. Lett.*, 2010, **1**, 1472–1480.
60. J. D. Eshelby, *J. Appl. Phys.*, 1953, **24**, 176–179.
61. S. A. Morin, M. J. Bierman, J. Tong and S. Jin, *Science*, 2010, **328**, 476–480.
62. D. Maestre, D. Häussler, A. Cremades, W. Jäger and J. Piqueras, *Cryst. Growth Des.*, 2011, **11**, 1117–1121.
63. D. Maestre, D. Häussler, A. Cremades, W. Jäger and J. Piqueras, *J. Phys. Chem. C*, 2011, **115**, 18083–18087.
64. H. Wu, F. Meng, L. Li, S. Jin and G. Zheng, *ACS Nano*, 2012, **6**, 4461–4468.
65. F. Meng, S. A. Morin, A. Forticaux and S. Jin, *Acc. Chem. Res.*, 2013, **46**, 1616–1626.
66. J. Yeo, S. Hong, M. Wanit, H. W. Kang, D. Lee, C. P. Grigoropoulos, H. J. Sung and S. H. Ko, *Adv. Funct. Mater.*, 2013, **23**, 2216–3323.
67. J. B. In, H.-J. Kwon, D. Lee, S. H. Ko and C. P. Grigoropoulos, *Small*, 2014, **10**, 741–749.

68. J. Y. Lao, J. G. Wen and Z. F. Ren, *Nano Lett.*, 2002, **2**, 1287–1291.
69. H. Shi, L. Qi, J. Ma, H. Cheng and B. Zhu, *Adv. Mater.*, 2003, **15**, 1647–1651.
70. C. Ye, L. Zhang, X. Fang, Y. Wang, P. Yan and J. Zhao, *Adv. Mater.*, 2004, **16**, 1019–1023.
71. A. Kargar, K. Sun, Y. Jing, C. Choi, H. Jeong, G. Y. Jung, S. Jin and D. Wang, *ACS Nano*, 2013, 7, 9407–9415.
72. B. Tan and Y. Wu, *J. Phys. Chem. B*, 2006, **110**, 15932–15938.
73. K. Shankar, G. K. Mor, H. E. Prakasam, S. Yoriya, M. Paulose, O. K. Varghese and C. A. Grimes, *Nanotechnology*, 2007, **18**, 065707.
74. O. K. Varghese, M. Paulose and C. A. Grimes, *Nat. Nanotechnol.*, 2009, **4**, 592–597.
75. S. Ito, T. N. Murakami, P. Comte, P. Liska, C. Grätzel, M. K. Nazeeruddin and M. Grätzel, *Thin Solid Films*, 2008, **516**, 4613–4619.
76. S. H. Kang, S. H. Choi, M. S. Kang, J. Y. Kim, H. S. Kim, T. Hyeon and Y. E. Sung, *Adv. Mater.*, 2008, **20**, 54–58.
77. G. Larramona, C. Choné, A. Jacob, D. Sakakura, B. Delatouche, D. Péré, X. Cieren, M. Nagino and R. Bayón, *Chem. Mater.*, 2006, **18**, 1688–1696.
78. W.-Q. Wu, B.-X. Lei, H.-S. Rao, Y.-F. Xu, Y.-F. Wang, C.-Y. Su and D.-B. Kuang, *Sci. Rep.*, 2013, **3**, 1892.
79. N.-G. Park, J. Van de Lagemaat and A. Frank, *J. Phys. Chem. B*, 2000, **104**, 8989–8994.
80. M. McCune, W. Zhang and Y. Deng, *Nano Lett.*, 2012, **12**, 3656–3662.
81. Y. Qiu, K. Yan, H. Deng and S. Yang, *Nano Lett.*, 2011, **12**, 407–413.
82. F. Xu, Y. Lu, Y. Xie and Y. Liu, *Mater. Des.*, 2009, **30**, 1704–1711.
83. F. Xu, M. Dai, Y. Lu and L. Sun, *J. Phys. Chem. C*, 2010, **114**, 2776–2782.
84. K. Sun, Y. Jing, C. Li, X. Zhang, R. Aguinaldo, A. Kargar, K. Madsen, K. Banu, Y. Zhou and Y. Bando, *Nanoscale*, 2012, **4**, 1515–1521.
85. N. Takahashi, R. Matsumoto, A. Koukitu and H. Seki, *J. Cryst. Growth*, 1998, **189–190**, 37–41.
86. T. Kuykendall, P. Ulrich, S. Aloni and P. Yang, *Nat. Mater.*, 2007, **6**, 951–956.
87. Y. J. Hwang, C. H. Wu, C. Hahn, H. E. Jeong and P. Yang, *Nano Lett.*, 2012, **12**, 1678–1682.
88. H. Kim, M. Seol, J. Lee and K. Yong, *J. Phys. Chem. C*, 2011, **115**, 25429–25436.
89. Y.-L. Lee, C.-F. Chi and S.-Y. Liau, *Chem. Mater.*, 2009, **22**, 922–927.
90. S. Park, C. W. Lee, I. S. Cho, S. Kim, J. H. Park, H. J. Kim, D.-W. Kim, S. Lee and K. S. Hong, *Int. J. Hydrogen Energy*, 2013, DOI: 10.1016/j.ijhydene.2013.10.030.
91. S. Park, D. Kim, C. W. Lee, S.-D. Seo, H. J. Kim, H. S. Han, K. S. Hong and D.-W. Kim, *Nano Res.*, 2013, 7, 1–10.
92. A. Kargar, Y. Jing, S. J. Kim, C. T. Riley, X. Pan and D. Wang, *ACS Nano*, 2013, 7, 11112–11120.
93. Z. Yin, Z. Wang, Y. Du, X. Qi, Y. Huang, C. Xue and H. Zhang, *Adv. Mater.*, 2012, **24**, 5374–5378.

CHAPTER 9

Hierarchical Nanostructures: Application to Supercapacitors

JUNG BIN IN* AND COSTAS P. GRIGOROPOULOS

Mechanical Engineering, University of California, Berkeley, California 94720, U.S.A
*Email: jbin@berkeley.edu

9.1 Introduction

In conjunction with the energy crisis, environmental issues accompanying the utilization of energy resources have driven the development of new technologies that can realize highly efficient as well as environmentally friendly energy generation. The rise of green technology represents such an effort. In this respect, we are facing a new era of energy science: nanotechnology, a possible game changer. A recent endeavor to enhance the energy conversion efficiency of nanomaterial-based solar energy systems exemplifies the paradigm change.

Along with efficient harvesting, conversion and consumption of energy, the development of novel energy storage systems constitutes the main body of technical approaches to improve the overall efficiency of energy utilization. While the production of energy is mainly based on the large-scale plant industry, the energy produced is delivered to various consumer units ranging from residential houses to energy-intensive facilities. Since consumption of energy hardly occurs synchronously with its production, the development of energy storage systems is needed for stable use of the energy. Moreover, explosively growing demands for mobile electronics is

RSC Nanoscience & Nanotechnology No. 35
Hierarchical Nanostructures for Energy Devices
Edited by Seung Hwan Ko and Costas P Grigoropoulos

Published by the Royal Society of Chemistry, www.rsc.org

also spurring development of energy storage systems that can accommodate high energy density in a compact device.

Electricity is the most valuable form of energy. The electricity can be converted in a well-regulated manner to different forms of energy such as thermal, radiant, chemical, and mechanical energies. As the promise of zero-emission vehicles presents, utilization of the electricity can be free of emissive pollutants. Furthermore, electrical energy powers popular electronics that are indispensable in our daily life.

The development of high performance electrical energy storage systems has been a major concern in industry as well as in research societies. Recently, modern mobile electronics are rapidly extending their capabilities in order to accommodate large-data transfer and processing. Indeed, state-of-the-art smart phones are small energy-intensive units with multicore processors and high-resolution displays, and the performance requirement continues to grow. Wearable computers are promising futuristic applications and are expected to require additional functionalities amenable to mechanical deformation of the energy storage system.

Related to the development of advanced energy storage systems, two types of energy storage systems have attracted particular interest: electrochemical capacitors (so-called supercapacitors) and Li-ion batteries. The performance of these devices has been evaluated routinely based on specific power *vs.* specific energy relation and mapped in the so-called Ragone plot (Figure 9.1). As clearly indicated in Figure 9.1, batteries have extended discharge time with larger energy capacity. Thus batteries have been dominantly used to power mobile electronics such as cell phones and laptops.

In comparison with the batteries, supercapacitors feature higher specific power but relatively lower specific energy. The performance of supercapacitors culminates in various situations where a burst of high power output is required or relatively higher power should be supplied within a short period of time (less than a minute). Likewise, supercapacitors can be charged up in a comparably short time. Supercapacitors can also allow a very large number of charge–discharge cycles (>500 000) with extended lifetime compared with batteries and other conventional dielectric capacitors.[1,2] Owing to these characteristics, supercapacitors have been applied to backup power devices, heavy hybrid electric vehicles for cyclic energy capture, electric screwdrivers, and so on.[3]

In this chapter, we focus on the working principles of supercapacitors and application of nanomaterials, in particular those in the form of hierarchical nanostructures. As will be discussed in the following sections, the energy storage mechanism of supercapacitors is mostly based on surface phenomena. In this regard, using nanomaterials that are distinguished by extremely high surface to volume ratios offers a fundamental ground for realizing unusually high capacitance energy storage systems. Moreover, supercapacitor devices composed of nanomaterials can be light in weight, optically transparent,[4] and flexible,[5] which are promising characteristics for futuristic portable and wearable device applications. In the next section, we

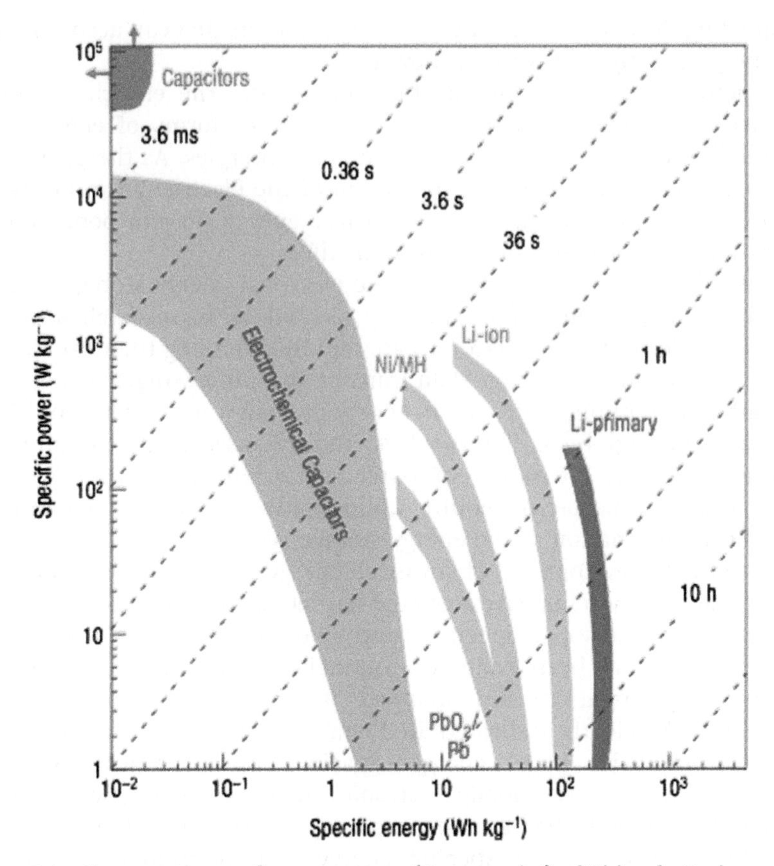

Figure 9.1 Comparative performance map (Ragone plot) of Li-ion batteries, supercapacitors, and traditional solid state and electrolytic capacitors. The dashed lines indicate the discharge time of the stored electricity. Reprinted by permission from Macmillan Publishers Ltd: *Nature Materials* ref. 1, copyright 2008.

start our discussion from a brief review of the energy storage mechanisms of supercapacitors.

9.2 Electrical Double-Layer Capacitors and Pseudocapacitors

9.2.1 Electrical Double-Layer Capacitors (EDLC)

Since the first development of a practical form of electrochemical capacitors (ECs) in 1966 by the Standard Oil Company of Ohio (SOHIO),[6] various concepts for high capacitance ECs have been proposed. Most of the electrochemical capacitors consist of three main components: electrodes, electrolytes, and separators. Figure 9.2(a) shows a simplified schematic of an

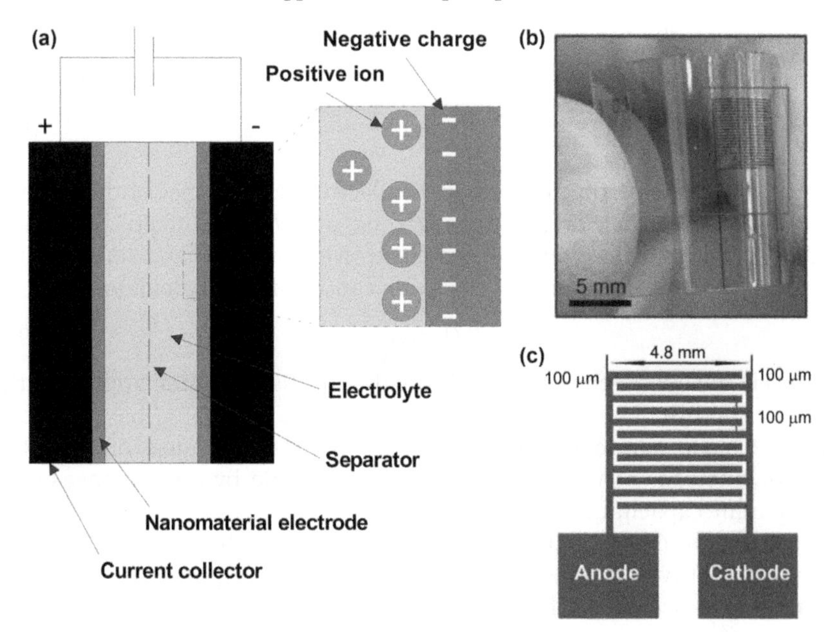

Figure 9.2 Structures of supercapacitors. (a) Conventional structure of electrochemical double layer supercapacitors. (b–c) Typical electrode structure of micro-supercapacitors. (b) Photo image of a real flexible micro-supercapacitor. (c) Schematic of typical interdigitated electrodes of micro-supercapacitors. (b–c) Reproduced from ref. 7.

EC consisting of these basic components. A differential potential externally imposed on the electrodes induces an electric field across the electrolyte. The electrodes provide electrical channels for electrons to be delivered to the interface with the electrolyte. To minimize the overall resistance the electrodes are connected to conductive (generally metallic) current collectors that lead to the external terminals. Electrolytes contain ionic molecules in liquid- or solid-state media. Separators are porous insulators that allow the transport of ions but prevent electrical short. However, the separator is not required with solid-state electrolytes since the electrolyte itself blocks possible contact between the anode and cathode. The separator is also not needed in micro-supercapacitors because of the separate patterns of interdigitated electrodes (Figure 9.2(b),(c)).

In EDLC, ionic molecules of the electrolyte are accumulated by coulombic attraction at the electrode surface. At the very surface of the electrode, they form a thin layer: the so-called Helmoltz layer. Along with this layer, an ion distribution can be further developed toward the electrolyte side, due to thermal fluctuations. This relatively thick layer is called the diffuse layer. These two layers constitute the essence of the charge storage mechanism of EDLC and play a major role in determining the electrochemical performance of the capacitors.

In principle, the capacitance of EDLC can be expressed by the following basic relation:

$$C = \varepsilon_r \varepsilon_0 \frac{A}{d} \tag{9.1}$$

where ε_r is the dielectric constant of the electrolyte, ε_0 is that of the vacuum ($\varepsilon_0 = 1$), A is the surface area of the electrode, and d is the effective thickness of the double layer developed in the electrolyte. Given a capacitance C, the amount of charge stored in the electrode can be simply calculated by:

$$Q = CV \tag{9.2}$$

where Q is the stored charge and V is the differential potential applied to the electrodes. While the capacitance of EDLC can be estimated based on eqn (9.1), the basic parameters lack sufficient accuracy because of non-ideal behavior. Instead, practical evaluation can be made by directly measuring the apparent capacitance of EDLC.

The capacitance of EDLC can be obtained from galvanostatic charging–discharging curves (CDCs). In CDC, a potential response to charging–discharging processes with a constant current is recorded as a V–t curve. From eqn (9.2), the capacitance can be expressed by the following relation:

$$C = \frac{I}{dV/dt} \tag{9.3}$$

According to eqn (9.3), the slope of the CDC (dV/dt) is directly correlated with the capacitance. In general, discharging curves are used for this purpose. Alternatively, the capacitance can be calculated from cyclic voltammetry (CV). In CV curves, the rate of potential change (dV/dt) is kept constant, and the current response is recorded with respect to the transient voltage (I–V curves). In contrast to an ideal capacitor that shows a rectangular response in CV characterization, general EDLCs show curved close-loops. In this case, the capacitance value is obtained by integrating and then normalizing the current response.

As discussed in section 9.1, specific energy and specific output power are most important figures of merit for supercapacitors. In conjunction with the above equations, the total energy (E) and the power (P) of EDLC can be expressed as:

$$E = \frac{1}{2} CV^2 \tag{9.4}$$

$$P = \frac{V^2}{4R_{ESR}} \tag{9.5}$$

where R_{ESR} indicates equivalent series resistance (ESR). ESR originates from (1) the contact resistance of the electrode and current collector; (2) the internal resistance of the electrode itself; (3) the resistance of electrolyte (bulk + pore);[8] (4) the resistance of the external lead contact. ESR can be

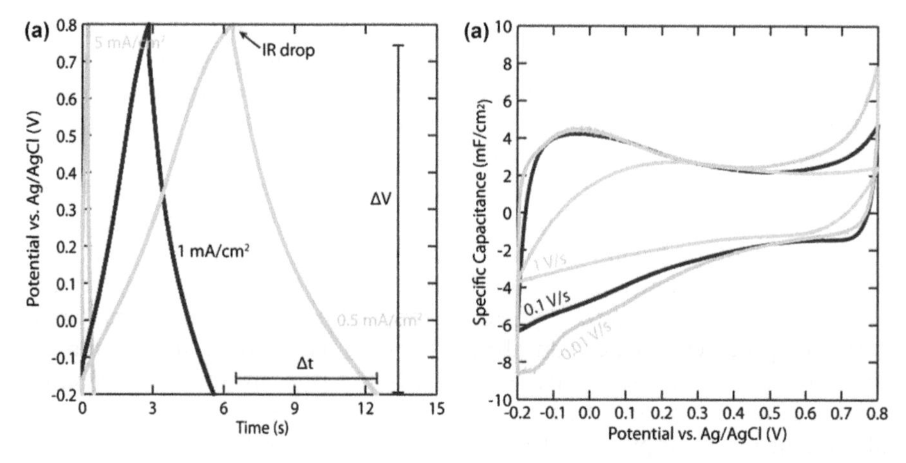

Figure 9.3 Measurement of electrochemical performance. (a) Charging–discharging curves of a pyrolyzed photoresist electrode at different current densities (0.5, 1, 5 mA cm^{-2}). (b) Cyclic voltammograms at different scanning rates (0.01, 0.1, 1 V s^{-1}). The specific capacitance was calculated from the corresponding values of current density based on eqn 9.3. Reprinted from ref. 9 with permission from Elsevier.

obtained from an abrupt potential drop (Figure 9.3(a)) in the discharging response of CDC, or from the real part of the high frequency response in electrochemical impedance spectroscopy (EIS).

Figure 9.3 shows examples of CDC and CV curves reported by Hsia *et al.*[9] Their carbon-based supercapacitor electrodes were produced from a photoresist by high-temperature pyrolysis. Indeed, the CDC curve shows an abrupt drop at the very start of discharging, which is indicative of ESR of the electrode. The CV responses are obviously non-ideal, and Hsia *et al.* suggested that this behavior could be accounted for by redox reactions of the aqueous electrolyte and possibly by pseudocapacitive contribution of surface groups formed during the pyrolysis process. (Pseudocapacitive charge storage will be discussed in the next section.)

For reasonable and practical evaluation, the stored energy and the power output are normalized with respect to extensive parameters of the supercapacitors. Unfortunately, the evaluation standard lacks consensus. In literature reports, researchers are still using scattered metrics, such as gravimetric (per gram), volumetric (per cm^3), and areal (per cm^2) densities. The quantity is calculated based on the single electrode or sometimes based on the final device. For realistic performance evaluation, proper metrics should be defined in consideration of device characteristics and performance scalability of electrode materials.[10]

From eqn (9.4) and (9.5), it is obvious that specific energy and specific power can be improved by increasing the specific capacitance and the operation voltage window, as well as by reducing the ESR. Since the specific capacitance is proportional to the specific surface area (SSA) (eqn (9.1)), using high SSA electrodes should be a straightforward approach to enhance capacitance. As will be discussed in section 9.3, various carbon nanomaterials, such as carbon nanotubes and graphene, have very high SSA. In this regard, porous assemblies of these carbon nanomaterials can be promising electrodes for high capacitance EDLCs.

In addition, the pore size of the supercapacitor electrode should be carefully tuned to achieve higher specific capacitance. While eqn (9.1) accounts for the capacitance of conventional planar electrodes, it is not directly applicable for calculation of the capacitance of nanoscale pores.[11] As revealed by Chmiola *et al.*,[12,13] when normalized by SSA, the capacitance of nanoscale pores experiences an abrupt increase as the pore size decreases below the size of solvated ions. They attributed this unexpected dramatic increase to desolvation of the ions in the extremely confined pore space and subsequent closer approach of the ions to the pore wall. As the pore size decreases further and approaches the size of electrolyte ions, size exclusion starts hindering the ions from accessing the surface inside the pore. As a result, the capacitance diminishes quickly. A similar trend was reported from a study on EDLC with an ionic liquid electrolyte, which is free of solvation/desolvation effects (Figure 9.4).[14]

It should also be considered that increasing specific capacitance can cause degradation of power performance. For instance, increased porosity of EDLC electrodes can compromise electrical conductance, resulting in an increase

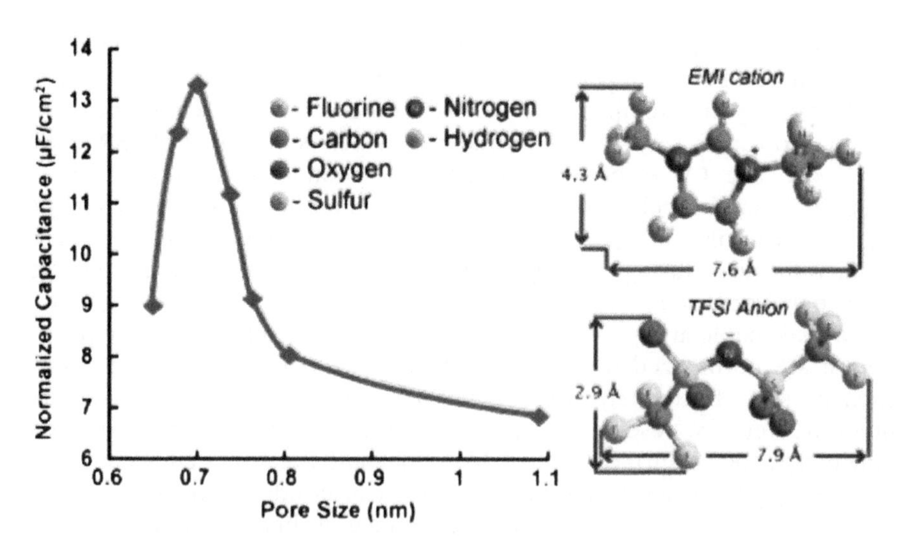

Figure 9.4 Capacitance change by variation of the pore size of carbide-derived carbon. The capacitance was normalized by the SSA of the electrode. An ethyl-methylimidazolium/bis(trifluoro-methane-sulfonyl)imide ionic liquid (EMI/TFSI) electrolyte was applied to exclude effects of solvation/desolvation of ions.
Reprinted with permission from ref. 14. Copyright 2008 American Chemical Society.

of ESR and hence decrease of specific power (eqn (9.5)). Therefore, a technical trade-off between porosity and conductance should be reasonably made based on an understanding of the ion transport mechanism through nanoscale pores.

The potential window is another important parameter that determines the specific energy and power densities. As eqn (9.4) and (9.5) indicate, the potential window has an influence on the performance by the second order. Thus increasing the potential window is a very effective way to enhance the energy/power density of EDLC. In the case of aqueous electrolytes, the potential window is restricted below 1.23 V, above which electrolysis of water molecules limits normal operation. In spite of the toxicity, organic electrolytes allow a larger potential (*i.e.* acetonitrile: 2.0 V). Recently, ionic liquid electrolytes are being actively investigated, due to their even larger voltage window ($\sim$4.0 V).

There exists a new type of EDLC: a micro-supercapacitor. In contrast to the conventional structure of EDLC, the electrodes are not facing each other, but are placed in the same plane. To enhance the specific capacitance, micro-supercapacitors adopt a characteristic interdigitated finger structure. Figure 9.2(c) shows a typical electrode structure of micro-supercapacitors. Calculating the capacitance of micro-supercapacitors is not as straightforward as that of traditional capacitors (eqn (9.1)). Geometric parameters of the finger structure should be comprehensively considered in conjunction with the conventional parameters. Readers are encouraged to refer to ref. 15 for a detailed discussion.

EDLC is distinguished from batteries in that no electron transfer occurs across the electrode–electrolyte interface (non-faradaic). As discussed above, the nature of energy storage is purely electrostatic. Thus, fast separation and recombination of charge (high specific power) are enabled, whereas charging–discharging rates of batteries are limited by relatively slow electrochemical redox reaction kinetics. Electrodes of EDLCs are relatively conductive and thereby generate minimal heat compared with batteries. EDLCs are also free of electrode dissolution, so their lifetime can be extended to a million cycles.

Notwithstanding the high power delivery and other merits, the performance of EDLC is still limited in the amount of stored energy (Figure 9.1). Therefore, a new form of supercapacitors has been investigated: the so-called pseudocapacitors. In the next section, the characteristics and energy storage mechanism of pseudocapacitors are discussed.

9.2.2 Pseudocapacitor

Pseudocapacitors inherit many useful features of EDLCs. However, the electrode of pseudocapacitors is not chemically inert. In pseudocapacitors, reversible redox reactions with the electrolyte occur near the electrode surface (faradaic). However, the reaction domain is not limited to the very surface, proceeding into the bulk of the material. Thus, pseudocapacitors feature increased capacitance that can be significantly higher than the general capacitance of EDLC. The effective capacitance of pseudocapacitors can be expressed in an additive manner as

$$C = C_E + C_P \tag{9.6}$$

where C_E and C_P indicate non-faradaic and faradaic contribution to the capacitance, respectively.

Metal oxides and conductive polymers are widely used as electroactive materials for pseudocapacitor electrodes. The pseudocapacitive metal oxides should be not only reversibly electroactive for redox reactions but also electrically conductive to deliver electrons that are essential for continuing the reactions. Ruthenium oxide (RuO_2) and manganese oxide (MnO_2) are the most representative metal oxides for pseudocapacitors. Other oxides such as vanadium oxide (V_2O_5),[16] nickel oxide (NiO),[17] cobalt oxide (Co_3O_4),[18] and tin oxide (SnO_2)[19] can be potentially used as economic materials for pseudocapacitor electrodes.

Ruthenium oxide has been widely investigated due to its superior performance compared with other metal oxides. Ruthenium oxide has three distinct redox reactions that are active within 1.2 V:

$$RuO_2 + xH^+ + xe^- \leftrightarrow RuO_{2-x}(OH)_x \tag{9.7}$$

where $0<x<2$.[19] The specific capacitance of ruthenium oxide could reach up to 720 F g^{-1}.[20] Ruthenium oxide also has high thermal stability, a relatively long lifetime, and good electrical and proton conductivities. In contrast, the use of ruthenium oxide is restricted because of its high price and toxicity.

Manganese oxide (MnO_2) is a promising material alternative to ruthenium oxide especially in terms of cost. From 2013, the price (per gram) of elemental manganese is almost 800 times lower than that of ruthenium. Moreover, manganese oxide is environmentally benign. The relevant redox reactions of manganese oxide can be expressed as:

$$MnO_2 + E^+ + e^- \leftrightarrow MnO_2E \tag{9.8}$$

where E^+ indicates the cation of the electrolyte.[19]

Notwithstanding the listed promising attributes of manganese oxide, it has a very poor electrical conductivity (10^{-5}–10^{-6} S cm^{-1}).[21] When independently used as an electrode, therefore, manganese oxide suffers a high ESR and poor power output. In addition, the ion intercalation–deintercalation can cause mechanical instability under cyclic charging–discharging loads. Moreover, chemical corrosion of the metal oxide, especially in aqueous electrolytes, can shorten the lifetime of the electrode. Recently, conductive polymer wrapping has been suggested to compensate for the low conductivity and protect MnO_2 from corrosion.[22,23]

Conductive polymers such as polypyrrole (PPy), polyaniline (PANI), and polythiophene (PT) are also promising candidates for pseudocapacitors.

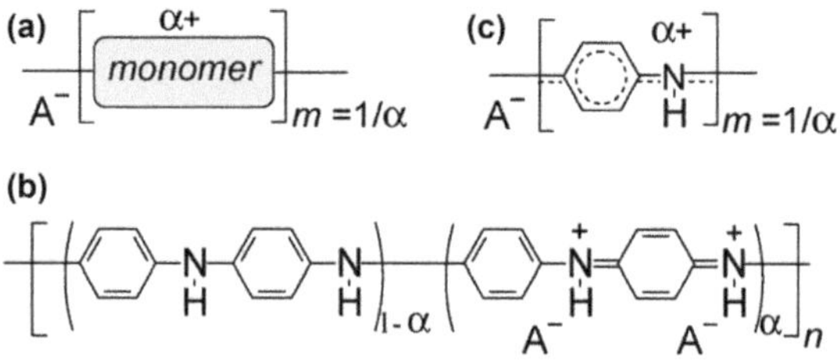

Figure 9.5 Chemical scheme of faradaic charge storage. (a) General conductive polymer, (b,c) polyaniline (PANI) ($0 \leq \alpha \leq 1$).
Reproduced from ref. 24.

The synthesis of conductive polymers is cheap and highly scalable. These materials have similar electrochemical characteristics to pseudocapacitive metal oxides. During charging–discharging, electrolyte ions move into or out of the backbone of the polymer *via* reversible redox reactions (Figure 9.5).[24] As is the case with metal oxides, the reaction domain proceeds into the bulk of the material as the electrode is charged. The intercalation and deintercalation of ions in the polymer electrode can cause cyclic swelling and shrinking, resulting in mechanical failure of the electrode. This stability issue limits the lifetimes of conductive polymer electrodes.

Pseudocapacitors resemble batteries in terms of their charge storage mechanism, and their performance is between those of EDLCs and batteries. The most valued feature of pseudocapacitors is their increased capacitance and thereby the increased energy density compared with EDLC. As expected, however, the relatively slow redox reaction kinetics and poor conductivity of electroactive materials can undermine the power performance of pseudocapacitors. To achieve high-energy storage and high-power output simultaneously, advanced material engineering has pursued the development of hybrid electrode systems. In the following sections, we will discuss various nanomaterials for electrodes of supercapacitors and their integration into novel hybrid electrode systems.

9.3 Nanomaterials for Supercapacitor Electrodes

As mentioned, assembled porous nanomaterials have an extraordinarily high specific surface area (SSA). They can be a very attractive choice for the electrodes of supercapacitors. A large degree of freedom in the design of the electrode and the selection of the material has spurred comprehensive research and performance evaluation of a variety of nanostructured electrodes. In the following sections, various nanomaterials especially used for electrodes of supercapacitors are introduced and their technical importance is discussed. Figure 9.6 shows a brief summary of the specific capacitance of supercapacitor electrodes of various materials. It should be especially noted that the performance of supercapacitors (capacitance, specific power, specific energy) of various nanomaterials can be quite scattered, depending

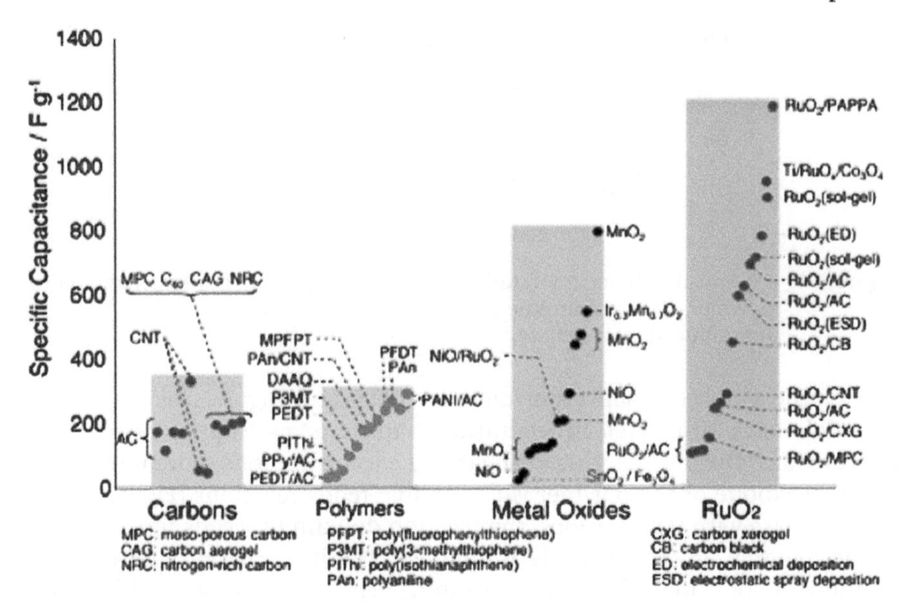

Figure 9.6 Specific capacitance of various electrode materials for supercapacitors. Reproduced from ref. 25 by permission of The Electrochemical Society.

on the choice of the electrolyte and the experimental conditions including scan rate, current load, and amount of material loading.[10] Ranges of values are provided in this section, and readers are encouraged to refer to the individual studies for more details.

9.3.1 Activated Carbon

Activated carbon has been a conventional material for carbon-based EDLCs. It can be produced from carbon-rich organic precursors *via* high-temperature carbonization in an inert atmosphere. A post-oxidation process in the presence of CO_2, water, and KOH can further enhance the surface properties. Activated carbon can be derived from abundant natural resources (wood, lignin, coconut shell, *etc.*) or other types of carbon such as coal and petroleum coke. In addition to the cheap carbon resources, the large-scale manufacturing process has enabled activated carbon to be commercially available.

Activated carbon consists of a myriad of micropores (<2 nm) and the SSA can reach 1000–2000 $m^2 g^{-1}$.[26] However, a lack of mesoporous ion channels in activated carbon limits the accessibility of electrolyte ions and thereby its capacitance is restricted to only a few tens of F g^{-1}. Moreover, activated carbon has a poor electrical conductivity, due to the low graphitic quality and the low connectivity between activated carbon agglomerates. The ESR of the activated carbon electrode is therefore higher than those of other graphitic carbon materials that will be discussed in the next section.

For the above reasons, the energy and power densities of the EDLCs composed of activated carbon are not as high as expected based on their high SSA (only 4–5 Wh kg^{-1} in energy density and 1–2 kW kg^{-1} in power density).[26] As a result, the development of new carbon-based electrode materials that can overcome the listed limits of activated carbon has been pursued. In the next section, the new carbon nanomaterials for EDLC electrodes are introduced.

9.3.2 Carbon-based Nanomaterials for EDLC

The high SSA and corrosion resistive properties of carbon-based nanomaterials rendered these materials most valuable for EDLC electrodes. Furthermore, carbon is an abundant element on earth and therefore, carbon-based nanomaterials can potentially be the most cost-effective competitor among other nanomaterials. Carbon-based nanomaterials for EDLC range from carbon nanotubes (CNTs), and graphene[1] to other graphitic or partially graphitic carbons such as carbon fibers,[27] carbon onions,[28] carbon nanohorns,[29] *etc.* In this section, we discuss the application of the most widely studied carbon nanomaterials: carbon nanotubes and graphene.

9.3.2.1 Carbon Nanotubes

The structure of a carbon nanotube can be described as rolled graphitic carbon sheets. The number of the sheets varies. Carbon nanotubes comprised of a single carbon sheet are called single-walled carbon nanotubes (SWNT). In the same manner, nanotubes with multiple walls are called multi-walled carbon nanotubes (MWCNT). The SSA of CNTs can reach about 1600 $m^2\ g^{-1}$ for SWNTs[30] and about 430 $m^2\ g^{-1}$ for MWNTs.[31] The graphitic carbon atoms are strongly bound to each other and are almost inert to foreign chemicals under the operation conditions of EDLC. The electrical conductivity of individual CNTs is excellent, extending to the order of 10^5 S cm^{-1}.[32] CNTs also have high strength (tensile strength: 10 GPa of SWNT) and outstanding mechanical resilience. Thus electrodes of CNTs are readily applicable to the electrodes of flexible energy storage systems.

Several large-scale manufacturing methods for CNTs can enable their commercial use. SWNTs can be produced by a gas-phase growth mechanism, the so-called HiPco process.[33] Catalytic chemical vapor deposition (CCVD) is a versatile method that can produce SWNTs and MWNTs selectively by controlling the size of the catalyst.[34] The CCVD growth can produce vertically aligned CNTs (VACNTs) or a CNT forest. In the CNT forest, the growth of individual nanotubes is directed vertically to the substrate. This structure is highly promising as the electrode of supercapacitors since the electrical and ionic transport path is straight to the current collector with minimal scattering by neighboring CNTs. Owing to this unique structure, the CNT forest can be used not only as a direct EDLC electrode but also as a mesoporous conductive template that can be decorated with other electroactive nanomaterials. (In section 9.3.3, the performance of this hybrid structure will be discussed.)

Capacitance of CNT electrodes ranges from 50–200 F g^{-1} in general aqueous electrolytes, and the energy density and the power density span 8–20 kW kg^{-1} and 1–10 Wh kg^{-1},[26] respectively. As reported by Lu *et al.*,[35] additional surface treatment and pore opening can enhance the capacitance. Indeed, they obtained an extraordinarily high capacitance of VACNTs (440 F g^{-1}) with an ionic liquid. The energy and power densities (148 Wh kg^{-1}, 315 kW kg^{-1}) were also high, compared with pristine VACNTs.

To apply CNTs to flexible supercapacitors, the nanotubes should be integrated on flexible materials such as polymers. However, direct growth of CNTs on polymeric substrates is highly challenging, due to the high temperature that CVD growth of CNTs requires. Instead of direct growth of CNTs on flexible substrates, CNTs can be transferred and assembled on flexible receptor substrates. This transfer process can incorporate CNTs into various flexible materials such as carbon micro-fibers,[36] cotton paper,[37] metal foils,[38] and plastic sheets.[39] Wet transfer techniques with solubilized CNTs are versatile and easy to apply for this purpose, but the CNTs become entangled and bundled. Performance of the entangled CNTs is significantly undermined, due to their high inter-CNT and CNT-current collector contact resistances.

In this regard, dry transfer of VACNTs to polymer sheets that does not affect the arrangement of the aligned nanotubes can be a plausible approach to realize high-performance CNT-based flexible capacitors. However, VACNTs have relatively poor in-plane conductivity in spite of good axial conductivity of the individual CNTs. Common flexible polymers are insulating and therefore the lateral electrical resistance of VACNTs can become even higher after the transfer. Recently, Marschewski *et al.*[40] reported a significant decrease in lateral resistance of VACNT electrodes by placing a nickel layer coat on top of the VACNTs. They used a laser welding technique to simultaneously transfer and pattern the nickel-coated VACNTs into interdigitated electrodes on polycarbonate substrates[41] (Figure 9.7). The fast conduction path provided by the nickel film enabled significant decrease of ESR, and the micro-supercapacitor that used the laser-welded VACNT electrodes showed excellent bendability.[42]

Despite the superior performance compared with the conventional activated carbon, use of CNT electrodes is limited by the following reasons. Firstly, commercialized CNTs still have a much higher price than activated carbon. Secondly, CNT growth requires catalyst particles, so these metal elements can act as impurities unless the CNTs are properly purified. As a result, unexpected pseudo-capacitance can appear. Purifying treatments with acidic solutions can readily remove the metal catalysts, but the wet post-processing can induce defects and surface groups on CNTs that can cause unwanted electrochemical behavior. Thirdly, despite the aforementioned effort, the in-plane (lateral) resistance of the CNT forest is still high. Metallic growth substrates can enhance the in-plane conductivity. However, direct

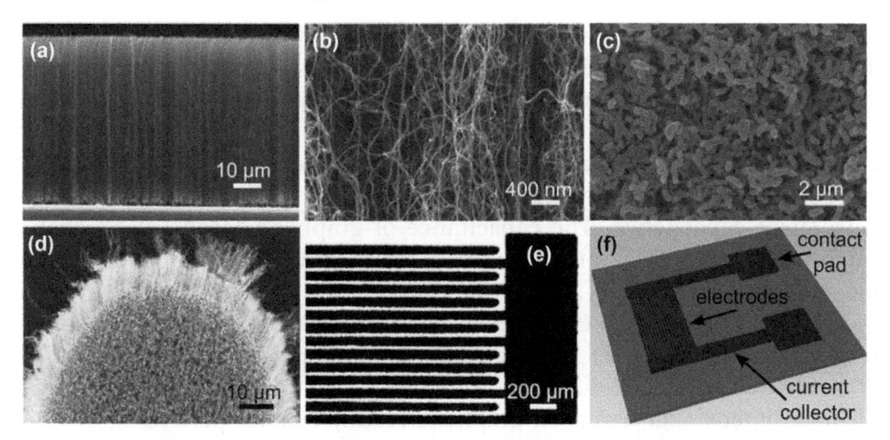

Figure 9.7 Laser-assisted patterned VACNTs for micro-supercapacitor electrodes. (a) SEM (side view) of Ni-coated VACNTs transferred on a polycarbonate substrate. (b) Enlarged SEM view of the lateral morphology of VACNTs. (c) SEM (top view) of Ni-coated VACNTs. (d) SEM (top view) of a line extremity of a laser-assisted transferred VACNT pattern. (e) Optical microscopy (top view) of the interdigitated VACNT pattern. (f) Illustration of the micro-supercapacitor device. © IOP Publishing. Reproduced from ref. 42 by permission of IOP Publishing. All rights reserved.

growth of CNTs on common metallic substrates is challenging since the growth catalyst can be deactivated by reaction with metals.[17]

9.3.2.2 Graphene

Graphene is a single layer of planar graphitic carbons that are sp^2-bonded to each other in a hexagonal honeycomb configuration. The resistivity of graphene can ideally be as low as 10^{-6} Ω cm^{-1},[43] and mechanical strength is over 100 GPa with a very high elastic modulus ($\sim$1 TPa).[44] The specific surface of graphene approaches 1520 m^2 g^{-1}.[45] In contrast to CNTs, the 2-dimensional structure of graphene enables improved electrical contact. These outstanding properties prove that graphene is a promising electrode material, equivalent or possibly superior to CNTs.

Similar to the CVD growth of CNTs, large-area graphene can be produced by CVD techniques. Copper foils are commonly used as catalytic growth substrates. A single or a few layers of graphene can grow on the copper surface with a methane feedstock gas at high temperatures (900–1100 °C).[46] While the CVD method is advantageous in growing large-grained graphene, the growth domain is restricted by the size of the catalytic copper foil (or the size of the growth system). Graphene films can be also extracted by exfoliation from a bulk form of graphite such as highly oriented pyrolytic graphite.[47] Wet exfoliation processes are highly promising and economic techniques for separating soluble graphene oxide (GO) layers from graphite or graphite oxide in a scalable manner.[48] To further enhance the electrochemical performance, the produced GO layers can be reduced with chemical agents such as hydrazine hydrate,[49] or by laser irradiation.[45]

The capacitance of graphene-based EDLCs can ideally reach about 550 F g^{-1}.[50] In practice, however, aggregation by stacking of graphene layers and degradation of the physical properties by chemical processing limit the consequent capacitance. The capacitance of graphene-based EDLCs spans roughly 100–200 F g^{-1},[51] and the corresponding specific energy and specific power generally reach ~30 Wh kg^{-1} and ~10 kW kg^{-1} with aqueous electrolytes. Colloidal GO can be easily coated on various substrates, and, as in the case of flexible CNT supercapacitors, graphene can equally be used as a flexible electrode.

The 2-dimensional nature of graphene readily enables patterning of thin GO films on various substrates for flexible in-plane micro-supercapacitors.[52] Laser scanning can be used to write an interdigitated pattern of reduced graphene oxide (rGO) with the pristine GO acting as both a separator and an efficient electrolyte.[53] Recently, El-Kady *et al.*[54] demonstrated DVD laser scribing on GO films for patterning of rGO on plastic substrates. Their all solid-state EDLC micro-supercapacitors with a poly(vinyl alcohol) (PVA)-H_2SO_4 polymer electrolyte showed an excellent areal capacitance (2.32 mF cm^{-2}).

9.3.3 Other Carbonaceous Materials

3-Dimensional macro-/meso-/micro- porous monolithic carbon structures are promising electrode materials for EDLCs. They can also serve as promising templates for building hierarchical nanostructures (see section 9.3.5). Their manufacturing processes are highly scalable. For their synthesis, organic precursors are introduced to a porous template, followed by high-temperature carbonization (or deposition of carbon) and removal of the template.[9,55–58] Natural porous carbon sources such as hemp can also be used for high-temperature carbonization to produce a 3D graphitic carbon monolith (Figure 9.8).[59] In this regard, these carbon structures are comparable to activated carbon. However, they are distinguished by larger pore sizes, relatively higher graphitic quality and better connectivity of the structure that leads to lower ESR and enhanced power performance.

Composites of carbonaceous materials can provide a new way of improving electrode performance. Carbon nanomaterials can be physically or chemically combined with another carbon nanomaterial of a different dimensionality,[60,61] pseudocapacitive polymers and metal oxides,[62] or precious metals. For instance, Lin *et al.*[62] reported that the capacitance of the horizontally aligned multi-walled carbon nanotubes (MWNTs) electrodeposited with PANI can be as high as 233 F g^{-1} @ 1A g^{-1}, which is 36 times higher than that of bare MWNTs. Zhang *et al.*[63] obtained even higher capacitance (1030 F g^{-1} @ 5.9 A g^{-1}) with vertically aligned MWNTs coated with PANI.

In the case of metals, despite the excellent electrical conductivity, instability of metals in electrolytes invalidates the use of metal nanomaterials for

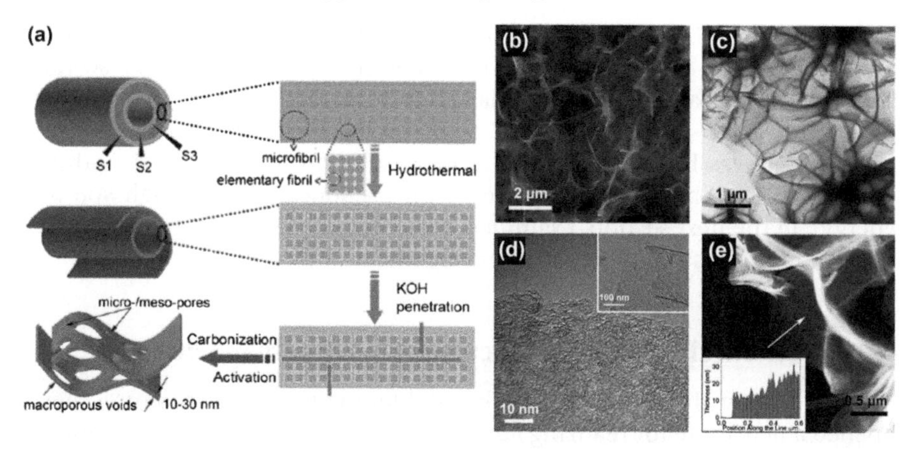

Figure 9.8 Interconnected graphitic carbon structure derived from hemp. (a) Schematic of the fabrication process. (b) SEM of the interconnected porous carbon structure. (c) TEM of the carbon structure. (d) High-resolution TEM of the corresponding graphitic carbon. (e) Annular dark field (ADF) TEM image and electron energy loss spectroscopy (EELS) thickness profile (inset) of the carbon nanosheet.
Reprinted with permission from ref. 59. Copyright 2013 American Chemical Society.

supercapacitor electrodes. Nanomaterials of precious metal elements could be candidates. However, the very high price prevents the direct use of precious metals as electrodes. Instead, precious metal nanomaterials can be used as an additive to significantly enhance the conductivity of carbon-based electrodes.[64]

In the composites of nanomaterials, multiple heterogeneous nanomaterials are dispersed uniformly or stacked layer by layer. In other cases, the component nanomaterials can be assembled into an intriguing hierarchical structure. Further details of the hybrid hierarchical nanostructure will be discussed in section 9.3.5.

9.3.4 Electroactive Nanomaterials for Pseudocapacitors

Electroactive materials for pseudocapacitors (section 9.2.2) can be also synthesized in the form of nanomaterials such as nanoparticles,[65] nanowires,[66] and nanosheets.[16] These structures can be fabricated by hydrothermal methods[67] or electrodeposition on conductive supports.

Consistently, supercapacitor electrodes of these materials benefit from significantly increased porosity and SSA and high pseudocapacitance. A decrease in the size can enhance the mechanical stability under cyclic charge–discharge processes.[19] Nanostructures of electroactive materials can also potentially overcome the intrinsic low conductivity owing to their short conduction path. For instance, nanoparticles of electroactive metal oxides, which feature high SSA, can provide a sufficient electrical conductance when supported on a conductive current collector.

Nevertheless, the conduction properties and the mechanical strength do not match those of carbon-based nanomaterials. As discussed in section 9.3.3, the electroactive nanomaterials can be loaded on foreign materials that can compensate for the listed problems. The concept of a *hybrid electrode system* was born as a novel technical solution to combine the advantages of EDLCs (high specific power) with pseudocapacitors (high specific energy). In the next section, examples of the hybrid electrode are introduced with an emphasis on hierarchical arrangement of electrode materials.

9.3.5 Hierarchical Nanostructures for Supercapacitors

As consistently discussed in this book, hierarchical nanostructures have opened a new route for realizing advanced energy materials. The constituent nanomaterials can be a single (unary system) or different (multi-component system: binary, ternary, and more) material(s) with *different morphological attributes* at each hierarchical level. Branched nanowires exemplify hierarchical nanostructures.[68] An intriguing expression, *nanoforest*, could be an analogical name of a hierarchical nanostructure.[69]

Hierarchical structures for supercapacitors are synthesized mostly based on the bottom-up approach because of fine control of the follow-up structure and better electrical contact between different hierarchical levels. For instance, a totally graphitic structure can be obtained by growing vertically aligned carbon nanotube arrays on graphene (Figure 9.9). The seamless contact at the interface leads to efficient electron collection from the

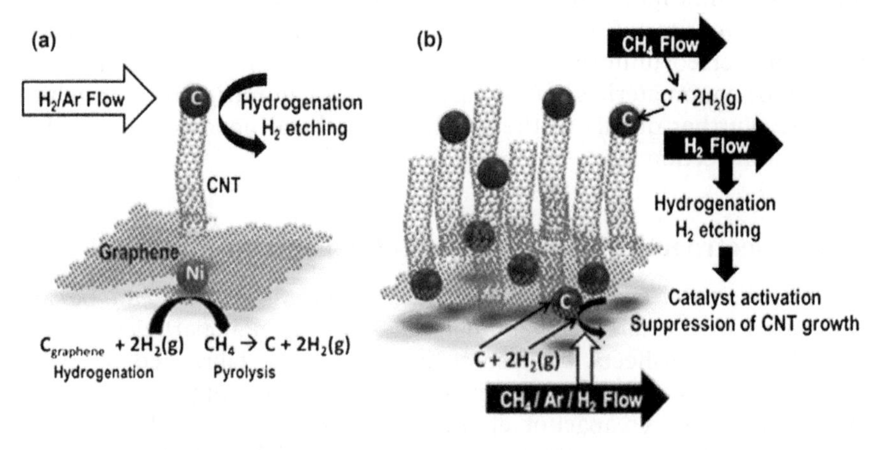

Figure 9.9 Illustration of growth processes in the catalytic CVD synthesis of a CNT–graphene structure. (a) Ni catalyst-assisted hydrogenation of graphitic carbon caused by hydrogen abundant gases (or absence of carbon feedstock). (b) Seamless connection of CNTs and the underlying graphene obtained by successful catalytic growth of CNTs in a methane carbon feedstock gas. © IOP Publishing.
Reproduced from ref. 71 by permission of IOP Publishing. All rights reserved.

nanotubes to the underlying graphene.[70,71] Related to supercapacitors, only a few exceptions of self-assembly approach for producing hierarchical nanostructure have been reported.[72]

The bottom-up synthesis processes consist of separated growth steps for each hierarchical level. The growth of the first step mainly produces a nanomaterial template that has larger porosity and better electrical conductance. In the following steps, nanostructures of a smaller feature with high SSA are produced on the template under equivalent or milder conditions so as not to degrade the structures grown in the previous stages. In the case of a unary system, the same precursor can be used, but in general, the growth parameters are modified in the follow-up stages to induce different morphologies. In the case of a hybrid system, growth precursors and conditions can be dramatically changed.

It has been reported that hierarchical nanostructures can overcome the restrictions of nanomaterials produced by single-mode growth and achieve unusual physical properties. Indeed, hierarchical nanostructures can improve the performance of supercapacitors (*i.e.* higher energy density, higher power density). A straightforward example is an increase of porosity or SSA by addition of branch materials that expose a surface area larger than the occupied sites of the mother material grown in the previous step.

As mentioned before, however, a one-dimensional approach with the aim of increasing SSA does not necessarily lead to the performance improvement of supercapacitors. The porosity of electrode materials can be classified into the following categories based on the diameter of pores: micropores (<2 nm), mesopores (2–50 nm), and macropores (>50 nm), according to IUPAC. Fuertes *et al.*[73] suggested that micro- and meso-pores contribute to the performance of supercapacitors in different manners: micropores are mainly responsible for capacitance, whereas small mesopores (2–4 nm) facilitate propagation of charges for high current loads. Larger pores such as macropores can act as accessible reservoirs of electrolyte ions. In this respect, the pore structure of the electrode should be tightly controlled to achieve optimal performance.

Hierarchical nanostructures potentially provide a larger degree of freedom for structure control by means of the multi-stage synthesis processes. Due to the separated growth for each hierarchical level, the porosity of hierarchical nanomaterials can be more actively tailored. In general, the nanomaterial in the first hierarchical level contains meso- or macro-pores, and its relatively large size enables relatively fast electrical conductance. For instance, mesoporous carbon nanomaterials, such as vertically aligned carbon nanotube arrays, can be a good candidate for the first level material. Nanostructures of the next hierarchy have finer pores and play a central role in the surface phenomena related to the charge storage. As a result, the resulting porosity can follow a multi-modal distribution.

In addition to the tunable porosity, hybrid hierarchical nanostructures consisting of different kinds of capacitive nanomaterials can benefit from specialized functionalities of the constituent nanomaterials. Representative hybrid hierarchical nanostructures for supercapacitors are carbon-based nanostructures (electrode material for EDLC) decorated with electroactive nanomaterials (electrode material for pseudocapacitors). This novel combination can take advantage of the high power performance of EDLC and high-energy capacity of pseudocapacitors.

For the past few years, many papers have reported the electrochemical performance of supercapacitors composed of hierarchical nanomaterials. Various combinations of nanomaterials have been examined to build the hierarchy. We can list them with examples in terms of the dimensionality of the constituent materials:[†]

(1) 1D + 0D: CNT^{PNP} (PNP: RuO_2,[74–76] MnO_2,[77] NiO)[17]
(2) 2D + 0D: GS^{PNP} (PNP: Co_3O_4)[78]; $GO^{NiO,79}$; GS–carbon sphere[72]
(3) 3D + 0D: MCT^{PNP} (PNP: MnO_2)[80]
(4) 1D + 1D: $CF^{CNT,81}$; CF^{CPNW} (CPNW: PANI)[82]
(5) 2D + 1D: $GS^{CNT,83,84}$; $CNT^{GS,85}$; $CNT^{MnO2\ nanosheet,86}$; GO^{CPNW} (CPNW: PANI)[87]
(6) 3D + 1D: $MCT^{CNT,88}$; MCT^{PNW} (Co_3O_4,[18] PANI)[89]
(7) 3D + 2D: $MCT^{MnO2\ nanosheet,58}$

When supported on other materials of larger dimensions, pseudocapacitive nanoparticles (0D) can avoid aggregation and the consequent loss of SSA. In most of the listed cases, the supporting nanomaterials (of higher hierarchy) are electrically more conductive and act as efficient current collectors and mesoporous templates that facilitate fast transport of ions. For instance, vertically aligned carbon nanotube arrays (VACNTs) decorated with pseudocapacitive nanoparticles are a promising hierarchical hybrid system. The interspace between the aligned nanotubes is on the order of tens of nanometres, and in contrast to activated carbon, the tortuosity of the space is relatively low (straighter transport path of ions).[90]

Recently, Jiang *et al.*[17] demonstrated the enhanced energy storage of the nickel oxide–VACNT system (Figure 9.10(a)–(e)) that benefits from the mesoporosity and fast electron transfer of VACNTs and the high capacitance of the pseudocapacitive nanoparticles. The nickel nanoparticles were deposited on the sidewalls of VACNTs by electrodeposition. They reported a specific capacitance (1.26 F cm^{-3}) that is 5.7 times higher than pure VACNTs and 94.2% was retained during 10 000 cycles. Zhang *et al.*[77] demonstrated the use of a MnO_2–VACNT hybrid electrode system for supercapacitors and reported an even higher capacitance of 199 F g^{-1} (305 F cm^{-3}) at a scan rate of 100 mV s^{-1} with 97% retention after 20 000 cycles.

[†]GS: graphene sheet; PNP: pseudocapacitive nanoparticles; PNW: pseudocapacitive nanowires; CPNW: conductive polymer nanowires; CF: carbon fiber; MCT: 3D monolithic macro-/mesoporous carbon template. The superscript indicates the nanomaterial at the lower level of the hierarchy.

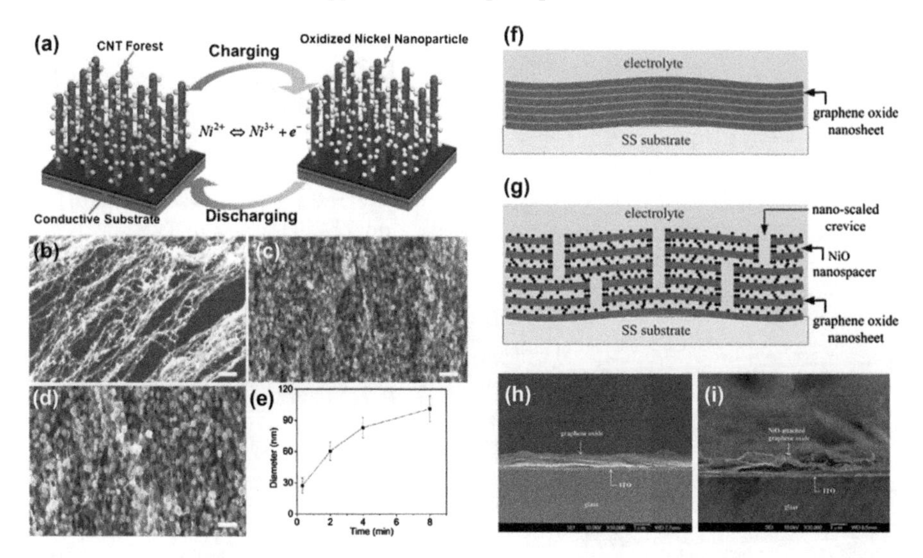

Figure 9.10 (a–e) CNT–NiO nanoparticle hybrid system. (a) Schematic of the charging–discharging processes of the hybrid electrode system. SEM images (side view) of the Ni-decorated VACNTs after various electrodeposition times: (b) 20 seconds, (c) 2 minutes, (d) 8 minutes. (e) Size (diameter) variation of Ni nanoparticles with increasing electrodeposition time. Reprinted with permission from ref. 17. Copyright 2013 American Chemical Society. (f–i) graphene oxide–NiO nanoparticle hybrid system. Schematics of stacked (f) pristine graphene oxide nanosheets and (g) NiO nanoparticle-decorated graphene oxide nanosheets in an electrolyte. SEM images (side view) of stacked (h) pristine graphene oxide and (i) NiO nanoparticle-decorated graphene oxide nanosheets. Reproduced from ref. 79.

Likewise, pseudocapacitive nanoparticles can be loaded onto 2D nanomaterials such as graphene as well as on 3D mesoporous carbon templates. Especially in the case of graphene-supported nanoparticles, the addition of intercalating nanomaterials to the multi-layers of graphene contributes not only to a straightforward increase of SSA and pseudocapacitance but they also act as *spacers* to suppress stacked aggregation of graphene layers (Figure 9.10(f)–(i)).[79,91]

Equivalently, 1D capacitive nanomaterials such as CNTs and pseudocapacitive nanowires can also be applied to other nanomaterials of higher hierarchy. For instance, Fan *et al.*[83] reported a significant increase of SSA (from 202 $m^2\ g^{-1}$ to 612 $m^2\ g^{-1}$) by growing CNTs on graphene nanosheets (Figure 9.11(a)–(e)). Interestingly, the cobalt nanoparticles used for catalytic growth of the CNTs contributed to pseudocapacitance in the form of cobalt hydroxide. As is the case of intercalating nanoparticles in graphene layers, the CNTs helped the graphene layers avoid aggregation during the wet process. They reported 385 F g^{-1} at a scanning rate of 10 mV s^{-1}.

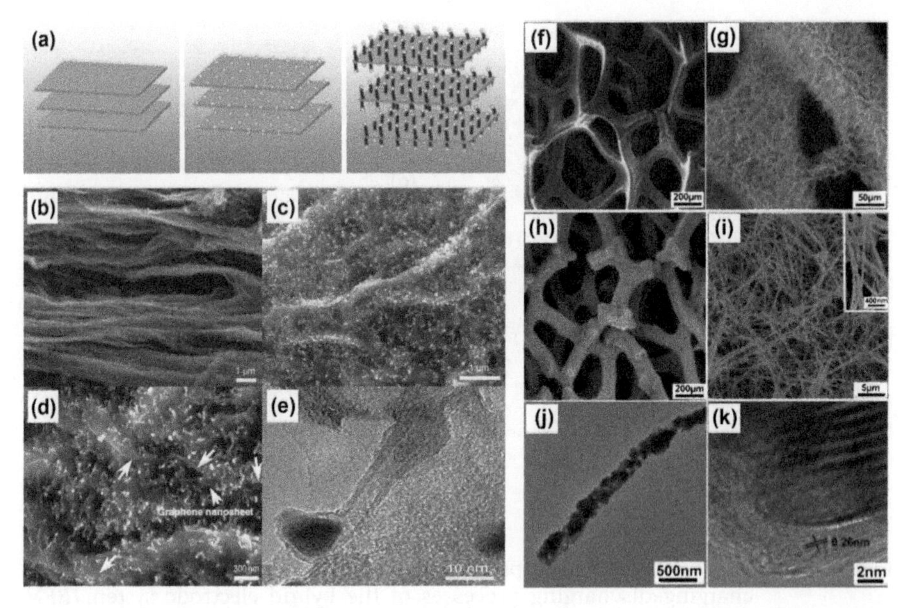

Figure 9.11 (a–e) CNT–graphene hybrid system. (a) Schematic of CNT growth from catalyst nanoparticles supported on graphene nanosheets. (b–e) SEM images of the CNT–graphene structure. (e) TEM image highlighting catalytic growth of the CNTs with the Co catalyst nanoparticle embedded at the tip. Reprinted from ref. 83 with permission. Copyright 2010, John Wiley and Sons. (f–k) Co_3O_4 NW–graphene foam hybrid system. (f) SEM image of the 3D interconnected graphene foam. (g, h, i) SEM image of the hierarchical CoO_3 NW–graphene foam structure. (j, k) TEM images of the Co_3O_4 nanowires grown on the graphene surface. Reprinted with permission from ref. 18. Copyright 2013 American Chemical Society.

Chemical vapor deposition of CNTs on graphene layers requires well-dispersed catalytic seeds to avoid undesirable agglomeration of the catalyst nanoparticles at high temperatrues.[83] In contrast, seedless hydrothermal synthesis of capacitive nanowires can enable the growth of dense nanowire arrays at relatively low temperatures. Due to the low diffusivity of solution precursors, supply of the precursor molecules should be facilitated *via* relatively large transport channels for uniform growth of the nanowires. In this regard, macro-/meso-porous carbon templates are plausible substrates to build a 3D structure of capacitive nanowires based on the hydrothermal method.

Dong *et al.*'s study exemplifies this capacitive nanowire (1D)–macro-/meso-porous carbon (3D) hybrid system (Figure 9.11(f)–(k)).[18] They synthesized pseudocapacitive Co_3O_4 nanowires by a hydrothermal method on a 3D graphene foam prepared from a Ni foam template by CVD. The produced graphene foam was conductive and light in weight, and served as an excellent 3D scaffold for the growth of the Co_3O_4 nanowires. The reported capacitance was 768 F g^{-1} at 50 mV s^{-1} (10 A g^{-1}), and it could increase up to ~1100 F g^{-1} after 500 cycles.

In addition to the introduced cases, similar approaches are also applicable with other combinations of nanomaterials. Furthermore, hierarchical nanostructures consisting of more than two levels can be applied to supercapacitors. Investigated ternary electrode systems are relatively rare, but recently it has been reported that conductive polymer coating on binary systems containing pseudocapacitive metal oxides can enhance the performance of the supercapacitors by providing additional electron conduction paths and protecting the metal oxides from corrosive electrolytes.[23,92] For instance, Yu *et al.*[23] demonstrated conductive wrapping of the hybrid graphene–MnO_2 nanostructure system to enhance cyclic performance as well as specific capacitance of the electrode. They reported a ~20% and ~45% increase of specific capacitance by a coating of poly(3,4-ethylenedioxythiophene)-poly(styrenesulfonate) (PEDOT:PSS) and single-walled carbon nanotubes, respectively. The capacitance reached a maximum (~380 F g^{-1}) at a scanning rate of 100 mV s^{-1} (current density: 0.1 mA cm^{-2}) with 95% retention after 3000 cycles.

9.4 Summary and Outlook

In this chapter, we have discussed two kinds of supercapacitors: EDLCs and pseudocapacitors. EDLCs have significantly higher power density compared with lithium ion batteries, but their energy density is limited, due to the restricted domain responsible for charge storage. The development of high specific area electrodes is pursued to improve the capacitance of EDLCs. Various carbon-based electrode materials of extremely high specific area include activated carbon, carbon nanotubes, and graphene. Pseudocapacitors based on electroactive metal oxides and conductive polymers can be promising replacements for EDLCs because of their battery-like charge storage mechanism. However, the increase of energy density is compromised by the decrease of power density.

Increasing demands for high energy and high power storage systems have led to extraordinary interest in nanostructured electrodes for supercapacitors. Capacitive nanomaterials are excellent candidates in themselves for improving the performance of supercapacitors. They can have physical and electrochemical properties that overcome the performance limit of conventional bulk materials. Hierarchical hybrid nanomaterials are a most advanced form of multi-functioning nanomaterial in that this novel material system can synergistically combine the strengths of constituent nanomaterials. Therefore, capacitive hierarchical hybrid nanomaterials are promising electrode materials for next-generation supercapacitors.

Nevertheless, there are issues to consider before we fully adopt the hierarchical material system for advanced supercapacitors. Firstly, the cost of the sophisticated fabrication processes should not be underestimated. Secondly, the resulting power performance (kW kg^{-1}) and energy storage (Wh kg^{-1})

should be well balanced so as not to lose the technical merits of high-power supercapacitors over lithium ion batteries.[1] Thirdly, it should be noted that the performance of nanomaterials may not be linearly proportional to the amount of the electrode material.[10] In the case of large storage systems, therefore, a rational architecture of the nanostructured electrode should be developed.

Acknowledgements

Parts of this work were supported by the U. S. National Science Foundation through the SINAM NSEC.

References

1. P. Simon and Y. Gogotsi, *Nat. Mater.*, 2008, 7, 845–854.
2. G. Wang, L. Zhang and J. Zhang, *Chem. Soc. Rev.*, 2012, **41**, 797–828.
3. J. R. Miller and A. F. Burke, *Electrochem. Soc. Interface*, 2008, **17**, 53.
4. H. Y. Jung, M. B. Karimi, M. G. Hahm, P. M. Ajayan and Y. J. Jung, *Sci. Rep.*, 2012, **2**, 773.
5. Y. He, W. Chen, C. Gao, J. Zhou, X. Li and E. Xie, *Nanoscale*, 2013, **5**, 8799–8820.
6. R. A. Rightmire, *Electrical Energy Storage Apparatus*, U.S. patent, 1966, **3**, 288, 641.
7. W. Si, C. Yan, Y. Chen, S. Oswald, L. Han and O. G. Schmidt, *Energy Environ. Sci.*, 2013, **6**, 3218–3223.
8. S. Yoon, C. W. Lee and S. M. Oh, *J. Power Sources*, 2010, **195**, 4391–4399.
9. B. Hsia, M. S. Kim, M. Vincent, C. Carraro and R. Maboudian, *Carbon*, 2013, **57**, 395–400.
10. Y. Gogotsi and P. Simon, *Science*, 2011, **334**, 917–918.
11. J. Huang, B. G. Sumpter and V. Meunier, *Chem.–Eur. J.*, 2008, **14**, 6614–6626.
12. J. Chmiola, G. Yushin, Y. Gogotsi, C. Portet, P. Simon and P. L. Taberna, *Science*, 2006, **313**, 1760–1763.
13. J. Chmiola, C. Largeot, P.-L. Taberna, P. Simon and Y. Gogotsi, *Angew. Chem., Int Ed.*, 2008, **47**, 3392–3395.
14. C. Largeot, C. Portet, J. Chmiola, P.-L. Taberna, Y. Gogotsi and P. Simon, *J. Am. Chem. Soc.*, 2008, **130**, 2730–2731.
15. D. Pech, M. Brunet, T. M. Dinh, K. Armstrong, J. Gaudet and D. Guay, *J. Power Sources*, 2013, **230**, 230–235.
16. J. Zhu, L. Cao, Y. Wu, Y. Gong, Z. Liu, H. E. Hoster, Y. Zhang, S. Zhang, S. H. Yang, Q. Yan, P. M. Ajayan and R. Vajtai, *Nano Lett.*, 2013, **13**, 5408–5413.
17. Y. Jiang, P. Wang, X. Zang, Y. Yang, A. Kozinda and L. Lin, *Nano Lett.*, 2013, **13**, 3524–3530.
18. X.-C. Dong, H. Xu, X.-W. Wang, Y.-X. Huang, M. B. Chan-Park, H. Zhang, L.-H. Wang, W. Huang and P. Chen, *ACS Nano*, 2012, **6**, 3206–3213.

19. M. Zhi, C. Xiang, J. Li, M. Li and N. Wu, *Nanoscale*, 2013, **5**, 72–88.
20. J. Zheng, P. Cygan and T. Jow, *J. Electrochem. Soc.*, 1995, **142**, 2699–2703.
21. D. Bélanger, L. Brousse and J. W. Long, *Electrochem. Soc. Interface*, 2008, **17**, 49.
22. J. Tao, N. Liu, W. Ma, L. Ding, L. Li, J. Su and Y. Gao, *Sci. Rep.*, 2013, **3**, 2286.
23. G. Yu, L. Hu, N. Liu, H. Wang, M. Vosgueritchian, Y. Yang, Y. Cui and Z. Bao, *Nano Lett.*, 2011, **11**, 4438–4442.
24. C. Peng, D. Hu and G. Z. Chen, *Chem. Commun.*, 2011, **47**, 4105–4107.
25. K. Naoi and P. Simon, *Electrochem. Soc. Interface*, 2008, **17**, 34–37.
26. T. Chen and L. Dai, *Materials Today*, 2013, **16**, 272–280.
27. L.-F. Chen, X.-D. Zhang, H.-W. Liang, M. Kong, Q.-F. Guan, P. Chen, Z.-Y. Wu and S.-H. Yu, *ACS Nano*, 2012, **6**, 7092–7102.
28. M. E. Plonska-Brzezinska and L. Echegoyen, *J. Mater. Chem. A*, 2013, **1**, 13703–3714.
29. S. Zhu and G. Xu, *Nanoscale*, 2010, **2**, 2538–2549.
30. M. Cinke, J. Li, B. Chen, A. Cassell, L. Delzeit, J. Han and M. Meyyappan, *Chem. Phys. Lett.*, 2002, **365**, 69–74.
31. C. Niu, E. K. Sichel, R. Hoch, D. Moy and H. Tennent, *Appl. Phys. Lett.*, 1997, **70**, 1480–1482.
32. T. W. Ebbesen, H. J. Lezec, H. Hiura, J. W. Bennett, H. F. Ghaemi and T. Thio, *Nature*, 1996, **382**, 54–56.
33. M. J. Bronikowski, P. A. Willis, D. T. Colbert, K. A. Smith and R. E. Smalley, *J. Vac. Sci. Technol., A*, 2001, **19**, 1800–1805.
34. T. Yamada, T. Namai, K. Hata, D. N. Futaba, K. Mizuno, J. Fan, M. Yudasaka, M. Yumura and S. Iijima, *Nat. Nanotechnol.*, 2006, **1**, 131–136.
35. W. Lu, L. T. Qu, K. Henry and L. M. Dai, *J. Power Sources*, 2009, **189**, 1270–1277.
36. L. Viet Thong, H. Kim, A. Ghosh, J. Kim, J. Chang, V. Quoc, An, P. Duy Tho, J.-H. Lee, S.-W. Kim and Y. H. Lee, *ACS Nano*, 2013, **7**, 5940–5947.
37. S. Hu, R. Rajamani and X. Yu, *Appl. Phys. Lett.*, 2012, **100**, 104103.
38. Y. W. Zhu, X. D. Lim, M. C. Sim, C. T. Lim and C. H. Sow, *Nanotechnology*, 2008, **19**, 325304.
39. M. Kaempgen, C. K. Chan, J. Ma, Y. Cui and G. Gruner, *Nano Lett.*, 2009, **9**, 1872–1876.
40. J. Marschewski, J. B. In, D. Poulikakos and C. P. Grigoropoulos, *Carbon*, 2014, **68**, 308–318.
41. J. B. In, D. Lee, F. Fornasiero, A. Noy and C. P. Grigoropoulos, *ACS Nano*, 2012, **6**, 7858–7866.
42. B. Hsia, J. Marschewski, S. Wang, J. B. In, C. Carraro, D. Poulikakos, C. P. Grigoropoulos and R. Maboudian, *Nanotechnology*, 2014, **25**, 055401.
43. J.-H. Chen, C. Jang, S. Xiao, M. Ishigami and M. S. Fuhrer, *Nat. Nanotechnol.*, 2008, **3**, 206–209.
44. C. Lee, X. Wei, J. W. Kysar and J. Hone, *Science*, 2008, **321**, 385–388.

45. M. F. El-Kady, V. Strong, S. Dubin and R. B. Kaner, *Science*, 2012, **335**, 1326–1330.
46. S. Bhaviripudi, X. T. Jia, M. S. Dresselhaus and J. Kong, *Nano Lett.*, 2010, **10**, 4128–4133.
47. K. S. Novoselov, A. K. Geim, S. V. Morozov, D. Jiang, Y. Zhang, S. V. Dubonos, I. V. Grigorieva and A. A. Firsov, *Science*, 2004, **306**, 666–669.
48. S. Park and R. S. Ruoff, *Nat. Nanotechnol.*, 2009, **4**, 217–224.
49. S. Stankovich, D. A. Dikin, R. D. Piner, K. A. Kohlhaas, A. Kleinhammes, Y. Jia, Y. Wu, S. T. Nguyen and R. S. Ruoff, *Carbon*, 2007, **45**, 1558–565.
50. J. Xia, F. Chen, J. Li and N. Tao, *Nat. Nanotechnol.*, 2009, **4**, 505–509.
51. M. D. Stoller, S. Park, Y. Zhu, J. An and R. S. Ruoff, *Nano Lett.*, 2008, **8**, 3498–3502.
52. Z. Niu, L. Zhang, L. Liu, B. Zhu, H. Dong and X. Chen, *Adv. Mater.*, 2013, **25**, 4035–4042.
53. W. Gao, N. Singh, L. Song, Z. Liu, A. L. M. Reddy, L. Ci, R. Vajtai, Q. Zhang, B. Wei and P. M. Ajayan, *Nat. Nanotechnol.*, 2011, **6**, 496–500.
54. M. F. El-Kady and R. B. Kaner, *Nat. Commun.*, 2013, **4**, 1475.
55. Z. Lausevic, P. Y. Apel, J. B. Krstic and I. V. Blonskaya, *Carbon*, 2013, **64**, 456–463.
56. J.-C. Yoon, J.-S. Lee, S.-I. Kim, K.-H. Kim and J.-H. Jang, *Sci. Rep.*, 2013, **3**, 1788.
57. D.-W. Wang, F. Li, M. Liu, G. Q. Lu and H.-M. Cheng, *Angew. Chem., Int. Ed.*, 2008, **47**, 373–376.
58. Y. He, W. Chen, X. Li, Z. Zhang, J. Fu, C. Zhao and E. Xie, *ACS Nano*, 2013, 7, 174–182.
59. H. Wang, Z. W. Xu, A. Kohandehghan, Z. Li, K. Cui, X. H. Tan, T. J. Stephenson, C. K. King'ondu, C. M. B. Holt, B. C. Olsen, J. K. Tak, D. Harfield, A. O. Anyia and D. Mitlin, *ACS Nano*, 2013, 7, 5131–5141.
60. Q. Cheng, J. Tang, J. Ma, H. Zhang, N. Shinya and L.-C. Qin, *Phys. Chem. Chem. Phys.*, 2011, **13**, 17615–17624.
61. N. Jung, S. Kwon, D. Lee, D.-M. Yoon, Y. M. Park, A. Benayad, J.-Y. Choi and J. S. Park, *Adv. Mater.*, 2013, **25**, 6854–6858.
62. H. Lin, L. Li, J. Ren, Z. Cai, L. Qiu, Z. Yang and H. Peng, *Sci. Rep.*, 2013, **3**, 1353.
63. H. Zhang, G. Cao, Z. Wang, Y. Yang, Z. Shi and Z. Gu, *Electrochem. Commun.*, 2008, **10**, 1056–1059.
64. W. Z. Jian Zhi, Xiangye Liu, Angran Chen, Zhanqiang Liu and Fuqiang Huang, *Adv. Funct. Mater.*, 2013, **24**, 2013–2019.
65. W. Chen, R. B. Rakhi, L. Hu, X. Xie, Y. Cui and H. N. Alshareef, *Nano Lett.*, 2011, **11**, 5165–5172.
66. G. Wang, X. Lu, Y. Ling, T. Zhai, H. Wang, Y. Tong and Y. Li, *ACS Nano*, 2012, **6**, 10296–10302.
67. P. Yu, X. Zhang, D. L. Wang, L. Wang and Y. W. Ma, *Cryst. Growth Des.*, 2009, **9**, 528–533.
68. C. Cheng and H. J. Fan, *Nano Today*, 2012, 7, 327–343.

69. S. H. Ko, D. Lee, H. W. Kang, K. H. Nam, J. Y. Yeo, S. J. Hong, C. P. Grigoropoulos and H. J. Sung, *Nano Lett.*, 2011, **11**, 666–671.
70. J. Lin, C. Zhang, Z. Yan, Y. Zhu, Z. Peng, R. H. Hauge, D. Natelson and J. M. Tour, *Nano Lett.*, 2013, **13**, 72–78.
71. Y.-S. Kim, K. Kumar, F. T. Fisher and E.-H. Yang, *Nanotechnology*, 2012, **23**, 015301.
72. C. X. Guo and C. M. Li, *Energy Environ. Sci.*, 2011, **4**, 4504–4507.
73. A. B. Fuertes, G. Lota, T. A. Centeno and E. Frackowiak, *Electrochim. Acta*, 2005, **50**, 2799–2805.
74. Y.-T. Kim and T. Mitani, *Appl. Phys. Lett.*, 2006, **89**, 033107.
75. Y. T. Kim, K. Tadai and T. Mitani, *J. Mater. Chem.*, 2005, **15**, 4914–4921.
76. J.-K. Lee, H. M. Pathan, K.-D. Jung and O.-S. Joo, *J. Power Sources*, 2006, **159**, 1527–1531.
77. H. Zhang, G. Cao, Z. Wang, Y. Yang, Z. Shi and Z. Gu, *Nano Lett.*, 2008, **8**, 2664–2668.
78. G. He, J. Li, H. Chen, J. Shi, X. Sun, S. Chen and X. Wang, *Mater. Lett.*, 2012, **82**, 61–63.
79. M.-S. Wu, Y.-P. Lin, C.-H. Lin and J.-T. Lee, *J. Mater. Chem.*, 2012, **22**, 2442–2448.
80. X. Zhao, L. Zhang, S. Murali, M. D. Stoller, Q. Zhang, Y. Zhu and R. S. Ruoff, *ACS Nano*, 2012, **6**, 5404–5412.
81. Z. Zhou, X.-F. Wu and H. Fong, *Appl. Phys. Lett.*, 2012, **100**, 023115.
82. P. Yu, Y. Li, X. Yu, X. Zhao, L. Wu and Q. Zhang, *Langmuir*, 2013, **29**, 12051–12058.
83. Z. Fan, J. Yan, L. Zhi, Q. Zhang, T. Wei, J. Feng, M. Zhang, W. Qian and F. Wei, *Adv. Mater.*, 2010, **22**, 3723–3728.
84. S.-Y. Yang, K.-H. Chang, H.-W. Tien, Y.-F. Lee, S.-M. Li, Y.-S. Wang, J.-Y. Wang, C.-C. M. Ma and C.-C. Hu, *J. Mater. Chem.*, 2011, **21**, 2374–2380.
85. H.-C. Hsu, C.-H. Wang, S. K. Nataraj, H.-C. Huang, H.-Y. Du, S.-T. Chang, L.-C. Chen and K.-H. Chen, *Diamond Relat. Mater.*, 2012, **25**, 176–179.
86. H. Xia, Y. Wang, J. Lin and L. Lu, *Nanoscale Res. Lett.*, 2012, 7, 1–10.
87. J. Xu, K. Wang, S.-Z. Zu, B.-H. Han and Z. Wei, *ACS Nano*, 2010, **4**, 5019–5026.
88. M. G. Hahm, A. L. M. Reddy, D. P. Cole, M. Rivera, J. A. Vento, J. Nam, H. Y. Jung, Y. L. Kim, N. T. Narayanan, D. P. Hashim, C. Galande, Y. J. Jung, M. Bundy, S. Karna, P. M. Ajayan and R. Vajtai, *Nano Lett.*, 2012, **12**, 5616–5621.
89. Y. Yan, Q. Cheng, G. Wang and C. Li, *J. Power Sources*, 2011, **196**, 7835–7840.
90. D. N. Futaba, K. Hata, T. Yamada, T. Hiraoka, Y. Hayamizu, Y. Kakudate, O. Tanaike, H. Hatori, M. Yumura and S. Iijima, *Nat. Mater.*, 2006, **5**, 987–994.
91. Y. Si and E. T. Samulski, *Chem. Mater.*, 2008, **20**, 6792–6797.
92. R. B. Rakhi, W. Chen, D. Cha and H. N. Alshareef, *Adv. Energy Mater.*, 2012, **2**, 381–389.

CHAPTER 10

Hierarchical Field Emission Devices

E. STRATAKIS

Institute of Electronic Structure and Laser, Foundation for Research & Technology—Hellas, (IESL-FORTH), P.O. Box 1527, Heraklion 711 10, Greece
Email: stratak@iesl.forth.gr

10.1 Introduction to Field Electron Emission Technology

Field-electron emission (FE) refers to the emission of electrons from a solid cathode (generally in the shape of a sharp tip) into a vacuum under the influence of a strong electric field of the order of 10^6–10^7 V cm^{-1}, otherwise known as cold cathode electron emission. The pointed or conical shape of the cathode strongly enhances the electric field at the tip, originating from biasing the cathode negatively with respect to a nearby anode. FE is a form of quantum mechanical tunneling which occurs when electrons pass through the potential energy barrier at the interface between the cathode and the vacuum. Field emitting cathodes can be integrated into small, lightweight devices and can be operated at high repetition rates due to their rapid on/off switching speed. Potential applications include vacuum microelectronic devices, such as electron guns and microwave power amplifiers, electron microscopy and FE electronic devices, such as flat panel FE displays (FEDs).

One of the first-studied applications of FE was developed in surface science for the study of surface diffusion of adsorbates over clean metal

RSC Nanoscience & Nanotechnology No. 35
Hierarchical Nanostructures for Energy Devices
Edited by Seung Hwan Ko and Costas P Grigoropoulos

Published by the Royal Society of Chemistry, www.rsc.org

surfaces.[1] Later on, several technological applications were demonstrated. In particular, field emission sources are routinely used in scanning electron microscopy and e-beam lithography because of their small optical size. Furthermore, a high density two-dimensional array of emitters can be fabricated by a variety of techniques resulting in cathode elements suitable for flat panel displays.[2] Fabrication of FEDs is considered to be the most promising application of field electron emission. A FED is a thin, flat cathode ray tube comprising two-dimensional arrays of electron cathodes (the pixels) and gating elements that drive the emitting electrons towards a patterned phospor screen (Figure 10.1a). Each pixel comprises thousands of micro or nanotips (typical density of 10^6 cm^{-2}). Among the advantages of this technology, are the low power consumption and cost and the potential to provide displays that are thin and have low weight. Although FEDs have the optical quality of a cathode ray tube, they do not produce X-rays, nor are they sensitive to magnetic fields or large temperature variations. For all of these reasons, FEDs can find a wide range of applications in industrial and automobile applications or as a monitor for scientific and medical instruments (such as mass spectrometers and oscilloscopes).

Owing to the potential unique properties of FEDs, an intense research effort has been devoted to the design and fabrication of cold cathode electron emitters exhibiting low operation voltage, high current emissivity, and increased durability under poor vacuum conditions. For this purpose, a wide range of materials have been considered as cold cathode field emitters including metallic, metal–insulator–vacuum structures, thin films, diamond and diamond films, graphite and graphite pastes, polymers, carbon fibers, nanowires and nanotubes. Research and technology in this area indicate that the cathodes developed to date suffer from several factors that limit the operating performance and stability such as degradation due to Joule heating and chemical instability at high emission currents, as well as ion

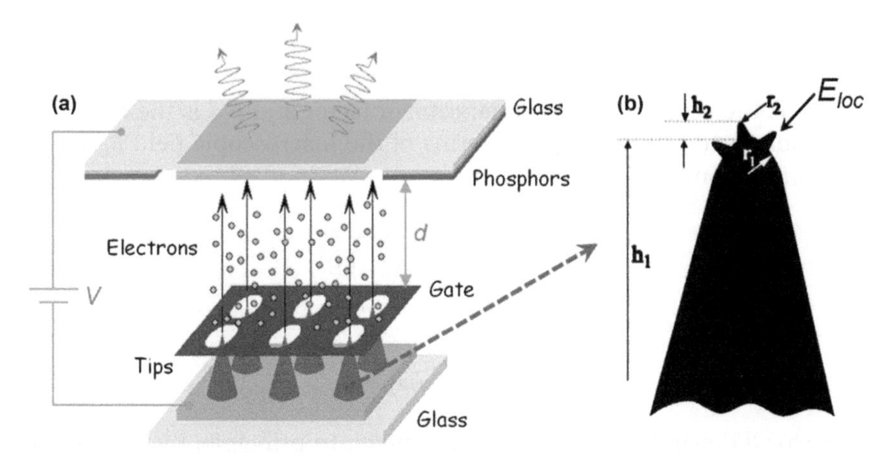

Figure 10.1 Schematic of: a flat panel FE display (a); a hierarchical field emitting tip (b).

bombardment causing erosion and blunting of the emitting tips. Despite the great progress, obtaining sharp, robust and chemically stable cathodic materials and architectures that will enable the establishment of a FED industry and boost FEDs to be the dominant display technology is an ongoing effort.

The basic formulation of quantum mechanical tunneling of conduction electrons from a flat metallic surface into a vacuum was introduced by Fowler and Nordheim in 1928[3a] resulting in the Fowler–Nordheim (FN) equation. This equation provides the emission current density, J, corresponding to electron tunneling from an exact triangular barrier without any correction factors. Later, Gomer[3b] derived the image potential corrected FN equation which can be written as

$$J = A\varphi^{-1}E^2 exp\left(-\frac{B\varphi^{\frac{3}{2}}}{\mathrm{E}}\right), \tag{10.1}$$

where φ is the work function of the emitter and A, B are constants with $A = 1.54 \times 10^{-6}$ A eV V^{-2} and $B = 6.83$ V$^{-3/2}$ V nm^{-1}. $E = V/d$ is the electric field at the emitting surface, where V is the applied voltage bias between the emitting cathode and a collecting anode and d is the spacing between them. J is given by the measured emission current in Ambers times the emitting area, S ($I = S \times J$). In the case of semiconductors, the field electron emission theory is complex. Various parameters including the contribution of the effective electron mass, field penetration, surface states, doping and band structure have to be taken into account. A detailed account of the theory field emission from semiconductors is available.[4]

To account for the FE behavior of non-planar structures with certain geometries, one should consider that the local field at the emitting point, E_{loc} (Figure 10.1b), is different from the average field E of a flat emitter. In a first approximation the relation between the two fields is:

$$E_{loc} = \beta E = \beta\frac{V}{d} \tag{10.2}$$

where β is the geometric field enhancement factor and β/d is the local field conversion factor. Accordingly, in terms of the macroscopic field E, the FN eqn (10.1) becomes

$$J = A\varphi^{-1}(\beta E)^2 exp\left(-\frac{B\varphi^{\frac{3}{2}}}{\beta\mathrm{E}}\right) \tag{10.3}$$

or

$$\ln\left(J/E^2\right) = \ln\left(A\beta^2/\varphi\right) - B\varphi^{\frac{3}{2}}/\beta E \tag{10.4}$$

The factor β depends on the emitter geometry. In principle, the calculation of β is complex and several models have been proposed for its estimation. Such models are not universally applicable and may vary for different types of materials and geometries. In the simplest model, for emitters of height h with a tip of radius of curvature r, it can be expressed as $\beta = h/r$, which is often called 'aspect ratio' (AR). Alternatively, the value of β can be determined from the current–voltage measurements, *via* calculation of the slope of the $\ln(J/E^2)$ *vs* $(1/E)$ plot (known as FN plot) in eqn (10.4). However, in most cases, the value obtained may be physically unrealistic in the sense that it is orders of magnitude higher than the emitter AR. As can be seen from eqn (10.3), the emission performance of an emitter can be enhanced by increasing the AR and in series the enhancement factor β.

The highest current density attainable, lower turn-on voltage (defined as the applied bias needed to obtain measurable current), increased β, stability of the FE current and lifetime are the figures of merit for FE cathodes. Among the important parameters affecting the current stability are the residual gas pressure and atom adsorption and/or desorption causing local work function variations. Ease of fabrication, high charge carrier mobility and electrical conductivity are also important for the applicability of FE sources in electronic devices.

10.2 Hierarchical Field Emission Cathodes

A critical parameter for a FE cathode performance is the field enhancement factor β which can be enhanced by increasing the 'AR', h/r of the emitters. Owing to their unique geometries of small curvature radius, one-dimensional (1D) nanostructures including carbon nanotubes (CNTs) nanowires and nanofibers have attracted significant interest for their potential field emission (FE) applications. Besides this, two-dimensional (2D) nanostructures such as carbon nanowalls, graphene and other 2D nanosheets should also give rise to high geometric field enhancement. To further increase β thus enhance the FE capability of various 1D or 2D nanostructures, different approaches have been proposed. Among the most promising, one that has been frequently employed recently is based on hierarchical development of the emitting material, for example *via* using a micron-sized structured substrate and subsequently growing nanostructures onto it (Figure 10.1b). This approach offers the advantages of not only further increasing the enhancement factor β of the emitters, but also significantly increasing the effective electron emitting area and therefore the density of the emitting sites.

The effect of regular and random variations in emitter morphology was first addressed by Atlan *et al.*[5] According to this work, the geometric mismatch among the protruding structures and thus the fractal dimension of the emitting surface determine the degree of field enhancement while the total emission current is determined by the geometry of protrusions. Zhirnov *et al.* considered an analytical model to address emission from small protrusions on the surface.[6] According to this model, the emitting surface may be thought of as a number of primary structures with a height of h_1 and sharpness of r_1, decorated by tiny emitters with a height of h_2 and sharpness of r_2 (Figure 10.1b). Assuming a 'two-step field enhancement (TSFE)' approach, the electric field on the primary tip is

$$E_1 = \beta_1(V/d) = (h_1/r_1)(V/d), \tag{10.5}$$

while that on the very end of the secondary protrusions is

$$E_2 = \beta_2 E_1 = (h_2/r_2)E_1 = \beta_1\beta_2(V/d). \tag{10.6}$$

Later on, Solntsev *et al.*[7] considered FE cathodes consisting of N distinct stages or substructures, which is defined as a multistage field enhancement effect; the field enhancement factor for such structures has been given as

$$E_2 = \prod_{i=0}^{N} \beta_i E_o \tag{10.7}$$

Later, Zhirnov pointed out that in most cases this analytical model overestimates the field enhancement factor,[8] indicating that the local surface structure and thus the dimensional interaction between the various secondary geometric protrusions of the hierarchical emitter is an important electrostatic factor that significantly affects the surface field and potential barrier. Such nano and mesoscale electrostatic perturbations are complex phenomena that are extremely difficult to treat analytically and without the use of sophisticated numerical methods that account for the multiscale nature of the secondary structures.

To date, a wide variety of hierarchical field emitters (HFE) have been demonstrated including, carbon-based, metallic, metal oxide, metal–organic and semiconducting, as summarized in Table 10.1. This chapter demonstrates the FE characteristics of those exhibiting the best performance and stability under device operation. It is concluded that multiscale emitters are superior compared with single-length scale ones, demonstrating the great potential of hierarchical FE technology for future electron device applications.

10.2.1 Carbon-based Field HFE Cathodes

Carbon-based nanostructures, including nanotubes, nanowalls and graphene, exhibit a unique combination of properties for superior field emission performance. Sato *et al.* were the first to indicate that the field emission of single-wall carbon nanotubes (SWCNTs) can be enhanced *via* depositing them on a micro-rough substrate.[9] For this purpose, using chemical vapor deposition (CVD), they had selectively grown SWCNTs on lithographically etched silicon microprotrusions. FE experiments showed that this approach was favourable towards the improvement of field emission characteristics including the FE threshold and enhancement factor. Improved FE properties had also been measured for SWCNTs grown, using an electrophoretic method, on Ag microparticles.[10] According to a TSFE process, the field enhancement of each emission site can be understood by a coupling of field enhancement of Ag particles and intrinsic field

Table 10.1 Critical performance parameters of hierarchical field emitters reported in the literature.

Hierarchical FE cathode	Turn-on field ($V\ \mu m^{-1}$) [at 10 $\mu A\ cm^{-2}$]	Enhancement factor, β	Stability: testing time - fluctuation	Ref.
Single-wall CNTs on Si microprotrusions	1.2	—	—	9
Single-wall CNTs on Ag microparticles	3.1	3641	—	10
Single-wall CNT hairpin arrays on Si	—	2091	—	11
Multi-wall CNTs on Si pyramids	1.95	6980	—	12
Carbon nanowalls on Si microcones	0.9	3533	90 h - 15%	13
Graphene on Si microcones	2.3	1604	90 h - 5%	14
Graphene nanostructures on carbon fibers	0.41	24 065	2 h - 5%	15
Graphene on ZnO nanowires	1.8	10 179	2 h - 4%	16
ZnO nanosleeve-fishes	1.3	3010	—	17
ZnO tips on ZnO nanorods	5.6	2300	—	18
ZnO nanoscrewdrivers	8	—	—	19
6-fold symmetrical hierarchical ZnO nanostructures	6.1	1351	—	20
ZnO hexagonal towers	2.37	2691	—	21
Zn/ZnO hierarchical arrays	8.5	3490	—	22
ZnO/MWCNT array	1.5	—	5 h - 2%	23
Branched TiO_2 hierarchical nanostructures	2.76	7440	—	24
TiO_2 nanostructures on carbon fibers	2.72	5438	2 h - 3%	25
MoO_2 nanowires with nanoprotrusions	0.2	42 991	1 h - 15%	26
SnO_2 sallow-like hierarchical nanostructures	1.88	3028	2 h - 5%	27
WO_3 nanowires on carbon fibers	1.8	6900	12 h - 5%	28
Fe_2O_3 hierarchical columnar arrays	17.8	463	3 h - 50%	29
AlN hierarchical nanostructures	2.5	—	2 h - 7%	30
AlN multitipped nanowires	3.8	950	—	31
WS_2 nanotubes on Si microcones	1.1	3526	18 h - 15%	32
Flowerlike In-doped SnS_2	7.5	—	—	33
AgTCNQ nanowire on carbon fibers	1.72	1165	2 h -10%	34

enhancement of each CNT. Finally, novel SWCNT micro tip arrays with a hairpin structure were fabricated on a patterned silicon wafer *via* tailoring cross-stacked SWCNT sheets with a focused laser beam.[11] By using an adhesive tape to modify the hairpin tip (Figure 10.2a), CNT field emitters can provide 150 μA intrinsic emission currents with low beam noise. Such a good field emission was ascribed to the Joule-heating-induced desorption adsorbents off the emitter surface, the high temperature annealing effect, and the surface morphology.

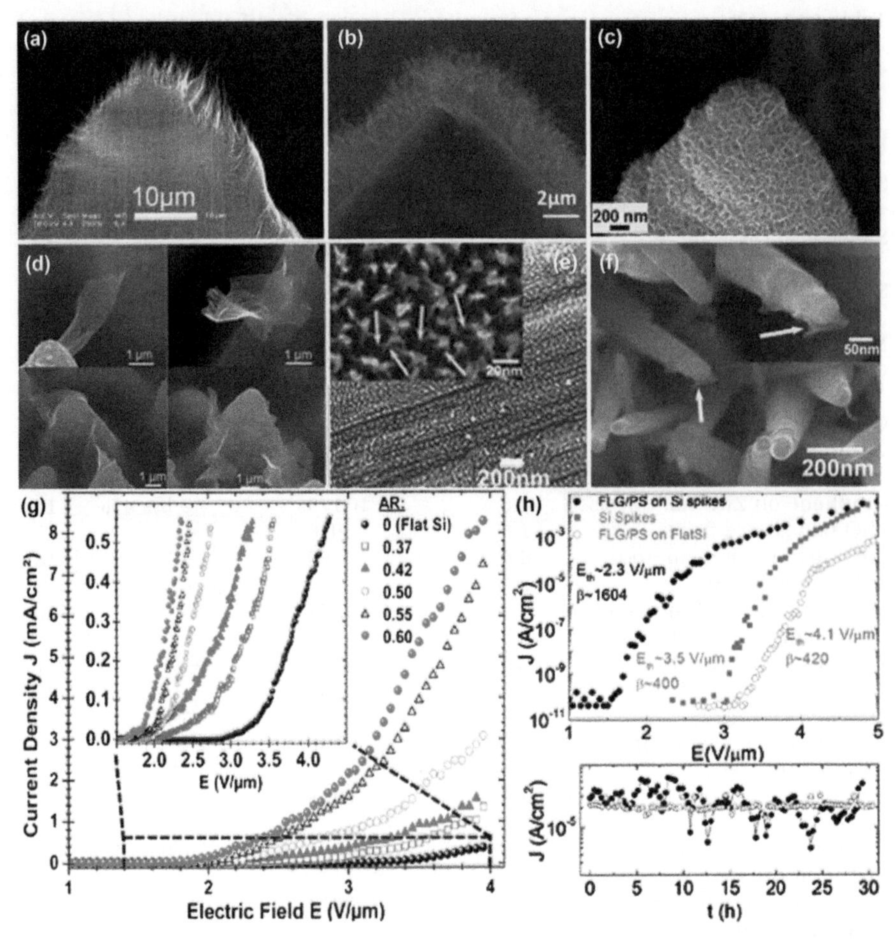

Figure 10.2 Carbon-based field HFE cathodes: SWCNT micro tip arrays with hairpin structure (a) – reproduced with permission from ref. 11; MWCNT on KOH-etched Si pyramids (b) – reproduced with permission from ref. 12; CNWs on micro-conical Si spikes (c) – reproduced with permission from ref. 13; free-standing graphene on conical microspikes (d) – reproduced with permission from ref. 14; graphene nanostructures on fabric (e) – reproduced with permission from ref. 15; graphene on ZnO nanowires (f) – reproduced with permission from ref. 16; typical J–E curves of the field electron emitting hierarchal MWCNT cathodes with various pyramid aspect ratio values along with that of flat Si reference substrate. The inset shows a zoomed-in part of the J–E curves (g) – reproduced with permission from ref. 12; J–E FE characteristics for different cathodes of free-standing graphene on conical microspikes (h-top) and emission current stability over time (h-bottom) – reproduced with permission from ref. 14.

Gautier *et al.* fabricated multiwall carbon nanotube (MWCNT) hierarchical emitting cathodes *via* a combination of KOH-etched Si pyramids and plasma-enhanced chemical vapor deposition (PECVD) growth of vertically aligned MWCNTs.[12] The resulting cathodes can be shown in Figure 10.2b. The FE properties of the MWCNTs grown on pyramids with various ARs were measured and compared with flat MWCNT mats grown on planar silicon substrates. It is shown that the higher the AR of the Si pyramids, the lower the threshold field of the respective cathodes. In particular, a threshold field value as low as 1.95 V μm^{-1} was achieved for the hierarchical MWCNT emitters with an AR value of 0.6, which was lower by more than 40% compared to planar ones (Figure 10.2g). Analysis of the respective FN curves revealed the existence of two distinct low-field (LF) and high-field FE regimes. In both regimes, the estimated β values for the hierarchical cathodes were found to be almost two times higher compared to those measured on flat cathodes.

Stratakis *et al.* reported an efficient methodology for the fabrication of large scale regular arrays of hierarchical carbon nanowall (CNW) field emitters.[13] The respective cathodes had been produced by CVD of CNWs on forests of micro-conical Si spikes (CNW/μSi) fabricated by ultrafast laser structuring. Figure 10.2c depicts the corresponding SEM images of CNW/μSi field emission cathodes after the CVD process. It is observed that CNWs follow the surface and decorate the microspikes forming a flower-like coating. The FE properties of CNWs grown on spikes with low and high AR were measured and compared with CNWs layers grown on planar silicon substrates. It is found that the FE performance of hierarchical CNW structures is far superior to that of planar CNW mats and comparable to that reported for optimized CNT-based emitters. The improved field emission properties of the fabricated arrays were attributed to the dual micro and nanomorphology of the emitters, involving a TSFE (eqn 10.7).

Owing to its inherent 2D geometry, graphene should give rise to high geometric field enhancement, allowing the extraction of electrons at low threshold electric fields. However, typical deposition methods provide graphene flakes that tend to lie parallel to or protruding at small angles from the substrate, thus limiting the geometrical field enhancement. In this respect, controlled deposition of free-standing graphene nanosheets on microstructured substrates could provide state- of-the art hierarchical FE cathodes. Stratakis *et al.* reported a simple and general solution-based approach for deposition of free-standing few-layer graphene (FLG) sheets.[14] The method is based on drop-casting an FLG–polymer solution on forests of conical micro-spikes engraved on Si (FLG/μSi) exhibiting different hydrophilicities. It is shown that, depending on the deposition conditions, the FLG flakes become free-standing with their edges anchored to the micro-spikes (Figure 10.2d). As shown in Figure 10.2h, the hierarchical FLG/μSi cathodes were found to exhibit excellent FE performance, with turn-on fields as low as 2.3 V μm^{-1} and a field enhancement of a few thousand, and a stability that was superior to that of thin-film-type FLG emitters prepared on planar substrates.

Maiti *et al.* developed a hierarchical FE cathode comprising graphene nanostructures over a flexible fabric substrate.[15] Nanostructuring was realized through plasma treatment of graphene, coaxially deposited over individual carbon fibers by means of an aqueous phase electrophoretic deposition technique. Upon variation of the duration of plasma etching, the planar graphene structure sheathed around carbon fiber was turned into a hierarchical geometry composed of vertical-wall or cone-like morphology (Figure 10.2e). The cathodes developed exhibited outstanding electron emission performance with a turn-on field as low as 0.41 V μm^{-1}, while the performance was preserved upon extreme bending of the flexible cathode. The same group has also demonstrated hierarchical FE cathodes based on spin coating of a solution-processed ZnO nanowire array with graphene and subsequent plasma modification (Figure 10.2f).[16] The superior field emission properties were attributed in part to the multistage geometrical field enhancement as well as to the facile electron transfer from nanowires to graphene due to band bending at the ZnO–graphene interface.

10.2.2 Metal Oxide HFE Cathodes

Metal oxides are proving to be the richest family of nanostructures developed to date, while a plethora of self-organized hierarchically structured metal oxide nanomaterials have been synthesized. Among these, ZnO-based HFEs are the most frequently tested as cathode field electron emitters. Figure 10.3 summarizes the best performing ZnO HFE architectures to date. In particular, Xu *et al.* fabricated, through thermal chemical vapor deposition, four different hierarchical ZnO nanostructures[17] including nanosleeve-fishes, radial nanowire arrays, nanocombs and nanoflowers (Figure 10.3a). The morphology can be controlled by both the growth temperature and the gas flow rate. The FE measurements performed showed that such structures exhibit high emission current densities and low turn-on fields of 1.3, 1.9, 2.5 and 3.4 V μm^{-1} respectively (Figure 10.3g). Later, the FE properties of well-aligned ZnO nanorod arrays with three kinds of tip morphology—abruptly sharpened (Figure 10.3b), tapered and plane—have been comparatively investigated.[18] It is found that the FE performance is strongly affected by the tip morphology and relies not only on the nanorod's radius of curvature and array density but also on the detailed tip morphology. Among the three kinds of samples, the nanorods with abruptly sharpened tips exhibit the best field emission property. This excellent performance was attributed to the unique two-stage geometrical configuration of the abruptly sharpened tip. Three different types of hierarchical ZnO nanostructures, *i.e.* nanocombs, nanoscrewdrivers and nanonails, have been through a simple thermal evaporation process and tested for their FE properties.[19] The successful fabrication of HFE cathodes comprising aligned arrays of single-crystalline, 6-fold-symmetrical, hierarchical ZnO nanostructures was also reported.[20] Meanwhile, Wang *et al.* successfully synthesized[21] by a chemical vapor deposition method hierarchical ZnO hexagonal towers. Kuan *et al.* studied the FE properties of six-fold core-shelled three-dimensional hierarchical Zn/ZnO micro/nano tip arrays,

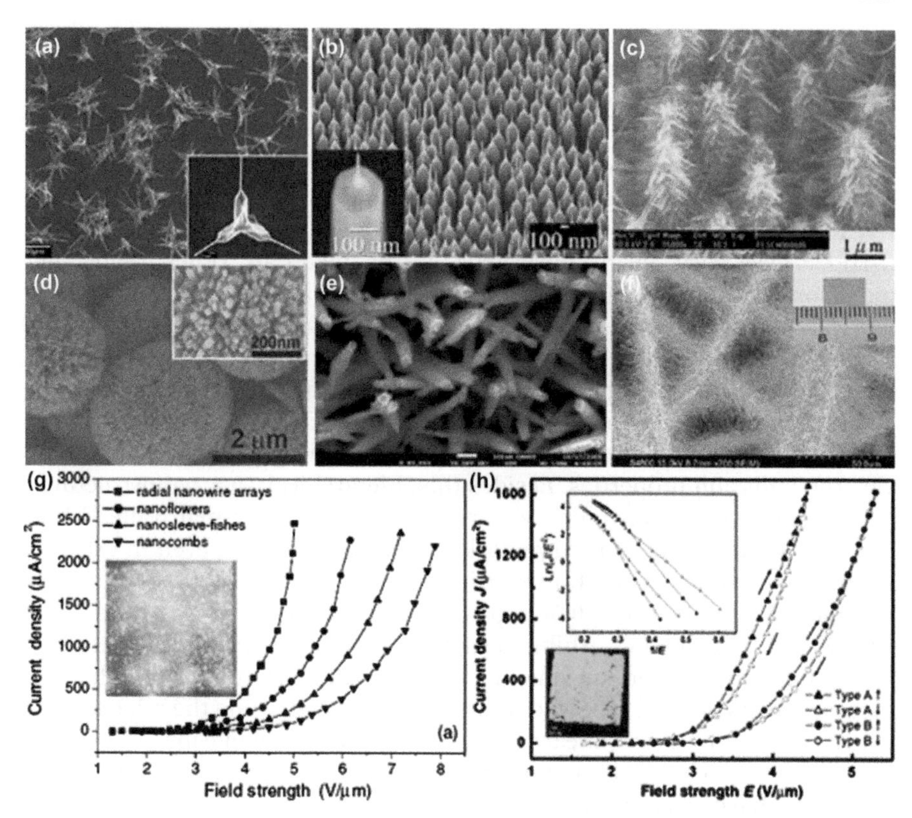

Figure 10.3 Metal oxide HFE cathodes: ZnO nanosleeve-fishes (a) – reproduced with permission from ref. 17; ZnO nanorod arrays with abruptly sharpened tips (b) – reproduced with permission from ref. 18; hierarchical Zn/ZnO micro/nano tip arrays (c) – reproduced with permission from ref. 22; cauliflower, rutile TiO_2 hierarchical nanostructures (d) – reproduced with permission from ref. 24; MoO_2 nanowires decorated with nano-protrusions (e) – reproduced with permission from ref. 26; WO_3 NWs onto carbon fibers (f) – reproduced with permission from ref. 28; J–E plots of the field emission from four types of hierarchical ZnO nanostructures (the transparent anode image of the electron emission from radial nanowire arrays at a field strength of 4.4 V μm^{-1} is shown in the inset (g) – reproduced with permission from ref. 17; J–E plots of the field emission from the two hierarchical SnO_2 structures (the corresponding F–N plots and the transparent anode image of the electron emission from sallow-like structures at field strength of 3.98 V μm^{-1} are shown in the two insets respectively), (h) – reproduced with permission from ref. 27.

fabricated by a direct annealing process.[22] Transmission electron microscopy analysis revealed that the oxide layer grows epitaxially from the Zn microtips and then the branches grow epitaxially from the oxide layer (Figure 10.3c). Finally, Nayeri *et al.* synthesized single-crystalline ZnO nanowires on vertically aligned MWCNT arrays. ZnO nanowires are grown on the base of individual CNTs through the low-temperature wet-chemical batch deposition technique, while the size and interspacing of the nanowires can be controlled by precursor concentration, growth temperature and time duration.[23] The special double-stage structure of the achieved nanostructures in addition to the high density of ZnO's tiny tips caused an enhancement in their field emission behaviour, in comparison with ZnO-NW and CNT arrays.

HFEs have also been fabricated out of materials other than ZnO metal oxide nanostructures. Using a hydrothermal process, Sarkar *et al.* fabricated[24] different rutile TiO_2-based hierarchical nanostructures (Figure 10.3d). In particular, cauliflowers, 3D microspheres, densely-packed nanorod arrays, step edge faceted nanorods and branched structures have been demonstrated. The FE characteristics of these architectures were studied showing that the multilevel branched cathodes exhibited the best electron emission properties. Besides this, hierarchical TiO_2 nanostructures over conducting carbon fabric was realized *via* hydrothermal route.[25] The resulting FE cathodes showed remarkable FE performance, even in extreme bending conditions. The FE properties of MoO_2 nanowires decorated with nanoprotrusions (Figure 10.3e), grown on a silicon substrate have been studied too.[26] Such HFEs exhibited superior emission with a turn-on field of 0.2 V μm^{-1} and an enhancement factor of 42991. It was found that FE measurements depend on cathode–anode distance, d, while the relationship between $1/\beta$ and $1/d$ was linear, suggesting that the FE behaviour followed a two-step FE model. Self-organized, sallow-like, hierarchical SnO_2 nanostructures have been successfully prepared by a vapor phase transport method.[27] Field emission measurements of these nanostructures showed a high emission current density and low turn-on field of 1.88 V μm^{-1} (Figure 10.3h). Li *et al.* studied the FE properties of single-crystalline WO_3 nanowires on carbon paper synthesized by a chemical vapor deposition process.[28] SEM imaging showed that the WO_3 nanowires were uniformly grown onto the carbon fibers (Figure 10.3f). The FE cathodes tested showed remarkable FE performance and current stability. The successful fabrication of well-ordered and aligned hematite (Fe_2O_3) hierarchical columnar arrays *via* pulsed laser deposition was demonstrated.[29] The hematite comprised an array of microcolumns decorated by nanoplates or nanoparticles that can be controlled by oxygen pressure during PLD. It was found that the FE properties of such SFE cathodes depend on cathode–anode distance, d, while the relationship between $1/\beta$ and $1/d$ was linear, suggesting that the FE behaviour followed a two-step FE model.

10.2.3 Semiconducting HFE Cathodes

Single-crystalline wurtzite AlN hierarchical nanoarchitectures were self-assembled through a chemical vapor transport and compensation process.[30] FE applications of AlN materials are of interest, considering their negative electron affinity; therefore electrons excited into the conduction band of

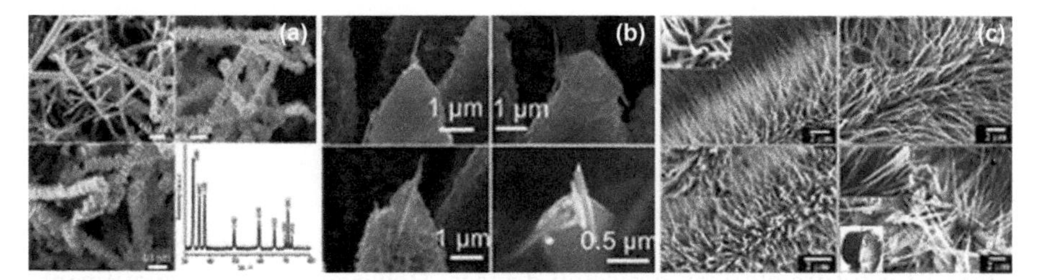

Figure 10.4 Semiconducting and metal–organic HFE cathodes: AlN hierarchical nanoarchitectures (a) – reproduced with permission from ref. 30; free-standing WS_2 nanotubes on Si microspikes – reproduced with permission from ref. 32; metal–organic charge transfer complex nanowires on carbon fabric fibres – reproduced with permission from ref. 34.

such materials can be freely emitted into a vacuum. The nanostructures consisted of stems growing along the <0001> direction decorated with comb-like nanotips growing perpendicularly out from both sides of the stems along the <0110> direction (Figure 10.4a). Such a hierarchical architecture exhibited remarkable FE properties. The intrinsically low electron affinity of AlN and the enhanced emission from sharp nanotips and thin edges are assumed to contribute to the high emission currents observed. Later, the growth of well-aligned AlN nanorods with hairy surfaces *via* a vapor–solid process was reported.[31] The overall hierarchical cathode had an aligned columnar structure, while each nanorod exhibited a hairy surface.

Hierarchical WS_2 nanotube (NT)-based cathodes have been fabricated *via* drop casting of WS_2 NTs/P3HT solutions onto low and high-aspect ratio Si microspike arrays fabricated by ultrafast laser processing.[32] Following drop-casting, the microspikes were partially covered by the nanocomposite, while some NTs protruded out of the P3HT matrix and decorated the top of the microspikes, forming a random array of free-standing NTs (Figure 10.4b). It was demonstrated that free-standing WS_2NT-based field emitters exhibit comparable threshold field but superior stability compared to optimized CNT mats.

Zhong *et al.* reported[33] on the hydrothermal synthesis of SnS_2 and In-doped SnS_2 exhibiting a flower-like hierarchical structure. It was found that both materials showed excellent FE properties, while the emitters fabricated by doped structures exhibited superior FE behavior.

10.2.4 Metal–organic-based HFE Cathodes

Maiti *et al.* have fabricated metal-based hierarchical FE arrays *via* the chemical growth of metal–organic charge transfer complex nanowires on conducting carbon fabric fibres.[34] Control over the growth parameters during the organic solid phase reaction gave rise to a modification in nanowire morphological properties, as well as their distribution atop the carbon fibers (Figure 10.4c). The FE performance of the fabricated cathode was measured to be comparable to that of the best organic field emitters reported.

10.3 Conclusion

HFEs based on various materials will continue to be an area of high research interest for future electron emission devices. However, although significant progress has been made for the realisation of efficient HFE cathode elements, more efforts are required for the development of a HFE technology suitable for large-scale industrial applications.

References

1. A. G. J. van Oostrom, *Philips Res. Rep.*, 1966, **1**, 11.
2. C. Spindt, C. E. Holland, A. Rosengreen and I. Brodie, Field emitter array development for gigahertz operation, in *Proceedings of IEEE Transactions on Electron Devices ED-38, San Francisco, 3. CA, May, 1991*, IEEE.
3. (a) R. H. Fowler and L. Nordheim, *Proc. R. Soc. London*, 1928, **A119**, 173–181; (b) R. Gomer, *Field Emission and Field Ionisation*, Harvard University Press, Cambridge, MA, 1961.
4. A. Modinos, *Field, Thermionic and Secondary Electron Emission Spectroscopy*, Plenum Press, New York, US, 1984.
5. D. Atlan, G. Gardet, V. T. Binh, N. Garcia and J. J. Saenz, *Ultramicroscopy*, 1992, **42**, 154 and later by O. B. Isayeva, M. V. Eliseev, A. G. Rozhnev and N. M. Ryskin, *Solid-State Electron.*, 2001, **45**, 871.
6. V. V. Zhirnov, E. I. Givargizov and P. S. Plekhanov, *J. Vac. Sci. Technol., B*, 1995, **13**, 418.
7. V. A. Solntsev and A. N. Rodionov, *Solid-State Electron.*, 2001, **45**, 853.
8. D. L. Jaeger, J. J. Hren and V. V. Zhirnov, *J. Appl. Phys.*, 2003, **93**, 691.
9. H. Sato, K. Hata, H. Miyake, K. Hiramatsu and Y. Saito, *J. Vac. Sci. Technol., B*, 2005, **23**, 754.
10. W. Lua, H. Songa, Y. Jina, H. Zhao, Z. Lia, H. Jiang and G. Miao, *Microelectron. J.*, 2008, **39**, 782.
11. Y. Wei, P Liu, F. Zhu, K. Jiang, Q. Li and S. Fan, *Nano Lett.*, 2012, **12**, 2071.
12. L.-A. Gautier, V. Le Borgne, S. Al Moussalami and M. A. El Khakani, *Nanoscale Res. Lett.*, 2014, **9**, 55.
13. E. Stratakis, R. Giorgi, M. Barberoglou, Th. Dikonimos, E. Salernitano, N. Lisi and E. Kymakis, *Appl. Phys. Lett.*, 2010, **96**, 043110.
14. E. Stratakis, G. Eda, H. Yamaguchi, E. Kymakis, C. Fotakis and M. Chhowalla, *Nanoscale*, 2012, **4**, 3069.
15. U. N. Maiti, S. Maiti, N. S. Das and K. K. Chattopadhyay, *Nanoscale*, 2011, **3**, 4135.
16. U. N. Maiti, S. Maiti, T. P. Majumder and K. K. Chattopadhyay, *Nanotechnology*, 2011, **22**, 505703.

17. F. Xu, K. Yu, G. Li, Q. Li and Z. Zhu, *Nanotechnology*, 2006, **17**, 2855.
18. N. Pan, H. Xue, M. Yu, X. Cui, X. Wang, J. G. Hou, J. Huang and S. Z. Deng, *Nanotechnology*, 2010, **21**, 225707.
19. D. Peng, Y. Huang, K. Yu, L. Li and Z. Zhu, *J. Nanomater.*, 2010, **2010**, 560409.
20. Z. Wang, J. Gong, Y. Su, Y. Jiang and S. Yang, *Cryst. Growth Des.*, 2010, **10**(6), 2455.
21. P. Wang, X. Zhang, J. Wen, L.-L. Wu, H. Gao, E. Zhang and G. Miao, *J. Alloys Compd.*, 2012, **533**, 88.
22. C.-Y. Kuan, M.-H. Hon, J.-M. Chou and I.-C. Leu, *Cryst. Growth Des.*, 2009, **9**(2), 813.
23. F. D. Nayeri, S. Darbari, E. A. Soleimani and S. Mohajerzadeh, *J. Phys. D: Appl. Phys.*, 2012, **45**, 285101.
24. D. Sarkar, C. K. Ghosh and K. K. Chattopadhyay, *CrystEngComm*, 2012, **14**, 2683.
25. S. Maiti, U. N. Maiti and K. K. Chattopadhyay, *Synth. React. Inorg., Met.-Org., Nano-Met. Chem.*, 2014, **44**, 1255.
26. A. Khademi, R. Azimirad, Y.-T. Nien and A. Z. Moshfegh, *J. Nanopart. Res.*, 2011, **13**, 115.
27. Q. Wang, K. Yu and F. Xu, *Solid State Commun.*, 2007, **143**, 260.
28. L. Li, Y. Zhang, X. Fang, T. Zhai, M. Liao, X. Sun, Y. Koide, Y. Bando and D. Golberg, *J. Mater. Chem.*, 2011, **21**, 6525.
29. L. Li and N. Koshizaki, *J. Mater. Chem.*, 2010, **20**, 2972.
30. L. W. Yin, Y. Bando, Y. C. Zhu, M. S. Li, Y. B. Li and D. Golberg, *Adv. Mater.*, 2005, **17**, 110.
31. J. H. He, R. Yang, Y. L. Chueh, L. J. Chou, L. J. Chen and Z. L. Wang, *Adv. Mater.*, 2006, **18**, 650.
32. G. Viskadouros, A. Zak, M. Stylianakis, E. Kymakis, R. Tenne and E. Stratakis, *Small*, 2014, **10**, 2398–2403.
33. H. X. Zhong and C. X. Wang, *Nano: Brief Reports and Reviews*, 2011, **6**, 489.
34. S. Maiti, U. N. Maiti, S. Pal and K. K. Chattopadhyay, *Nanotechnology*, 2013, **24**, 465601.

CHAPTER 11

Sensors

DONGJIN LEE

School of Mechanical Engineering, Konkuk University, Korea
Email: djlee@konkuk.ac.kr

11.1 Introduction

Sensors are analytical devices that quantify physical or chemical information. Recently, there has been an increasing demand for high performance sensors with high sensitivity, high resolution, short response/recovery time, long lifetime, multiplexity, disposability, *etc.* Nanomaterial-based sensors have been of great interest due to their high performance over the past two decades. This high performance is attributed to a high surface-to-volume ratio, size similarity to target macromolecules[1] and novel detection mechanisms.[2–4]

Among these, the high surface-to-volume ratio enhances the interaction between a sample analyte and the sensor's surface. However, it is not a good strategy to keep decreasing the size of nanomaterials since van der Waals forces become dominant at the nanoscale leading to aggregation of nanomaterials. Then, the free surface is screened by neighboring individual elements and the diffusion of target molecules into the inner part of aggregates is suppressed.[5] Therefore, the target molecule can interact only with the outer surface of aggregates and the change in material properties is limited on the outer surface, which makes it difficult to observe. The effect of such behavior in aggregates is to decrease the sensitivity and/or resolution of nanomaterial-based sensors. Moreover, the restrained diffusion of the target molecule yields a sluggish response/recovery.

RSC Nanoscience & Nanotechnology No. 35
Hierarchical Nanostructures for Energy Devices
Edited by Seung Hwan Ko and Costas P Grigoropoulos

Published by the Royal Society of Chemistry, www.rsc.org

As a further step, researchers have developed hierarchical nanostructures for sensing applications in order to retain a higher surface-to-volume ratio. Hierarchical nanostructures are termed as the higher dimensional structures that are composed of low dimensional nano building blocks such as nanoparticles, nanowires, nanorods, nanotubes and nanosheets, *etc.*[5] With hierarchical nanostructures, almost all of the surface is exposed to the environment, meaning an extremely high surface-to-volume ratio, so that a slight change in analyte concentration can induce a great change in output signal (high sensitivity/resolution). Such structures also facilitate the diffusion and mass transport of chemical species such as molecules and ions in bulk analyte samples to the sensor's surface, so that sensors do not demonstrate diffusion-limited behavior.[5] Therefore, target molecules quickly obtain access to the surface leading to fast changes in the property of sensing materials (fast response/recovery).

In this chapter, sensors that use hierarchical nanostructures are summarized and discussed in terms of gas sensors, biosensors and surface enhanced Raman spectroscopy (SERS) sensors. The classification of sensors is dependent on the kind of target and transducing mechanism. The targets are limited to gas phase molecules such as ethanol, methanol, acetone, liquid petroleum gas (LPG), *etc* in gas sensors whereas biosensors employ a special type of liquid analyte that plays an important role biologically or biochemically. Proteins, RNA/DNAs, cells and tissues may become other types of target substances for biosensors. However, the classification of SERS sensors is based on the transducing mechanism, especially using the optical technique. The targets are not limited in SERS sensors, since they use spontaneous adsorption of molecules to the surface, where the Raman signal is greatly enhanced. Nevertheless, SERS sensors are of interest in this chapter, since hierarchical nanostructures are a very promising platform for SERS, which will be discussed later. Firstly, the hierarchy and morphology suitable for sensing applications are summarized and discussed. Then, the preparation of hierarchical nanostructures for sensors is briefly presented in terms of the material type, which has its own application. The various sensors that use hierarchical nanostructures are summarized and discussed in terms of sensor performance. One will see the sensing performance significantly increases compared to dense and compact counterparts. Finally, concluding remarks and outlook are provided.

11.2 Hierarchy and Morphology Suitable for Sensors

Individual nano building blocks are assembled in such a way that more than two kinds of dimensionality are found, resulting in hierarchical nanostructures. Building blocks may or may not have the same dimensionality, shape and scale. Therefore, hierarchical nanostructures usually demonstrate regular patterns on the surface, which plays an important role in sensing applications. Many types of hierarchical nanostructures have been reported depending on hierarchy and morphology as shown in Table 11.1. The nano

Table 11.1 Hierarchical nanostructures that have been used for sensing applications: the background color represents the types of sensors for which hierarchical nanostructures were used (red: biosensors, green: SERS sensors, blue: gas sensors). All figures are re-used with permission.

Nano building block	Final shape of nanostructure		
0D nanoparticle nanocrystal 	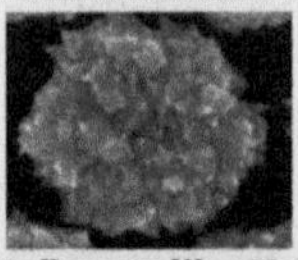 0 flower-like [14]		
1D nanowire nanorod nanotube nanobelt 	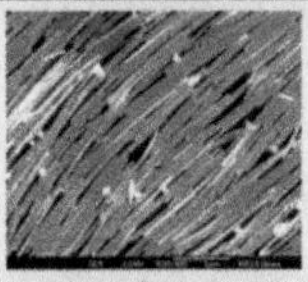 1 monolayer [11]	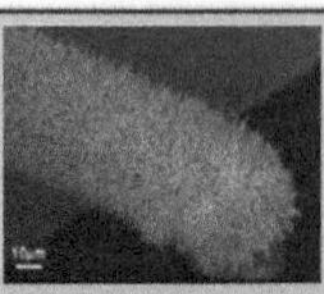 1 cactus-like [36]	 1 urchin-like [22]
	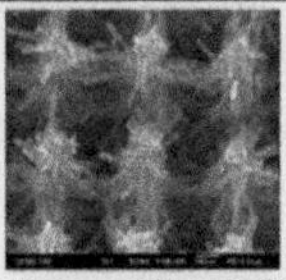 1 tree-like [12]	 1 dendrite [33]	 1 thread-like [18]
	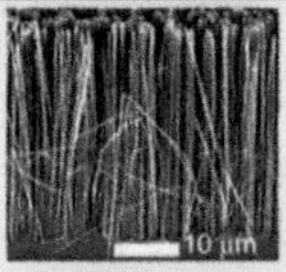 1 multilayer [13]	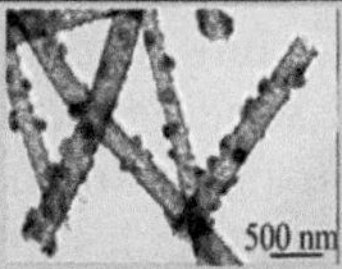 1 bamboo-like [29]	 1 thornbush-like [25]
2D nanosheet 	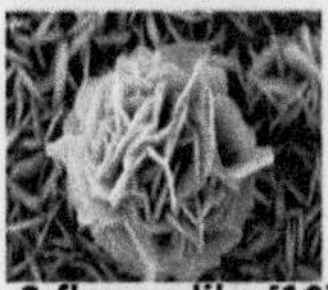 2 flower-like [10]	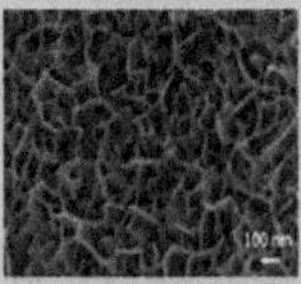 2 mesh-like [32]	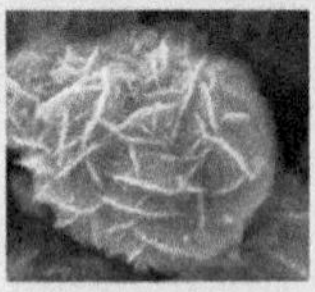 2 flower-like [16]
	 2 core-shell [9]	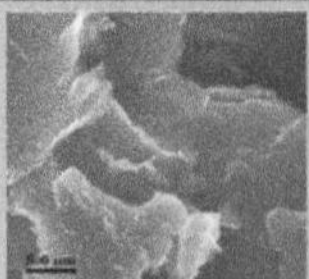 2 multi-sheet [34]	
		 2 wall-like [28]	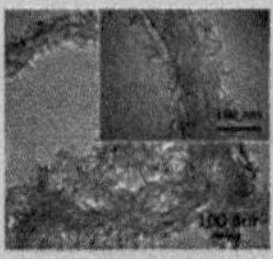 2 ivy-like[31]

building block is the unit to formulate the final structure. It may be classified into 0 (nanoparticle, nanocrystal), 1 (nanowire, nanorod) and 2 (nanosheet) dimensions.[5] The final structure might be in 2D or 3D with various types of morphology such as flower-like, urchin-like, cactus-like, *etc*. For example, a 1-3 flower-like hierarchical nanostructure is a three-dimensional flower-like structure composed of 1D nano building blocks such as nanowires or nanorods. Table 11.1 summarizes the hierarchical nanostructures that have been used in sensing applications. Particularly, structures with a red background were used as biosensors, a green one as SERS sensors, and a blue one as gas sensors. This will be discussed in detail in section 11.4.

Usually, a 3D hierarchical structure is more suitable for sensors than a 2D structure due to its higher surface-to-volume ratio. Particularly, in chemical sensors where the analyte sample is in either gas or liquid phase, 3D structures are preferable. In the case of SERS sensors, however, a 2D structure might be used if directionality is needed upon interaction with light. On the other hand, mechanical sensors such as force and strain sensors, even though they will not be considered in this chapter, prefer 1D or 2D types of final morphology. In this chapter, 3D hierarchical nanostructures are the focus for chemical sensing applications.

11.3 Preparation of Hierarchical Nanostructures for Sensors

11.3.1 Hierarchical Metal Oxide Nanostructures

Metal oxides have the electrical properties that are explained using a well-established semiconductor theory. Since metal oxides are intrinsically semiconductors, they can be applied to electronic devices such as field-effect transistors (FETs). Nanostructures allow more room to be utilized for sensors due to the electrical behavior that is sensitive to external physical or chemical quantities such as ultraviolet illumination,[6] oxidizing/reducing agents,[7] and charge carrier doing/trapping species.[2,3] For this reason, metal oxides are widely used for electrochemical sensing applications suitable for chemical sensors such as gas sensors and biosensors. In this section, methods that have been used for the preparation of hierarchical metal oxide nanostructures for sensing applications are summarized and discussed. Mostly they have been prepared either (a) by evaporation and condensation in the gas phase or (b) by hydrothermal and solvothermal growth in the liquid phase.

11.3.1.1 Evaporation and Condensation

This method is based on the vapor phase growth that has been extensively used for producing 1D nano building blocks such as nanowires or nanorods. The vaporized source elements condense resulting in directional nanostructures. Hierarchical metal oxide nanostructures were prepared by multi-step growth. The condensed power source material in a vessel is vaporized at the elevated temperature usually inside a quartz tube. The lowered pressure facilitates the sublimation of the solid source to vapor. The vapor is transported by carrier gas, usually inert argon gas, and condenses under certain conditions such as temperature, pressure, atmosphere, substrate, *etc.*[6] The detailed mechanisms are reviewed and summarized in reference.[5] If a mixture of several elements is used as the source material, doped random hierarchical nanostructures might be produced.[8]

11.3.1.2 Hydrothermal and Solvothermal Self-assembly

The hydrothermal/solvothermal method involves chemical reactions to prepare crystalline nano building blocks of oxides. The synthesis is conducted in a liquid. Usually, water or other solvents are used, so that the preparation method is termed as a hydrothermal/solvothermal growth process. This method is advantageous in that it can produce good quality crystalline metal oxide nanostructures while maintaining control over the composition, size and shape distribution. It is also known that crystalline nano building blocks form incident aggregates which serve as nuclei for further radial growth under appropriate conditions. Various hierarchical metal nanostructures were produced using this method as shown in Table 11.1.

11.3.2 Hierarchical Metal Nanostructures

Hierarchical metal nanostructures are widely used in SERS applications as shown in Figure 11.1. There are many reports that show that the enhancement factor increases greatly with the degree of hierarchy of the metal nanostructures resulting in higher resolution sensors. Metal nanostructures for SERS applications were mostly prepared by a top-down approach. This method is very advantageous and is more controllable, but is limited to a very narrow region and the fabrication cost is very high. Since a top-down approach is beyond of the scope of this book, readers are kindly encouraged to refer to other literature for the method. Due to a lack in the capacity of the top-down approach, bottom-up growth or a combination of the two approaches was widely used to prepare SERS active platforms that will be used for molecular sensing applications.

11.3.2.1 Bottom-up Approach

A bottom-up approach is widely used to fabricate nanostructures using chemical or physical forces. It is difficult to fabricate a structure with a size less than 50 nm using top-down approaches, so the bottom-up approach has been noticed as the alternative fabrication method. As shown in Figure 11.1(e), Zhang *et al.*[9] fabricated core–shell Ag wires for high sensitivity

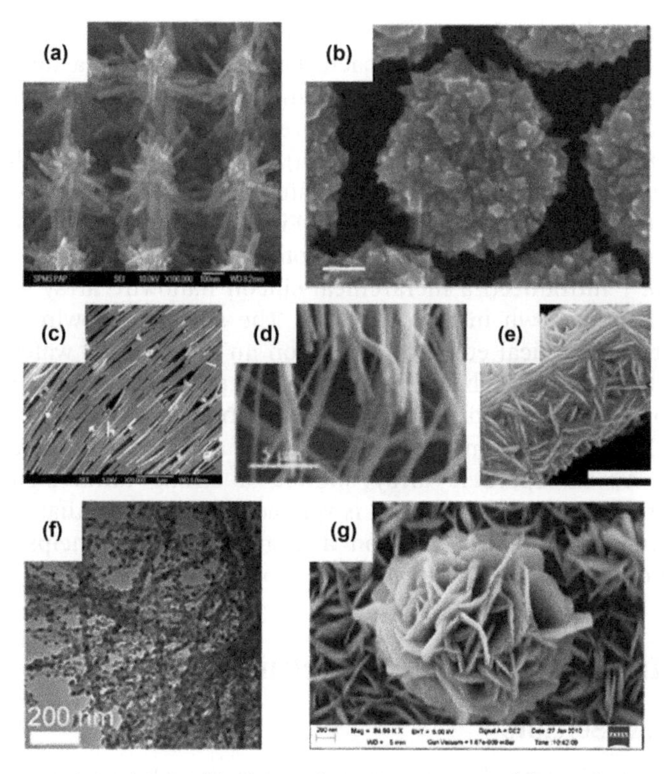

Figure 11.1 Various hierarchical metal nanostructures: (a) Ag decorated Si/ZnO nanotrees (adapted from ref. 12 with permission), (b) Au particle array deposited with hierarchical surface roughness on an ITO (adapted from ref. 14 with permission), (c) Ag nanowire array which was fabricated by using the Langmuir–Blodgett technique (adapted from ref. 11 with permission), (d) Cu microtube which was made by metal deposition (adapted from ref. 48 with permission), (e) core–shell Ag nanowires covered by nanosheets (adapted from ref. 9 with permission), (f) silicon nanowires coated with silver nanoparticles (adapted from ref. 13 with permission), (g) β-Ni(OH)2 nano-flowers (adapted from ref. 10 with permission).

SERS and an additional layer of Ag wire nanosheets. They showed the acid-directed self-assembled hierarchical nanostructure for bimolecular detection. Sarkar *et al.*[10] synthesized β-$Ni(OH)_2$ nano-flowers, which were made with ammonia and nickel acetate by a seedless hydrothermal method as shown in Figure 11.1(g). Tao *et al.*[11] adopted the Langmuir–Blodgett technique to fabricate an aligned Ag nanowire monolayer that had an excellent packing density as shown in Figure 11.1(c).

11.3.2.2 Combination of Top-down and Bottom-up Approaches

Recent studies on hierarchical metal nanostructures have used a combination of top-down and bottom-up approaches. Basically, large area patterns are defined by the top-down approach and smaller features are fabricated afterwards by the bottom-up approach. One could obtain a uniform and regular array of hierarchical structures on a large area by the combinative method. For example, Cheng *et al.*[12] prepared an Ag decorated Si/ZnO nano-tree array as shown in Figure 11.1(a). First of all, Si nanopillars were fabricated using photolithography on a wafer and ZnO nanorods were grown by a bottom-up hydrothermal method. Finally, silver nanoparticles were deposited on Si/ZnO nano-trees by photochemical reduction. Similarly, Zhang *et al.*[13] introduced a hierarchical silicon nanowire array with silver nanoparticles as shown in Figure 11.1(f). The silicon nanowire array was produced by a chemical etching process on an n-type (100) wafer, and Ag nanoparticles were grown by galvanic redox reaction. Duan *et al.*[14] proposed a novel fabrication method for uniformly distributed surface roughened Au nanoparticles as shown in Figure 11.1(b). As a result, the hierarchically roughened Au nanoparticle array showed a strong SERS signal. Consequently, the combinative method is very advantageous in that it can prepare hierarchical nanostructures on a large area, which helps with the realization of sensors for practical applications.

11.4 Hierarchical Nanostructure Sensors

11.4.1 Gas Sensors

The operation principle of metal oxide sensors is based on the change of the number of charge carriers (electrons or holes) in the material upon interaction with target molecules such as gases and chemical species. There are two types of metal oxides: n-type (ZnO, SnO_2, TiO_2, Fe_2O_3) and p-type (NiO, CoO). The n-type oxides respond to reducing gases such as H_2, CH_4, CO, C_2H_5OH and H_2S, while the p-type oxides react to oxidizing gases such as O_2, NO_2 and Cl_2. Figure 11.2 exemplarily depicts a scheme of the sensing mechanism in n-type semiconductor gas sensors in the case of CO as a target gas.[15] Oxygen in ambient air traps free electrons from the conduction band near the surface, resulting in an electron-depletion layer represented by the white region in Figure 11.2(a). However, once the sensor is exposed to a reducing gas which reacts with oxygen, the trapped electrons are released back to the bulk leading to a narrower depletion region thereby reduced resistance.

$$\frac{1}{2}O_2 + e^- \rightarrow O^-(s)$$

$$R(g) + O^-(s) \rightarrow RO(g) + e$$

where *e* is an electron from the oxide. *R(g)* is the reducing gas, and *g* and *s* are the gas and surface, respectively. The change of electrical resistance/conductance is usually measured by the microfabricated interdigitated electrodes in Figure 11.2(c). It is noted that these reactions take place at elevated temperatures, so that the sensing platform should possess an

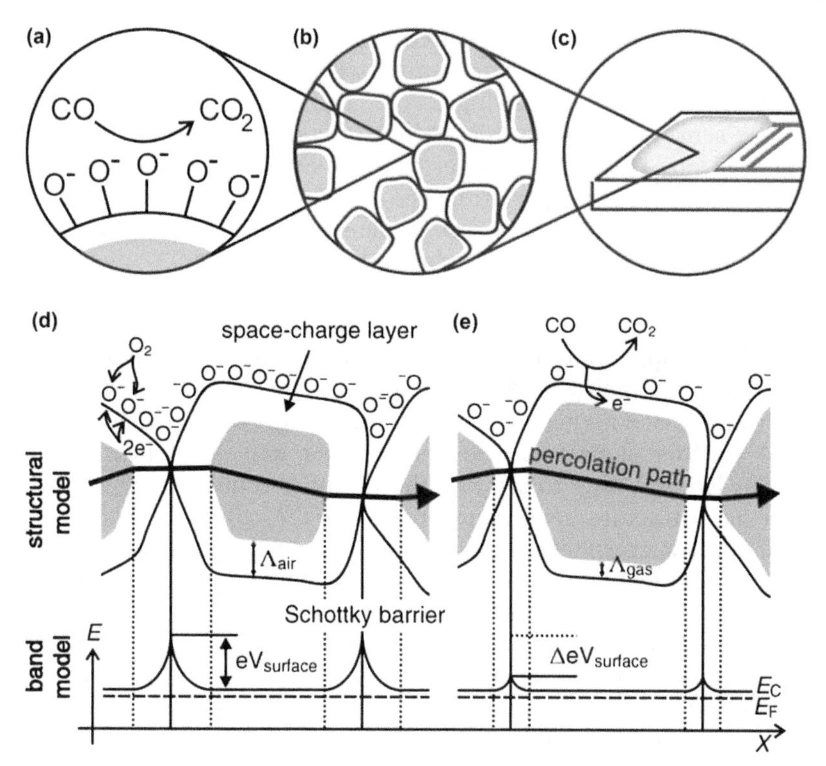

Figure 11.2 Electrical sensing mechanism of an n-type metal oxide nanostructure gas sensor: (a) surface adsorption of oxygen, (b) microstructure of the sensing layer, (c) measurement of output resistance, and band model showing (d) initial state and (e) effect of reducing gases on space-charge layer (Λ_{air}) and Schottky barrier ($eV_{surface}$) (adapted from ref. 15 with permission).

embedded heater. On the other hand, oxidizing gas removes electrons on the surface of p-type materials leaving holes which play the role of charge carrier, producing higher conductance. The electrical resistance/conductance is also dependent on the interparticle barrier, known as the 'Schottky barrier' as shown in Figure 11.2(d) and (e). As more electrons are trapped by ionosorbed oxygen, a wider space–charge layer and higher Schottky barrier are induced. The space–charge layer becomes thinner during the gas reactions above, so that a lower Schottky barrier along the percolation path exists.

In gas sensors, there are two main quantities to evaluate sensor performance. One is the gas response that is defined as the ratio of resistance when the sensor is exposed to ambient air (R_a) to that when it is subjected to the target gas (R_g). It is very similar to the sensitivity in that a higher gas response is expected at higher gas concentration unless the surface is not saturated producing stiffer behavior in a gas response *versus* concentration curve. The other performance is the response/recovery time, which was defined as the time required for reaching 90% of the final equilibrium state

($t_{90\%}$). The response time is measured when the sensor is exposed to gas whereas the recovery time is measured when the sensor is released to ambient air.

The gas sensors exploiting hierarchical metal oxide nanostructures are summarized in Table 11.2 in terms of material type, hierarchy and morphology, gas sensing performances, and operation temperatures. The 2-3 flower-like hierarchical SnO_2 microspheres composed of self-assembled nanosheets were prepared by hydrothermal synthesis of Sn_3O_4 microspheres followed by calcination.[16–18] The gas sensor made of hierarchical SnO_2 microspheres in Figure 11.3(a) showed a gas response of 7.7–33.1 in the range of 10–50 ppm ethanol (C_2H_5OH) at 400 °C with a very short response time of 1–3 s[16] as shown in Figure 11.3(b).

It is evident that a higher gas response was found at each concentration of gas and a stiffer slope was also observed for nanosheet microstructure sensors. It is interesting to see that the response time is constant in hierarchical nanostructure sensors regardless of the concentration of gases, while compact microspheres showed a decreased response time. Sun *et al.*[17] also found that the hierarchical nanostructure sensor had a gas response 3–4 times higher than dense microspheres to other gases such as acetone, butanone, methanol, *etc.* as well as C_2H_5OH by using the same hierarchical nanostructures as shown in Figure 11.4(a). Furthermore, the response/recovery time of sensors was also found to be notably shorter, 0.6 and 11 s, respectively, than the dense counterpart of a SnO_2 microsphere.[18] The ethanol selectivity of the sensor was tested with respect to 500 ppm of various gases (CO_2, NO_2, CH_4, H_2, Cl_2, C_2H_4, and CO) at optimal operation temperature, resulting in the highest gas response of the tested gases. Qin *et al.*[19] synthesized square-shaped single-crystalline SnO_2 nanowires and 1-3 urchin-like hierarchical nanostructures as shown in Figure 11.4(b), successfully using the bottom-up hydrothermal approach. The gas response to 20 ppm acetone (CH_3COCH_3) at 290 °C was 5.5 with very low response and recovery times, 7 and 10 s, respectively. Thong *et al.*[20] synthesized a 1-2 dendrite SnO_2 hierarchical nanostructure as shown in Figure 11.4(c) by a two-step thermal evaporation process and conducted gas sensing experiments for liquid petroleum gas (LPG) and ammonia (NH_3). The gas response to 2000 ppm LPG at 350 °C was 20.4, which was almost four times higher than that in SnO_2 nanowires. Furthermore, they showed operation temperatures of 350 and 200 °C for LPG and NH_3, respectively, which demonstrated a similar range of gas response. This suggests that the optimal operation temperature depends on gas species and the selectivity of sensors is tunable with operation temperature.

Wang *et al.*[21] synthesized a 2-3 flower-like α-Fe_2O_3 hierarchical nanostructure as shown in Figure 11.4(d) by a simple solvothermal method combined with a subsequent annealing process. The gas response to 100 ppm C_2H_5OH at 280 °C was 37.9 with very short response and recovery times of 1 and 0.5 s, respectively. They demonstrated that the flower-like α-Fe_2O_3 hierarchical nanostructure had a higher selectivity for ethanol than other

Table 11.2 Summary of gas sensors that use hierarchical nanostructures.

			Performance				
Sensor type	Structure	Target gas	Gas response	Response/ recovery time (s)	Operating temperature (°C)	Note	Ref.
α-Fe_2O_3	2-3 flower-like	C_2H_5OH	37.9	1/0.5	280	100 ppm C_2H_5OH	21
α-Fe_2O_3	1-3 thornbush-like	C_2H_5OH	15	1–3/4–8	150	5 ppm C_2H_5OH	23
α-Fe_2O_3	1-3 urchin-like	C_2H_5OH	12	2/40	260	100 ppm C_2H_5OH	22
SnO_2/α-Fe_2O_3	1-3 thornbush-like	C_2H_5OH	6	—	350	100 ppm C_2H_5OH	24
SnO_2	2-3 flower-like	C_2H_5OH	7.7–33.1	1–3	400	10–50 ppm C_2H_5OH	16
SnO_2	2-3 flower-like	C_2H_5OH	56.2	—	275	100 ppm C_2H_5OH	17
SnO_2	1-3 thread-like	C_2H_5OH	3–3.5 (approximately)	0.6/11	300	50 ppm C_2H_5OH	18
SnO_2	1-2 dendrite	LPG	20.4	~3/~9	350	liquid petroleum gas (LPG) 2000 ppm (gas response) 1000 ppm (response/ recovery time)	20
SnO_2	1-3 urchin-like	CH_3COCH_3	5.5	7/10	290	20 ppm CH_3COCH_3	19
ZnO	1-3 flower-like	C_2H_5OH	10.9	5/7	340	20 ppm C_2H_5OH	49
ZnO	1-3 thornbush-like	C_2H_5OH	10	~10 (both)	265	50 ppm C_2H_5OH	25
ZnO	2-3 flower-like	C_2H_5OH	4	20.7/25.3 (approximately)	280	200 ppm C_2H_5OH	50
ZnO	2-3 flower-like	C_2H_5OH	177.1	4/357	400	100 ppm C_2H_5OH	26
ZnO	1-3 dendrite	H_2S	17.3	15–20/30–50	30	100 ppm H_2S	27
WO_{3-x}	1-3 thornbush-like	NO_2	2.96 (approximately)	—	200	5 ppm NO_2	51
In_2O_3	1-3 urchin-like	CO	2.16–3.81	34–38	400	10–50 ppm CO	52

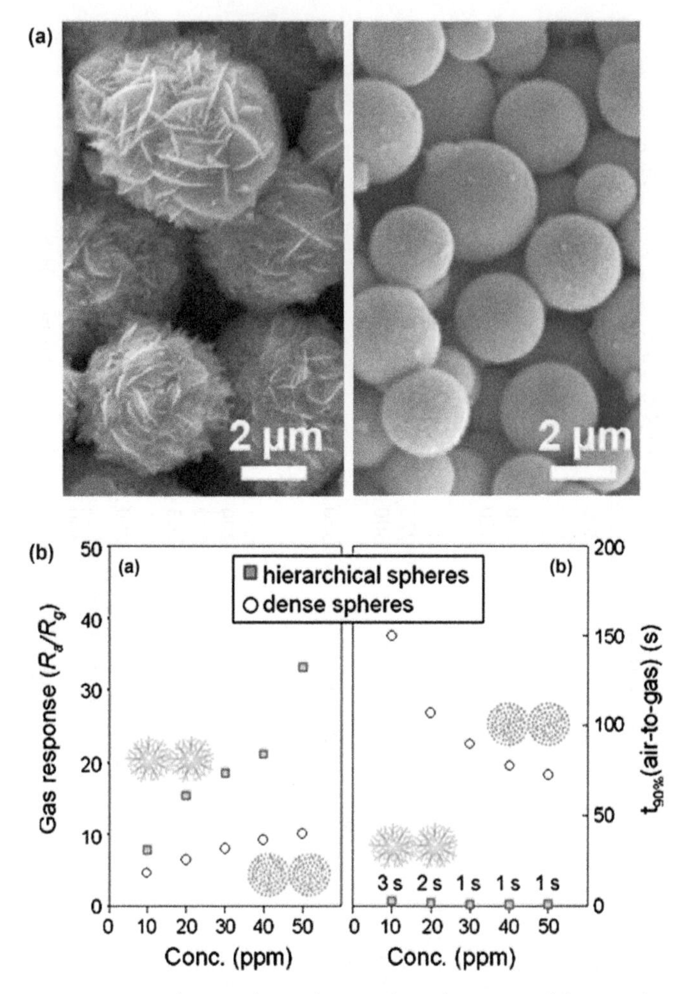

Figure 11.3 SnO_2 nanosheet microsphere ethanol sensor: (a) SEM image of the hierarchical nanostructure and its compact counterpart, (b) comparison of sensing performances (gas response and response time): a higher gas response was found at each concentration and a stiffer slope was also found (adapted from ref. 16 with permission).

gases. The gas response was eight times higher than compact α-Fe_2O_3 presumably due to an enhanced surface area. Sun *et al.*[22] synthesized 1-3 urchin-like α-Fe_2O_3 architectures by a simple one-step hydrothermal route combined with a subsequent calcination process. The gas response to 100 ppm C_2H_5OH at 260 °C was 12 with very short response and recovery times, 2 and 30 s, respectively. Gou *et al.*[23] prepared a 1-3 thornbush-like α-Fe_2O_3 hierarchical nanostructure through dehydration and recrystallisation of a hydrothermally synthesized β-FeOOH precursor. The gas response to 5 ppm C_2H_5OH at 150 °C was 15 and the response and recovery times were

Figure 11.4 Example of hierarchical metal oxide nanostructures used in gas sensors: (a) SnO_2 nanosheet microspheres (adapted from ref. 17 with permission), (b) 1-3 urchin-like SnO_2 hierarchical nanostructure (adapted from ref. 19 with permission), and (c) 1-2 dendrite SnO_2 hierarchical nanostructure (adapted from ref. 20 with permission), (d) 2-3 flower-like α-Fe_2O_3 hierarchical nanostructure (adapted from ref. 21 with permission).

about 2 and 5 s, respectively. The improvement was five-fold for gas response. Chen *et al.*[24] prepared 1-3 thornbush-like SnO_2/α-Fe_2O_3 hierarchical hetero-nanostructures through a hydrothermal method, showing a gas response to 100 ppm C_2H_5OH at 350 °C of 6. It was reported that the sensitivity and selectivity of sensors were tunable with composite materials.

Zhang *et al.*[25] prepared a 1-3 thornbush-like hierarchical ZnO nanostructure by a hydrothermal process. The gas response to 50 ppm C_2H_5OH at 265 °C was 10 and response and recovery times were very short (~10 s). Compared to simple nanowires, hierarchical ZnO nanostructures had higher selectivity for ethanol than other gases. Kim *et al.*[26] synthesized highly crystalline 2-3 flower-like ZnO hierarchical nanostructures at room temperature through the alkaline hydrolysis of zinc salt by the forced mixing of two immiscible solutions. The gas response to 100 ppm C_2H_5OH at 300 °C was 177.1 and the response and recovery times were 4 and 357 s, respectively. The gas response was eight times higher than the sensors made by agglomeration. Zhang *et al.*[27] synthesized hierarchical 1-2 dendrite ZnO in a conventional horizontal tube furnace. The gas response to 100 ppm H_2S gas at 30 °C was found as 17.3, the response time was 15–20 s and the recovery time was 30–50 s. They also studied the effect of relative humidity on gas response and response/recovery time. It turned out that both gas response and response/recovery time increased with humidity.

11.4.2 Biosensors

Electrochemical biosensors that use hierarchical nanostructures are the focus of this section. Chemical or biochemical reactions in a sample solution produce or consume ions or electrons, which induces a variation in electrical properties in sensing materials. In biosensors, there are several main quantities to evaluate sensor performance. The sensitivity is a value of the electrode response with respect to analyte concentration, meaning an intensity of response generated due to detection of a target molecule. The linear range is the section of concentration in which current output shows linear behavior with respect to target concentration. Furthermore, response time means the time to reach the steady-state of current output. The limit of detection (LOD) is the minimum amount of target analyte for a discernible current output signal. It is one of the key parameters in evaluating sensor performance.

Biosensors using hierarchical nanostructures are summarized in Table 11.3 in terms of material type, hierarchy and morphology, target biomolecules, and performances such as linear range, sensitivity, and response time. In hierarchical biosensors, a reaction can occur easily between the analyte and receptor because the amount of receptor is abundant on the sensing materials.

The 2-3 wall-like Cu-NiO nanostructure was prepared by calcination of a Cu-$Ni(OH)_2$ precursor at 400 °C for 2 h.[28] The nanostructure produced was used to modify a glassy carbon electrode (GCE) for the implementation of non-enzymatic amperometric biosensors. The sensing performance was compared with a sensor possessing a porous NiO/GCE electrode. As a result, the hierarchical nanostructure sensor did not show Michaelis–Menton kinetics. It was noted that the electron transfer in the oxidation of glucose was improved because of an increased electrocatalytic active area due to the hierarchical nanostructure. The sensitivity was 76.36 μA cm^{-2} $(mM)^{-1}$ on a linear range of 0.5 μM to 5 mM and the LOD was 0.5 μM with a fast response time below 5 s. Chen *et al.*[29] prepared a hierarchical nano-composite of CuO nanoplates attached to the surface of TiO_2 nanotubes by electrospinning. Fluorine-doped tin oxide (FTO) glass was covered with TiO_2 nanotubes decorated with CuO nanoplates for nonenzymatic amperometric sensors for glucose detection. The sensor demonstrated a sensitivity of 1321 μA cm^{-2} $(mM)^{-1}$ on a linear range of 0.01 to 2 mM, a response time of 10 s and an LOD of 390 nM at the potential voltage of 0.7 V. This was a large improvement on the CuO/TiO_2 nanotube arrays without a hierarchical nanostructure[30] that showed a sensitivity of 79.79 μA cm^{-2} $(mM)^{-1}$ and a detection limit of 1 μM. Si *et al.*[31] grew 2-3 ivy-like hierarchical nanostructured TiO_2 by a solvothermal method using a multi-walled carbon nanotube as a template. The subsequent calcination removed the carbon nanotubes producing hierarchical structures as shown in Figure 11.5(a) and (b). It was used to make a glucose sensor, which had a linear range up to 1.5 mM with a sensitivity of 9.9 μA cm^{-2} $(mM)^{-1}$, an LOD of 1.29 μM and a

Table 11.3 Summary of biosensors that use hierarchical nanostructures.

Material	Structure	Target	Linear range (mM)	Sensitivity (μA cm^{-2} (mM)$^{-1}$)	Response time (s)	LOD (μM)	Applied potential (V)	Ref.
Cu-NiO	2-3 wall-like	Glucose	0.5×10^{-3}–5	76.36	<5	0.5	0.4	28
Cu-TiO_2	1-3 bamboo-like		0.01–2	1321	<10	0.39	0.7	29
TiO_2	2-3 ivy-like		–1.5	9.9	<5	1.29	–0.45	31
Mn_3O_4/3DGF	2-3 mesh-like		0.1–8	360	<5	10	0.4	32
Cu-Co	1-2 dendrite		0.5×10^{-3}–14	—	<5	0.1	0.65	33
Mn_3O_4/3DGF	2-3 mesh-like	H_2O_2	2×10^{-3}–6.5	1030	<3	1	0	32
Cu-Co	1-2 dendrite		1×10^{-3}–11	—	—	0.75	–0.4	33
Graphene/HRP	2-3 multi-sheet		1×10^{-3}–2.6	220	<2	2.03	–0.08	34
Au	1-3 sphere		–0.8	—	<3.5	1.2	–0.2	35
IrO_2	1-3 cactus-like	NADH	—	2.9	<10	5	0.3	36

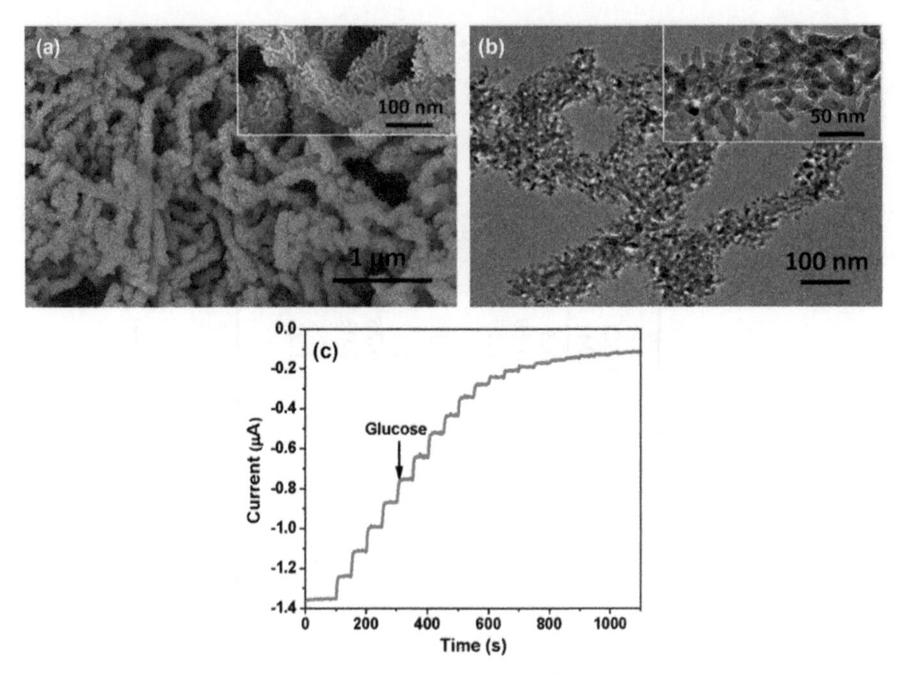

Figure 11.5 SEM image (a) and TEM image (b) of the hierarchical TiO_2 nanostructure and current–time curve (c) for successive addition of 0.15 mM of glucose aliquots at −0.45 V (adapted from ref. 31 with permission).

response time of 5 s at an applied potential of −0.45 V. The sensitivity was improved 2–3-fold compared to other TiO_2-based glucose sensors. Furthermore, it was interesting to observe a saturated current at a higher glucose concentration due to enzymatic behavior as shown in Figure 11.5(c).

The 2–3 mesh-like hierarchical structure of Mn_3O_4 was fabricated on 3D graphene foam using an electrochemical deposition method.[32] This flexible and freestanding hierarchical Mn_3O_4 composite was used for the non-enzymatic detection of glucose and hydrogen peroxide (H_2O_2) which are key analytes in the health care and food industries. The glucose sensor showed a sensitivity of 360 μA cm^{-2} $(mM)^{-1}$ on a linear range of 0.1 to 8 mM, a response time of less than 5 s and an LOD of 10 μM at an applied potential voltage of 0.4 V. On the other hand, the H_2O_2 sensor had a sensitivity of 1030 μA cm^{-2} $(mM)^{-1}$ on a linear range of 2 μM to 6.5 mM, a response time of 3 s and an LOD of 1 μM without applying a potential voltage. The authors suggested that the direct electrooxidation of glucose could be catalyzed by MnO_2, resulting in a positive current flow. However, MnO_2 experienced a series of redox mechanisms leading to a cathodic current peak, which was seen as a negative current. Noh *et al.*[33] reported a 1–2 dendrite hierarchical nanostructure of a Cu–Co alloy as shown in Figure 11.6(a) and (b) by electrochemical synthesis followed by application to a biosensor that measured either glucose or H_2O_2. The sensor for glucose showed a linear range of

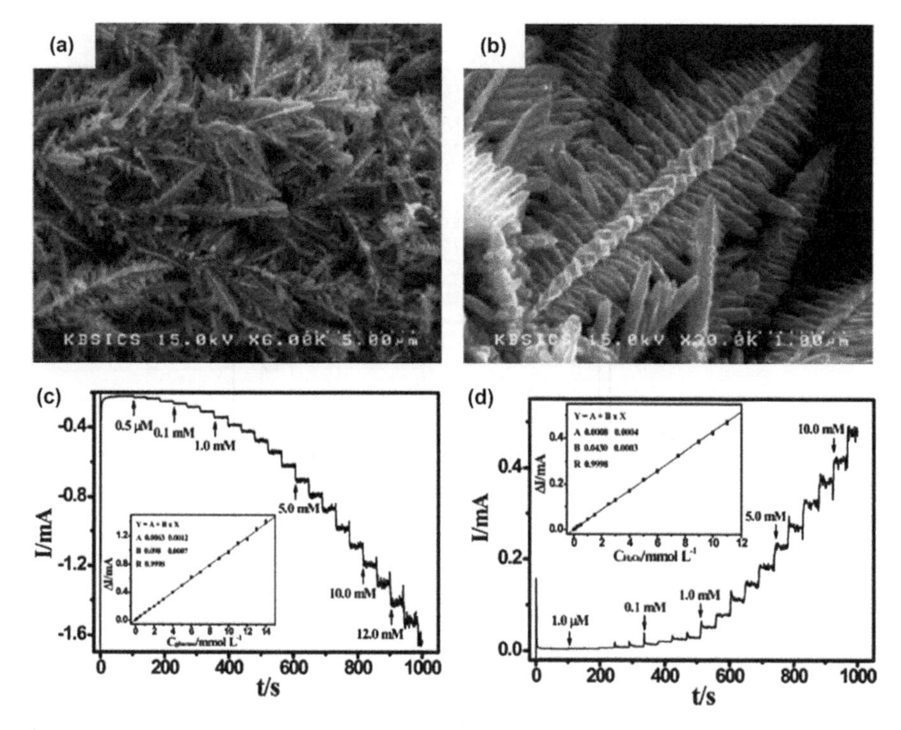

Figure 11.6 SEM image of a Cu–Co alloy dendrite nanostructure (magnification: (a) 6000× and (b) 30 000×) and chronoamperometric responses in (c) 0.1 M NaOH with glucose spikes (applied potential: +0.65 V) and in (d) 0.1 M PBS after successive spikes with H_2O_2 (applied potential: − 0.40 V) (adapted from ref. 33 with permission).

0.5 μM to 14 mM, a response time of 5 s and an LOD of 0.1 μM at an applied potential of 0.65 V. In the case of the H_2O_2 sensor, the linear range was from 1 μM to 11 mM with a sensing response time of 5 s and the LOD was 0.75 μM at an applied potential of −0.4 V as shown in Figure 11.6(c) and (d). It was noted that the Cu–Co alloy dendrite demonstrated a huge catalytic ability in both glucose oxidation and H_2O_2 reduction compared to traditional Cu and Co electrodes.

Zeng *et al.*[34] manufactured a graphene/HRP (horseradish peroxidase) 2–3 multi-sheet nanostructure by a layer-by-layer (LbL) self-assembled technique and used it for an enzymatic H_2O_2 sensor. The sensor had a sensitivity of 220 μA cm^{-2} $(mM)^{-1}$ on a linear range of 1 μM to 2.6 mM, a response time of less than 2 s and an LOD of 0.1 μM at an applied potential of −0.08 V. Guo *et al.*[35] made 1–3 sphere hierarchical Au nanowires by hydrothermal growth. The fabricated H_2O_2 sensor had a linear range up to 0.8 mM, a response time of 3.5 s and an LOD of 1.2 μM at an applied potential of −0.2 V. Furthermore, Shim *et al.*[36] synthesized 1–3 cactus-like IrO_2 nanowires on a platinum electrode by vapor phase growth as shown in Figure 11.7(a) and (b). It was used as a dihydronicotinamide adenine dinucleotide (NADH) sensor, which had a sensitivity

Figure 11.7 SEM image of IrO_2 nanowires grown on a Pt microwire (a–b) and current–time recordings (c) obtained at the 25 μm Pt wire (red) and IrO_2-Pt (blue) microelectrodes upon increasing the concentration of NADH in 0.005 M PBS solution (pH 7.4) at $E = 0.3$ V (c) (adapted from ref. 36 with permission).

of 2.9 μ A cm^{-2} $(mM)^{-1}$ and an LOD of 5 μM with a response time of less than 10 s at an applied potential of 0.3 V as shown in Figure 11.7(c).

In summary, the surface-to-volume ratio is greatly enhanced in hierarchical nanostructures compared to their dense and compact counterparts. It is believed that the improvement in sensing performance in electrochemical sensors exploiting hierarchical metal oxide nanostructures may be attributed to the increased surface area on which the sensing reaction occurs. Furthermore, the nano-sized feature guarantees that most of the material experiences a change in electrical property enabling a measurable signal at a slight change in analyte concentration, leading to higher sensitivity. In addition, the hierarchical nanostructure allows facile diffusion and mass transport for analyte molecules resulting in an immediate sensing reaction, which means an ultra-fast response/recovery of sensors.

11.4.3 SERS Sensors

As a different type of transduction mechanism from the two previously mentioned types of sensors, surface enhanced Raman spectroscopy or surface enhanced Raman scattering (SERS) sensors exploiting optical means of molecular detection are introduced and discussed in this section. Since its discovery 30 years ago,[37] SERS has been of great interest as a powerful high-throughput tool for the detection of molecules.[38,39] A laser induces a surface plasmon that leads to Raman scattering enhanced by means of an interaction with the adsorbed molecules on a metal surface. An enhancement factor of 10^{10} to 10^{11} was found,[40] which makes sensors suitable for trace-level molecule detection. In the past few years, much progress has been made in the fabrication of various SERS-active metal nanostructures for further signal enhancement.[12] Particularly, SERS platforms were usually constructed with the periodic metal nanostructures capable of sustaining reproducible field confinements.[38] In SERS, a high resolution with spectroscopic precision allows chemical and biomolecular sensing, which has huge potential for molecular fingerprinting. Moreover, a narrow spectral width endows a futuristic SERS sensing platform with multiplexed detection capability. In addition, the target molecules are not limited because the selectivity for certain molecules depends on physisorption or chemisorption determined by the composition and structures of substrates. However, it is disadvantageous in that SERS sensors cannot be designed as portable devices due to the complexity of the optical and optomechanical apparatus.

Recent studies[41–43] have demonstrated extremely strong local field enhancement in the gap among closely spaced metallic nano-elements. Highly ordered particle arrays,[44] micro-sized building blocks with nanoscale surface texture or roughness and well-aligned metal nanowires[45] have opened a new chapter in SERS applications due to narrow gaps between the adjacent features. Therefore, hierarchical nanostructures can play the role of a SERS-active platform if a huge amount of narrow gaps are obtained in the final structures. Hierarchical nanostructures were used as SERS platforms in the form of hierarchically branched silver nanostructures,[46] an electrochemically roughened silver surface,[47] *etc.* As a result, it turned out that the enhancement of the optical signal becomes extremely significant on a hierarchically nanostructured metal surface. Table 11.4 summarizes hierarchical nanostructures for SERS sensing applications.

Zhang *et al.*[9] fabricated an Ag core–shell structure using acid-directed self-assembly of metal nanostructures as shown in Figure 11.8(a) and (b). The structures constructed had an Ag nanowire core and shell made of Ag

Table 11.4 Summary of SERS sensors that use hierarchical nanostructures.

No.	Material	Structure	Test material	Resolution	Ref.
1	Si/ZnO/Ag	1-3 tree-like	10^{-5} M R6G	1×10^{-9} M	12
2	Au	2-3 flower-like	10^{-6} M R6G	1×10^{-8} M	14
3	Ag	1-2 monolayer	10^{-9} M R6G	×	11
4	Cu	1-3 multilayer	CV	1×10^{-6} M	48
5	Ag	2-3 core–shell	melamine	5 ppm	9
6	SiNWs/AgNPs	1-3 multilayer	4.5×10^{-13} SDI	4.5×10^{-13} M	13
7	β-$Ni(OH)_2$	2-3 flower-like	4-mercaptopyridine	×	10

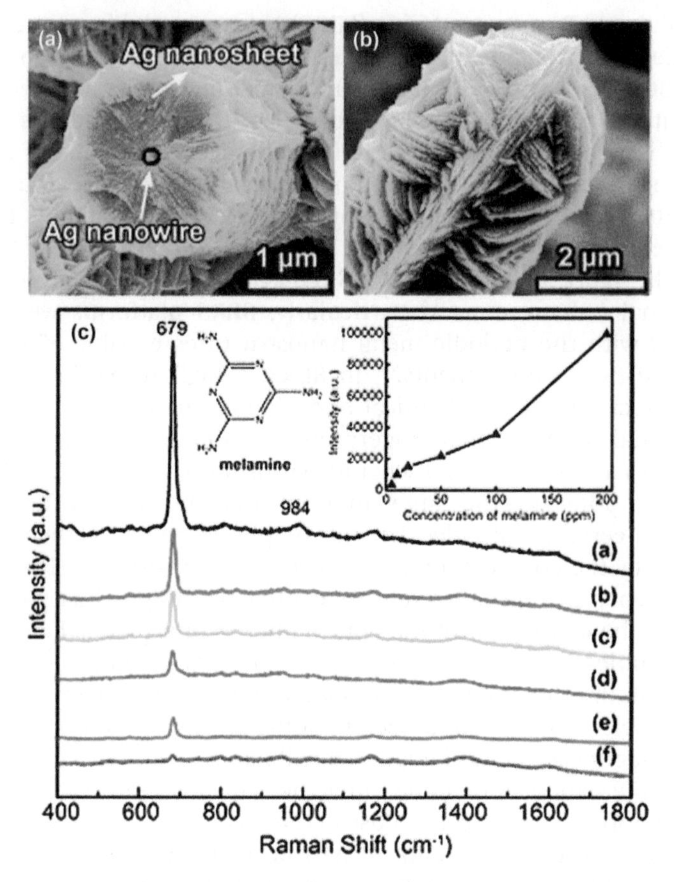

Figure 11.8 Ag core–shell hierarchical nanostructures for SERS sensing applications. (a) SEM image of the cross-sectional core–shell Ag wire, (b) the tip of a core–shell Ag wire covered by nanosheet assemblies, and (c) SERS spectra of melamine in different concentrations taken on core–shell Ag wire structures: a–f represent spectra from 200, 100, 50, 20, 10, and 5 ppm, respectively (adapted from ref. 9 with permission).

nanosheets, resulting in highly roughened surfaces that played a key role as a highly sensitive SERS platform. The fabricated Ag core–shell nanostructures showed especially high SERS sensitivity toward the analyte melamine, down to 5 ppm as shown in Figure 11.8(c), where a characteristic peak at 679 cm^{-1} was observed. It was reported that this hierarchical nanostructure had the highest sensitivity ever reported by means of fast detection in dairy products presumably due to the ample nanoscale gaps, which endow the metal substrate with applicability to highly-sensitive SERS sensors.

A silicon nanowire (SiNW) array decorated with silver nanoparticles (AgNP) was prepared and used as a SERS platform for the detection of Sudan dye.[13] The SiNW array was produced by chemical etching of a silicon wafer

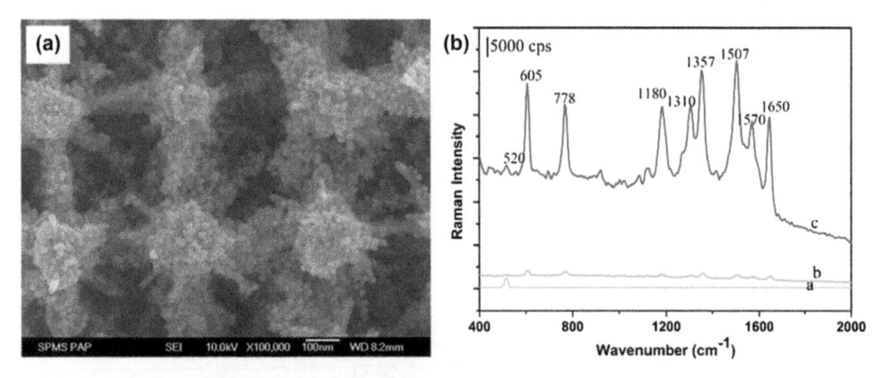

Figure 11.9 Hierarchical nanostructure of Si/ZnO nano-trees decorated with Ag nanoparticles for SERS applications. (a) SEM image (b) SERS spectra of R6G molecule with a concentration of 1×10^{-5} M. a–c represent spectra from Si/ZnO nano-trees, the 2D Ag nanoparticle substrate, and Si/ZnO nano-trees decorated with Ag nanoparticles (adapted from ref. 12 with permission).

followed by direct growth of AgNP using a galvanic redox reaction. It was demonstrated that the silicon nanowire array decorated with Ag nanoparticles had an enhancement factor of 10^8–10^{10}. Cheng *et al.*[12] prepared a higher degree hierarchical nanostructure of Ag decorated Si/ZnO nano-trees as shown in Figure 11.9(a) using a combination of top-down lithography for a Si nanopillar on a wafer scale and bottom-up hydrothermal growth of ZnO nano-trees followed by photochemical deposition of Ag nanoparticles. This hierarchical nanostructure was implemented for the detection of Rhodamin 6G (R6G) leading to the SERS spectra shown in Figure 11.9(b). The enhanced Raman signal was observed for the Ag-decorated Si/ZnO nano-tree structure (curve c) compared to control samples of a Si/ZnO nano-tree without Ag nanoparticles (curve a) and sputtered 2D Ag nanoparticles on a smooth surface (curve b). Furthermore, an enhanced SERS signal was found even for a 1×10^{-9} M concentration of R6G, which proved the excellence of the hierarchical nanostructures. The Si/ZnO nano-tree gives an ample surface area on which decorated Ag nanoparticles provide tremendous nanoscale gaps.

A highly porous Cu nanotube was prepared by Wu *et al.*[48] though electrodeposition of metals on electrospun fiber templates and subsequent wet etching as shown in Figure 11.10(a) and (b). The evenly distributed nanoscale pores on the sidewall played the role of 'hot spot' where SERS signals are much amplified. The target molecule used in this study was crystal violet (CV) solution. As a result, the hierarchical Cu porous nanotube had an enhanced SERS signal compared to a porous Cu substrate and smooth nanotubes as shown in Figure 11.10(c). This SERS platform could detect CV concentrations as low as 10^{-6} M.

In addition, the surface roughness of a metallic periodic structure is also important in SERS platforms. The surface roughness could be changed by

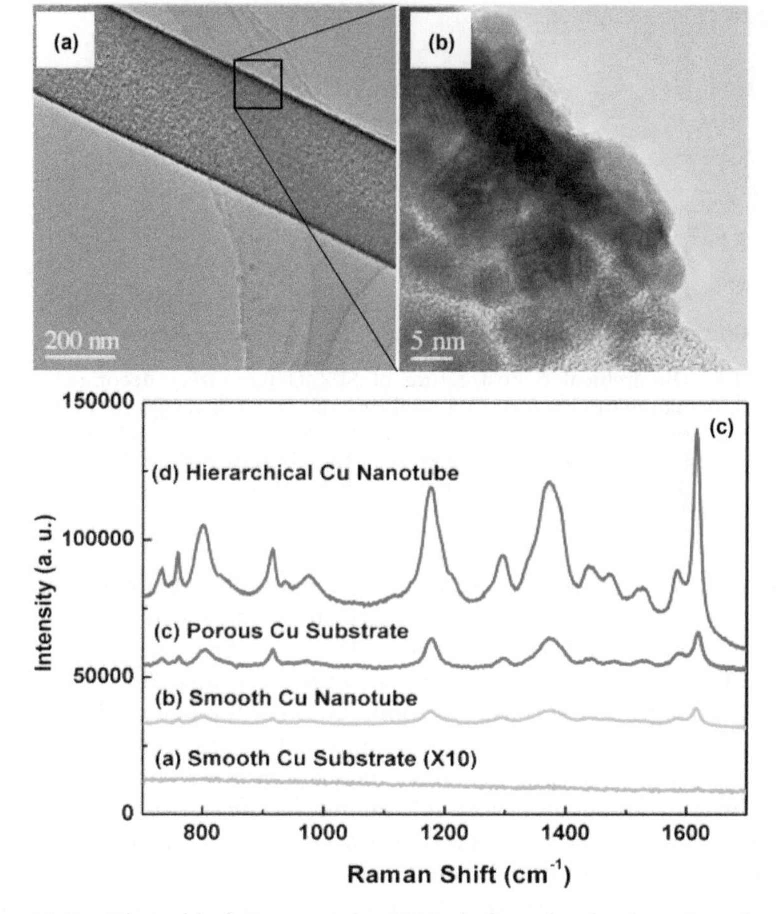

Figure 11.10 Hierarchical Cu nanotube SERS platform for the detection of crystal violet solution. (a) TEM images of Cu sub-microtubes with a porous sidewall, (b) Raman spectra of 1.0×10^{-6} M CV adsorbed on different substrates (adapted from ref. 48 with permission).

laser irradiation, demonstrating a lowered Raman signal for R6G on the periodic Au microparticle array.[14] The Langmuir–Blodgett (LB) technique was used to prepare a silver nanowire monolayer,[11] which was used as a SERS platform demonstrating an enhanced Raman signal for 10^{-9} M R6G. The silver nanowire monolayer provided nanoscale gaps between neighboring wires and an enhancement factor (EF) of 2×10^9 was found.

11.5 Concluding Remarks and Outlook

Hierarchically nanostructured sensors are very promising in terms of enhancing sensor performance: sensitivity, resolution and response/recovery time. In summary, hierarchical nanostructures have an extremely high surface-to-volume ratio enabling all surfaces to take part in a reaction simultaneously. Consequently, the sensitivity of the sensor increases greatly compared to the dense and compact counterparts of the same materials. Furthermore, the response/recovery time of the sensor decreases considerably since adsorption/desorption of gas onto the surface takes place very quickly due to the diffusion-free behavior of the target analyte in hierarchical nanostructures. By comparison, gas diffusion is very limited due to irregular distribution of pore size and the highly serpentine pathway in the agglomerates of nanomaterials. For this reason, hierarchical nanostructure sensors can be potent candidates for sensing biologically meaningful trace-level molecules, ions, and biomarkers. In addition, they might find applications in detecting toxins, hazardous gases and chemical and biological weapons that need acute action to relieve the damage to human beings. However, the mechanism by which hierarchical nanostructures are produced has not been understood fully so that the desired structures are not easy to realize at present. If the mechanism is clearly revealed, one can design appropriate hierarchical nanostructure sensor platforms in a controllable way for target analytes.

References

1. D. Lee, Y. Chander, S. M. Goyal and T. Cui, *Biosens. Bioelectron.*, 2011, **26**(8), 3482–3487.
2. Y. Cui, Q. Wei, H. Park and C. M. Lieber, *Science*, 2001, **293**(5533), 1289–1292.
3. D. Lee and T. Cui, *J. Vac. Sci. Technol., B*, 2009, **27**, 842–848.
4. D. Lee and T. Cui, *Langmuir*, 2011, **27**(7), 3348–3354.
5. J.-H. Lee, *Sens. Actuators B*, 2009, **140**(1), 319–336.
6. Z. R. Dai, Z. W. Pan and Z. L. Wang, *Adv. Funct. Mater.*, 2003, **13**(1), 9–24.
7. K. Arshak, E. Moore, G. M. Lyons, J. Harris and S. Clifford, *Sens. Rev.*, 2004, **24**(2), 181 –198.
8. J. Y. Lao, J. Y. Huang, D. Z. Wang and Z. F. Ren, *J. Mater. Chem.*, 2004, **14**(4), 770–773.
9. B. Zhang, P. Xu, X. Xie, H. Wei, Z. Li, N. H. Mack, X. Han, H. Xu and H.-L. Wang, *J. Mater. Chem.*, 2011, **21**(8), 2495–2501.
10. S. Sarkar, M. Pradhan, A. K. Sinha, M. Basu, Y. Negishi and T. Pal, *Inorg. Chem.*, 2010, **49**(19), 8813–8827.
11. A. Tao, F. Kim, C. Hess, J. Goldberger, R. He, Y. Sun, Y. Xia and P. Yang, *Nano Lett.*, 2003, **3**(9), 1229–1233.
12. C. Cheng, B. Yan, S. N. Wong, X. Li, W. Zhou, T. Yu, Z. Shen, H. Yu and H. J. Fan, *ACS Appl. Mater. Interfaces*, 2010, **2**(7), 1824–1828.
13. M.-L. Zhang, X. Fan, H.-W. Zhou, M.-W. Shao, J. A. Zapien, N.-B. Wong and S.-T. Lee, *J. Phys. Chem. C*, 2010, **114**(5), 1969–1975.
14. G. Duan, W. Cai, Y. Luo, Y. Li and Y. Lei, *Appl. Phys. Lett.*, 2006, **89**(18), 181918–181918-3.
15. M. E. Franke, T. J. Koplin and U. Simon, *Small*, 2006, **2**(1), 36–50.

16. H.-R. Kim, K.-I. Choi, J.-H. Lee and S. A. Akbar, *Sens. Actuators B*, 2009, **136**(1), 138–143.
17. P. Sun, W. Zhao, Y. Cao, Y. Guen, Y. Sun and G. Lu, *CrystEngComm.*, 2011, **13**(11), 3718–3724.
18. L. Wang, Z. Lou, T. Zhang, H. Fan and X. Xu, *Sens. Actuators B*, 2011, **155**(1), 285–289.
19. Q. Lipeng, X. Jiaqiang, P. Qingyi, C. Zhiwuan, X. Qun and L. Feng, *Nanotechnology*, 2008, **19**(18), 185705.
20. L. V. Thong, L. T. N. Loan and N. V. Hieu, *Sens. Actuators B*, 2010, **150**(1), 112–119.
21. L. Wang, T. Fei, Z. Lou and T. Zhang, *ACS Appl. Mater. Interfaces*, 2011, **3**(12), 4689–4694.
22. P. Sun, W. Wang, Y. Liu, Y. Sun, J. Ma and G. Lu, *Sens. Actuators B*, 2012, **173**(0), 52–57.
23. X. Gou, G. Wang, X. Kong, D. Wexler, J. Horvat, J. Yang and J. Park, *Chem.-Eur. J.*, 2008, **14**(19), 5996–6002.
24. C. Yujin, Z. Chunling, S. Xiaoling, C. Maosheng and J. Haibo, *Nanotechnology*, 2008, **19**(20), 205603.
25. Y. Zhang, J. Xu, Q. Xiang, H. Li, Q. Pan and P. Xu, *J. Phys. Chem. C*, 2009, **113**(9), 3430–3435.
26. K.-M. Kim, H.-R. Kim, K.-I. Choi, H.-J. Kim and J.-H. Lee, *Sens. Actuators B*, 2011, **155**(2), 745–751.
27. N. Zhang, K. Yu, Q. Li, Z. Q. Zhu and Q. Wan, *J. Appl. Phys.*, 2008, **103**(10), 104305–6.
28. X. Zhang, A. Gu, G. Wang, Y. Huang, H. Ji and B. Fang, *Analyst*, 2011, **136**(24), 5175–5180.
29. J. Chen, L. Xu, R. Xing, J. Song, H. Song, D. Liu and J. Zhou, *Electrochem. Commun.*, 2012, **20**(0), 75–78.
30. S. Luo, F. Su, C. Liu, J. Li, R. Liu, Y. Xiao, Y. Li, X. Liu and Q. Cai, *Talanta*, 2011, **86**(0), 157–163.
31. P. Si, S. Ding, J. Yuan, X. W. Lou and D.-H. Kim, *Acs Nano*, 2011, **5**(9), 7617–7626.
32. P. Si, X.-C. Dong, P. Chen and D.-H. Kim, *J. Mater. Chem. B*, 2013, **1**(1), 110–115.
33. H.-B. Noh, K.-S. Lee, P. Chandra, M.-S. Won and Y.-B. Shim, *Electrochim. Acta*, 2012, **61**(0), 36–43.
34. Q. Zeng, J. Cheng, L. Tang, X. Liu, Y. Liu, J. Li and J. Jiang, *Adv. Funct. Mater.*, 2010, **20**(19), 3366–3372.
35. S. Guo, D. Wen, S. Dong and E. Wang, *Talanta*, 2009, **77**(4), 1510–1517.
36. J. H. Shim, Y. Lee, M. Kang, J. Lee, J. M. Baik, Y. Lee, C. Lee and M. H. Kim, *Anal. Chem.*, 2012, **84**(8), 3827–3832.
37. M. Fleischmann, P. J. Hendra and A. J. McQuillan, *Chem. Phys. Lett.*, 1974, **26**(2), 163–166.
38. H. Ko, S. Singamaneni and V. V. Tsukruk, *Small*, 2008, **4**(10), 1576–1599.
39. W. E. Doering, M. E. Piotti, M. J. Natan and R. G. Freeman, *Adv. Mater.*, 2007, **19**(20), 3100–3108.

40. E. C. Le Ru, E. Blackie, M. Meyer and P. G. Etchegoin, *J. Phys. Chem. C*, 2007, **111**(37), 13794–13803.
41. W. Li, P. H. C. Camargo, X. Lu and Y. Xia, *Nano Lett.*, 2009, **9**(1), 485–490.
42. H. Wei, F. Hao, Y. Huang, W. Wang, P. Nordlander and H. Xu, *Nano Lett.*, 2008, **8**(8), 2497–2502.
43. P. H. C. Camargo, M. Rycenga, L. Au and Y. Xia, *Angew. Chem., Int. Ed.*, 2009, **48**(12), 2180–2184.
44. X. Li, Y. Zhang, Z. X. Shen and H. J. Fan, *Small*, 2012, **8**(16), 2548–2554.
45. A. R. Tao, J. Huang and P. Yang, *Acc. Chem. Res.*, 2008, **41**(12), 1662–1673.
46. S. H. Han, L. S. Park and J.-S. Lee, *J. Mater. Chem.*, 2012, **22**(38), 20223–20231.
47. M. Moskovits, Surface-Enhanced Raman Spectroscopy: a Brief Perspective, in *Surface-Enhanced Raman Scattering*, ed. K. Kneipp, M. Moskovits and H. Kneipp, Springer, Berlin Heidelberg, 2006, pp. 1–17.
48. H. Wu, D. Lin and W. Pan, *Langmuir*, 2010, **26**(10), 6865–6868.
49. K.-I. Choi, H.-R. Kim and J.-H. Lee, *Sens. Actuators B*, 2009, **138**, 497.
50. S. Ma, R. Li, C. Lv, W. Xu and X. Gou, *J. Hazard. Mater.*, 2011, **192**, 730.
51. J. Zhang, S. Wang, M. Xu, Y. Wang, B. Zhu, S. Zhang, W. Huang and S. Wu, *Cryst. Growth Des.*, 2009, **9**, 3532.
52. A. Ponzoni, E. Comini, G. Sberveglieri, J. Zhou, S. Z. Deng, N. S. Xu, Y. Ding and Z. L. Wang, *Appl. Phys. Lett.*, 2006, **88**, 203101.

CHAPTER 12

Other Applications

SUKJOON HONG* AND SEUNG HWAN KO

Applied Nano and Thermal Science (ANTS) Lab, Department of Mechanical Engineering, Seoul National University, 1 Gwanak-ro, Gwanak-gu, Seoul 151-742, Korea
*Email: solaninhsj@gmail.com

12.1 Introduction

In the preceding chapters, various hierarchical nanostructures for energy efficient electronics applications among energy devices have been discussed. Electronics are the final stage for energy cycles and basically consume energy that was generated by energy harvesting devices and energy storage devices. Among the biggest electronics using nanostructures are displays (Chapter 10) and sensors (Chapter 11). Additionally, the application of hierarchical nanostructures can be found in various research topics: (1) highly efficient energy consumption devices including transparent conductors, highly flexible and stretchable electrodes, light emitting diodes, and (2) mechanical hierarchical nanostructures, for example, for super-hydrophobic and self-cleaning surfaces, pool boiling enhancement for more efficient heat transfer, and gecko-inspired adhesives. As long as the surface area or surface characteristics are considered an important factor to enhance the functionality of the devices, applying hierarchical nanostructures is one of the most powerful and promising approaches with a very high degree of design freedom.

In this chapter, other research trends for hierarchical nanostructures for highly efficient energy consumption devices other than the electrical devices discussed in the previous chapters and mechanical devices will be

RSC Nanoscience & Nanotechnology No. 35
Hierarchical Nanostructures for Energy Devices
Edited by Seung Hwan Ko and Costas P Grigoropoulos

Published by the Royal Society of Chemistry, www.rsc.org

introduced together with a discussion on how hierarchical nanostructuring will lead to better functionality in the devices.

12.2 Other Applications of Hierarchical Nanostructures

Hierarchical nanostructures can enhance the efficiency of electronic and other devices by expanding the functionality. For example, a transparent conductor and light emitting diode can enhance electrical functionality while maintaining the high transparency; a highly flexible or stretchable electrode can gain high mechanical stability while maintaining good electrical conductivity under various mechanical deformations; modification of surface characteristics can tune the surface hydrophobicity with broad tunability; heat transfer can be increased with large surface area and pool boiling enhancement; adhesion force can be increased with gecko inspired adhesives.

12.2.1 Metal Nanowire Percolation Network for Transparent Conductor Applications

A transparent conductor is a core component in many optoelectronic devices such as touch screens, LCDs, OLEDs, and solar cells, which have recently been showing a tremendously rapid growth. Indium tin oxide (ITO) has been the most dominant transparent conductor material. However, the further development and application of ITO have receded due to two issues: its scarcity and fragile ceramic nature.[1,2] These two key issues are actively being addressed by the emergence of the next generation of flexible transparent materials, such as conducting polymers,[3,4] carbon nanotubes (CNTs),[5–8] graphene,[9–12] as well as metal nanostructures.[13,14–26] Due to the poor conductivity and instability of conducting polymers and relatively low transmittance and electrical conductivity of carbon-based nanomaterials, metal nanowire based hierarchical nanostructures such as the silver nanowire percolation network are getting much attention as alternative transparent conductors.[1,2,16] Metals are amongst the most conductive materials on earth due to their high free-electron density, which consequently also makes them highly reflective in the visible wavelength range and not very transparent. However, metals with very small dimensions (smaller than the visible wavelength) can be highly transparent while maintaining good electrical conductivity.[1]

In flexible and transparent conductors, inherently stretchable materials such as carbon nanomaterials have been dominant research topics. Despite the many advantages of using metals, as mentioned earlier, metal nanowires have shown very limited applications with limited performance in flexible transparent electrodes.[16–26] This is because current widely used nanowires are relatively short (1–20 μm),[1,2] which deteriorates the transmittance, and also because the high temperature annealing process to improve junction resistance is not compatible with heat sensitive flexible or stretchable substrates. To address this current issue in metal nanowire percolation network transparent conductors, Lee *et al.*[1,27] presented a series of combined novel approaches to develop a very long Ag NW synthesis method and applied it as a new type of high performance flexible and transparent metal conductor combined with a low temperature nano-welding process as shown in Figure 12.1. Very long metallic NWs are critical factors enabling the high transparency, high electrical conductivity and superior mechanical compliance and strength at the same time, which are usually hard to achieve simultaneously. We found that very long metallic nanowire percolation network conductors combined with a low temperature nano-welding process enabled highly transparent flexible conductors with high transmittance (90–96%) and high electrical conductivity (9–70 ohm sq^{-1}) with good mechanical strength. These values beat the other related class of reported flexible and transparent conductors including carbon-based nanomaterials (CNT and graphene) and other metallic nanomaterial-based devices. To demonstrate the feasibility of our approach for high performance transparent and flexible conductors, a fully functioning transparent touch panel and highly flexible LED circuits were demonstrated for the first time as shown in Figure 12.2.

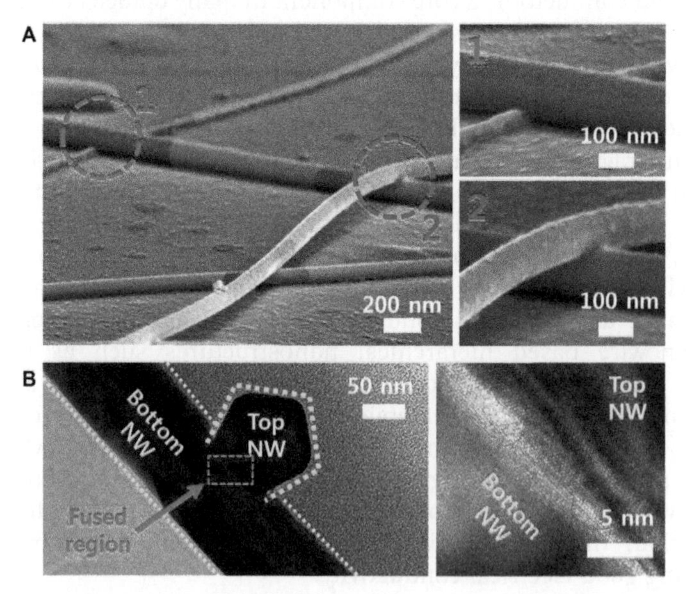

Figure 12.1 Nano-welding of a very long Ag NW network electrode for highly transparent and flexible electrodes. (A) Magnified SEM (pseudo-colored) and (B) HRTEM images of nano-welded spots between very long Ag NWs at optimum processing conditions. Note the high crystalline characteristics of each NW and the nano-welded spot between Ag NWs. The HRTEM sample was prepared by focused ion beam (FIB). Reproduced with permission from ref. 1.

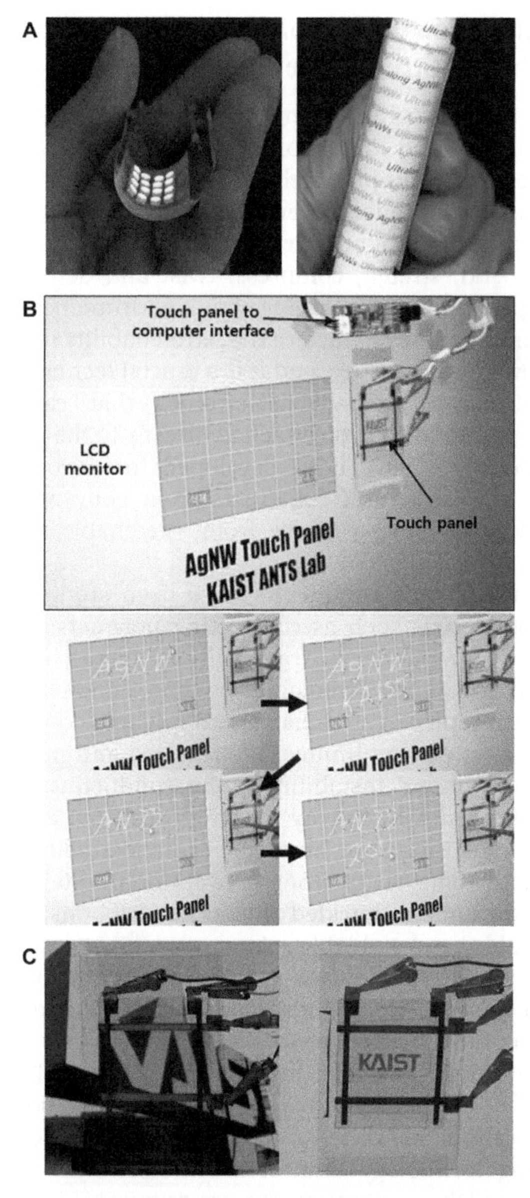

Figure 12.2 Applications of very long Ag NW network electrodes as flexible and transparent conductors. (A) Images of a flexible 4×4 array of LEDs with very long Ag NW network electrodes under bending. (B) Demonstration of applying a very long Ag NW transparent conductor for a touch panel. It was connected to a computer with control software. See the supplementary movie clip of its operation in Ref. 1. (C) Transparency of the touch panel using a Ag NW network transparent conductor on a commercial LCD monitor.
Reproduced with permission from ref. 1.

12.2.2 Metal Nanowire Percolation Network for Highly Flexible or Stretchable Electrodes

The recent dramatic progress in mechanically unconventional forms of electronics on foldable and stretchable substrates opened a new prospect for future electronics.[28–31] Examples include flexible and stretchable circuitries, flexible displays, flexible energy devices, smart skins, electronic eye-type imagers, soft and human-friendly devices and so on.[28–46] This new class of electronics can bend, stretch, compress, twist and deform into complex, non-planar shapes while maintaining good performance, reliability, and integration.[28] Among these characteristics, stretchability is the most difficult and advanced level of technology, and it is a crucial technology to realize the development of future wearable electronics that can be embedded into clothes and garments or even attached directly to the skin.[46] Stretchable and flexible electronics have been developed into two main categories: (1) engineering new structural constructs from conventional established materials or (2) assembling a device from stretchable new nanomaterial developments.[28–29,47]

Stretchable and foldable conductors have been studied in various new functional nanomaterials such as conducting polymers,[48,49] carbon nanotubes (CNTs),[50–54] graphene,[55–58] as well as hierarchical metal nanostructures.[47,59–75] Inherently stretchable carbon nanomaterials have been dominant research topics in stretchable and foldable conductors. However, they have demonstrated very limited applications and performance due to the poor conductivity and instability of the conducting polymer and the relatively low electrical conductivity and high material cost of the carbon-based nanomaterial. Inorganic hierarchical nanostructures with new engineered structural constructs from conventional established materials, such as metal thin films or buckled silicon nanoribbons, are getting attention as stretchable and foldable electronics.[28–29,47] Among the metal nanostructures, metallic nanowires may be very promising candidates for stretchable and foldable electronics. However, most research on metallic nanowires is focused on the transparent conductor as an alternative to ITO,[50,61] with very little research being done on the foldable electronics applications,[63–66,71,73] and there have been no demonstrations of stretchable electronics applications so far. This is because stretchable electronics is a far more challenging technology than foldable electronics and also mainly because current widely used nanowires are relatively short (1–20 μm),[28,50] which deteriorates the electrical conductance and mechanical compliance.

Recently, Lee *et al.*[28] presented novel routes to developing very long Ag nanowires and applying them as a new type of highly stretchable and highly conducting hierarchical metal electrode as shown in Figure 12.3. High aspect ratio metallic nanowires are critical factors enabling high electrical conductivity and mechanical compliance at the same time.[28] This is usually hard to achieve simultaneously and has been overlooked in most research on metallic nanowire conductors.[61–75] This mechanical compliance is very useful

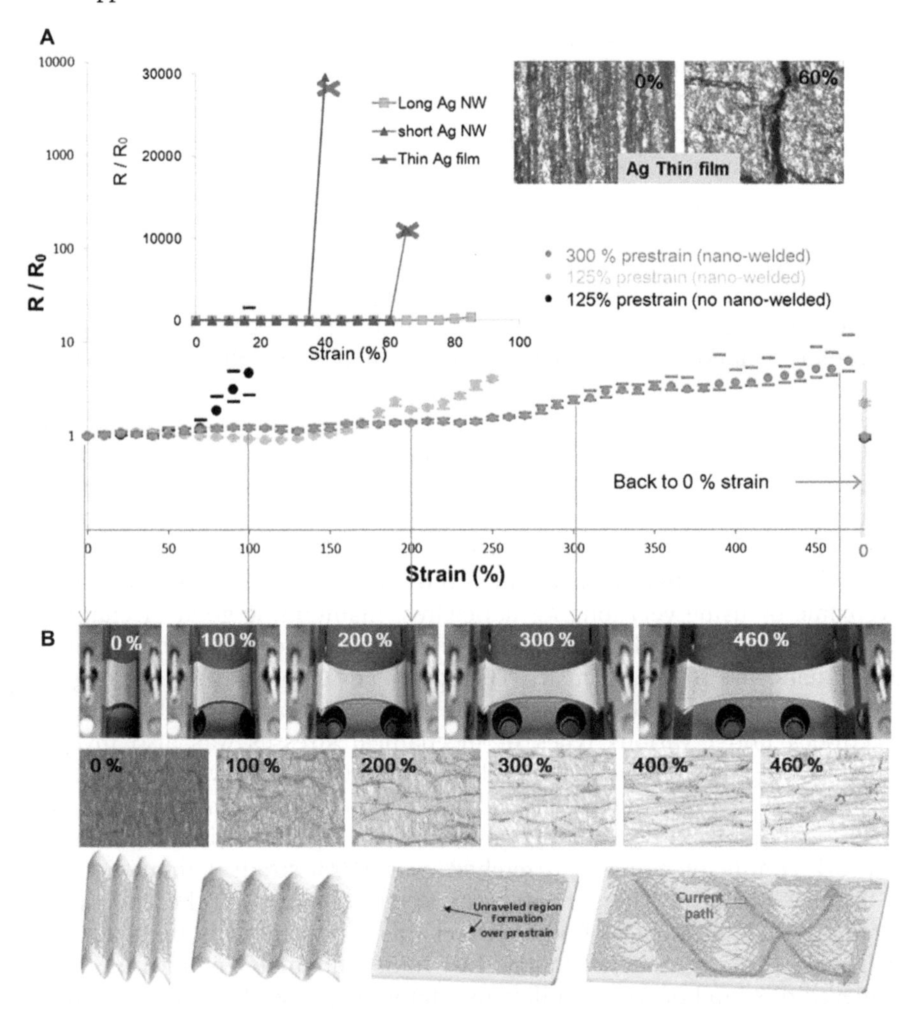

Figure 12.3 Metal nanowire hierarchical percolation network electrode for a highly stretchable metal electrode. (A) Relative resistance change measurement under various strains for a highly stretchable electrode on pre-strained Ecoflex substrates. The inset graph shows the relative resistance change for a Ag thin film (purple), short Ag NW percolation network electrode (blue) and electrode (green) during stretching on 50% pre-strained Ecoflex substrates. (B) Digital pictures of macroscopic (top row) and microscopic surface morphology (middle row) of an electrode on a pre-strained Ecoflex during 460% stretching process in Figure A (green circles). Schematic behavior of the VAgNPN electrode and an Ecoflex substrate during stretching (bottom row).
Reproduced with permission from ref. 28. Copyright 2013, Wiley-VCH.

and is even critical for the realization of high performance stretchable electronics. Lee *et al.* implemented the very long Ag nanowires to realize a high performance stretchable electrode with a world record strain above

460%, and low sheet resistance. These values beat the other related class of reported stretchable and transparent conductors including carbon-based nanomaterials (CNTs and graphene) and other metallic nanomaterial-based devices.[28] As well as a superior performance, their approach could offer further advantages in cost, due to the solution process, and easy scalability through a simple, fast, low temperature, non-vacuum process. Further, to demonstrate the feasibility of applying hierarchical nanostructures for high performance stretchable conductors, a fully functioning highly stretchable LED circuit was demonstrated as shown in Figure 12.4.[28]

The superior electrical conductance and mechanical compliance of hierarchical nano-networks of very long Ag nanowires allow major applications in highly stretchable and highly conductive metal conductors for stretchable electronics. High stretchability and high conductivity are hard to obtain simultaneously because the stretchability decreases sharply when the nanowire density is increased. When the nanowire density is increased to enhance the electrical conductivity, the Ag nanowire hierarchical percolation networks lose the good mechanical stretchability of an individual Ag nanowire and act like a metal thin film. However, this problem can be overcome by using very long Ag nanowires, maintaining good mechanical properties of individual metallic nanowires and good electrical conductivity at low nanowire number density. Taking full advantage of the very long Ag nanowires, the electrical property of stretchable hierarchical nanowire electrodes under various types of mechanical stresses was also investigated.[28] A highly stretchable hierarchical metal conductor was formed on a pre-strained, highly stretchable elastomer, which can be stretched up to 900% without failure. The highly stretchable Ag nanowire hierarchical network conductor could be stretched over 460% strain without notable resistance increase. There have been several reports of the individual strengths and advantages of other stretchable conductors, but no materials, including CNTs, graphene, metal thin films and metal nanowires, have achieved this level of stretchability and durability, as well as low resistance. The maximum strain was not limited by the failure of the nanowire electrode but by the failure of fixtures holding the nanowire electrode. Very long hierarchical nanowire networks showed much more robust and superior electrical and mechanical robustness under large strain than short nanowires and thin metal film electrodes. This proves the advantages of very long nanowires for high performance stretchable electronics. Ag thin films showed a dramatic resistance increase due to film rupture and short Ag nanowire percolation network electrodes showed a slightly better performance but was still much worse than very long hierarchical nanowire electrodes. The stretchable hierarchical metal conductor exhibits reversible mechanical and electrical changes after many stretching cycles indicating excellent electrical functionalities and stability under stretching.[28]

There are two major strategies to address the challenge in stretchable electronics, namely, "materials that stretch" and "structures that stretch". Lee *et al.*'s approach uses a hybrid of the two strategies, namely, "materials

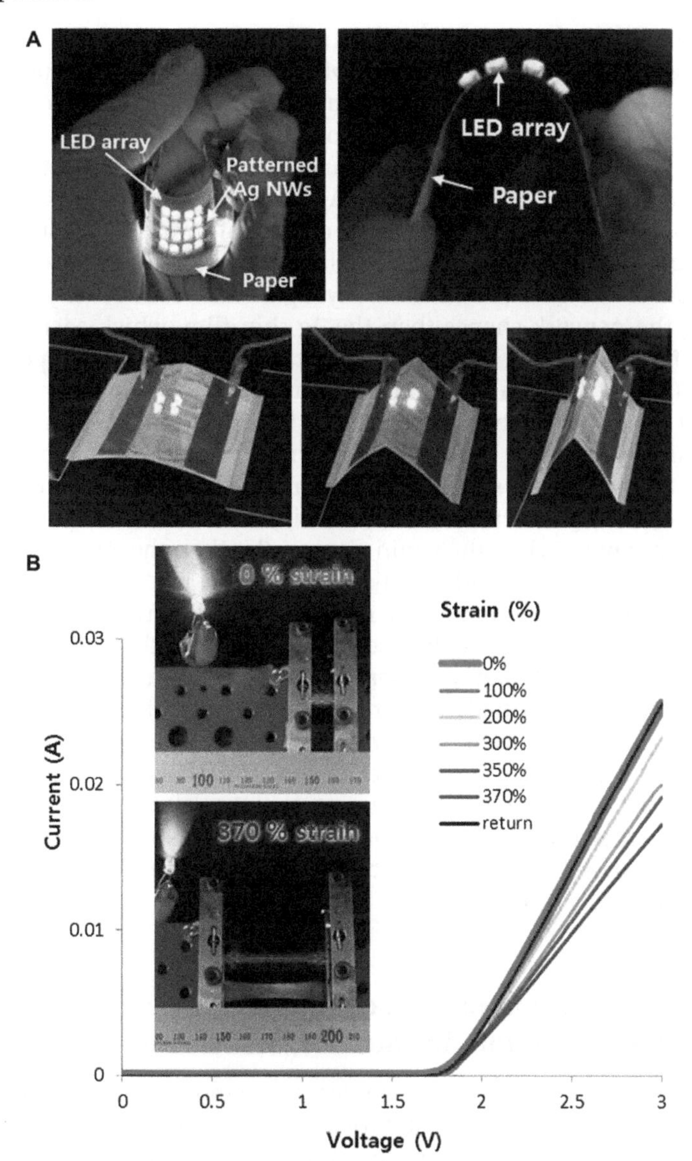

Figure 12.4 Applications of very long Ag nanowire hierarchical percolation network electrodes as highly stretchable and flexible electrodes. (A) Images of a flexible paper display with a 4×4 array of LEDs with electrodes (top pictures) and images of a flexible paper display under various bending radius. (B) Current–voltage measurement of the LED integrated stretchable electrode using a conductor on a highly elastic substrate at various tensile strains (over 370%) and after recovery from its stretching stage (black line). The insets show the digital images of the LED integrated circuits operating at 0 and 370% strain. A movie clip of a demonstration can be found in the supplementary information of Ref 28.
Reproduced with permission from ref. 28. Copyright 2013, Wiley-VCH.

and structures that stretch". Firstly, each Ag hierarchical nanowire possesses ideal material properties for a stretchable electrode. This is because bulk Ag itself is a one of the most ductile metals, so that Ag easily accommodates tensile stress and deforms. In addition, the pentagonal Ag nanowire is known to have superior yield strength and a size-dependent elastic property (higher Young's modulus than bulk Ag)[28] due to the enhanced stiffness because of the internal twin boundary of the nanowires.[28] Secondly, nano-welded Ag hierarchical nanowire percolation networks can more effectively accommodate deformation without any significant conductivity change by changing the network shapes than the Ag thin film, which easily ruptures into patches with electrical failure under a large strain. The percolation network made from longer hierarchical nanowires stretched more easily than that made from shorter nanowires for effective Ag mesh structures under the same stress. The percolation network with longer hierarchical nanowires experienced smaller maximum stress than that with shorter nanowires. Highly stretchable hierarchical nanowires experience two stretching regimes. When the strain was smaller than the Ecoflex pre-strain, the wavy substrate was gradually flattened. When the strain exceeded the pre-strain, the Ag hierarchical nanowire network structures themselves started to deform by aligning themselves with the direction of the strain. For some cases with a high nanowire density, the hierarchical nanowire behaved like a thin film and partially ruptured. However, the hierarchical nanowire percolation networks were still moderately good to yield robust high conductivity. The electrical and mechanical compliance characteristics of the conductor did not deteriorate over time. Furthermore, no changes in electrical characteristics or mechanical damages were observed after a lot of bending and stretching cycles, indicating excellent electrical functionalities and stability under stretching.

12.2.3 Modification of Surface Characteristics (Superhydrophobic Surface, Self-cleaning Surface)

Nature provides impressive examples of nanostructured and micro-structured systems with excellent optical and mechanical functionalities. In particular, many biological systems have evolved into multi-scale, hierarchical structures with sophisticated and smart functions.[76] The strong adhesion abilities of a gecko's foot,[77] superhydrophobic surface of a lotus leaf[78] and water strider leg,[79] photonic crystal structures in butterfly wings,[80] biomineralization of organic–inorganic structures[81] are just a few examples of hierarchical structures found in nature.[74] These unique structure–function relations have motivated researchers to fabricate hierarchical microstructures and nanostructures that mimic the unique functionalities of biological systems.[82,83] The wettability of solid surfaces is a very important property and is governed by both the chemical composition and the geometrical microstructure of the surface as shown in Figure 12.5.[76]

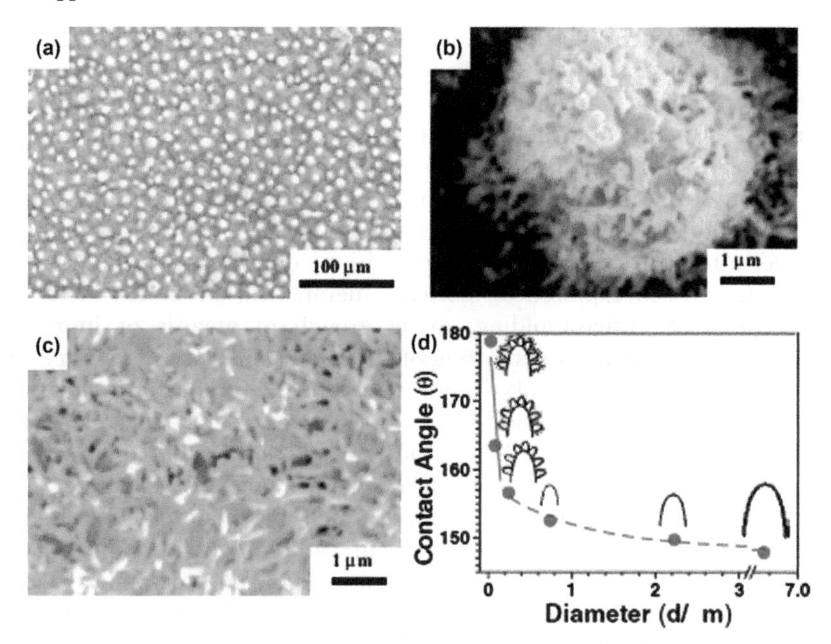

Figure 12.5 (A) Large area-SEM image of the surface of a lotus leaf. Every epidermal cell forms a papilla and has a dense layer of epicuticular waxes superimposed on it. (B) Enlarged view of a single papilla from (A). SEM images of the lower surface of the lotus leaf. (D) The fitted curve based on calculated data (contact angle, in degrees, against the mean outer diameter of protruding structures, in micrometres).

Conventionally, superhydrophobic surfaces have been produced in mainly two ways: (1) by creating a rough structure on a hydrophobic surface, and (2) by modifying a surface using materials with a low surface free energy. Until now, many methods have been developed to produce rough surfaces, including solidification of melted alkylketene dimer (AKD, a kind of wax),[84] plasma polymerization/etching of polypropylene (PP) in the presence of polytetrafluoroethylene (PTFE),[85] microwave plasma-enhanced chemical vapor deposition (MWPE-CVD) of trimethylmethoxysilane (TMMOS),[86] anodic oxidization of aluminium,[87] immersion of porous alumina gel films in boiling water,[88] mixing a sublimation material with silica or boehmite, phase separation,[88] and molding,[89] To obtain superhydrophobic surfaces, coating with low-surface-energy materials such as fluoroakylsilane (FAS) is often necessary.[87–90]

Hierarchical nanostructures can be used to enhance the superhydrophobicity of a surface. This idea came from the surface of a plant. The self-cleaning effect of some plant leaves (such as the lotus) is of great interest for practical applications in various fields.[76] The observation of hydrophobicity related to the topology of the surface of a plant leaf was previously reported by Barthlott and Neinhuis.[91] It was believed that this unique property is based on surface roughness caused by micrometre-scale papillae and epicuticular wax.[92] Zhai *et al.*[93] reported micro- and nanoscale hierarchical structures on the surface of a lotus leaf, *i.e.*, branch-like nanostructures on top of the micropapillae. These structures can induce superhydrophobic surfaces with a large contact angle and a small sliding angle.

Hierarchical nanostructures on a solid surface are very important for superhydrophobicity, which can induce a high contact angle. A number of methods have been explored to fabricate hierarchical structures including nanomolding and micromolding, direct growth of organic or inorganic structures, and lithography patterning of surfaces.[94] Li *et al.*[95,96] reported the preparation of a densely packed aligned carbon nanotube film with pure nanostructures. As an extension to this work, Ko *et al.*[74] introduced a simple and robust method to make hierarchical fibrillar arrays based on hybrid organic–inorganic material systems on mechanically flexible substrates. The hybrid structures are beneficial for multifunctional materials due to the mixed and synergetic functionalities of organic and inorganic components as shown in Figure 12.6.[74] The structures consist of polymer micropillar (μPLR) arrays decorated with ZnO nanowires. The polymer μPLR arrays are fabricated by replica molding on microfabricated silicon templates containing hexagonal micropore arrays. Subsequently, the ZnO nanowires are grown on the surface of μPLRs by a low-temperature, solution-based growth method, resulting in hierarchical microfibrillar and nanofibrillar arrays. The superhydrophobic surface properties of hierarchical μPLR/NW arrays with potential applications in self-cleaning smart surfaces, microfluidics, and biomedical devices, were demonstrated. For these applications, the mechanical flexibility of the polymeric support substrate is advantageous over traditional superhydrophobic surfaces with rigid silicon or glass support substrates.[74]

Guo *et al.*[75] reported the synthesis of a ZnO/CuO hetero-hierarchical nanotree array based on a CuO nanowire array *via* a simple hydrothermal approach combined with a thermal oxidation method as shown in Figure 12.7. Also, an as-prepared ZnO/CuO nanotree array after silanization presented remarkable superhydrophobic performance, which is attributed to the trapped air and hierarchical roughness. Furthermore, wettability of the ZnO/CuO hierarchical nanotree could be manipulated by the morphologies of hierarchical ZnO nanorods. By adjusting the reaction conditions, such as $Zn(NO_3)_2$ concentration and reaction time, the density, diameter and length of the hierarchical ZnO nanorod branch could be tuned. After silanization, the surface containing the ZnO/CuO hetero-hierarchical nanotree array exhibit impressive superhydrophobic behavior. Changing the structure from a CuO nanowire array to a ZnO/CuO nanotree array at the surface causes an obvious increase in static contact angle, which then decreases. These results demonstrate the switch between a Wenzel and Cassie–Baxter model. When in contact with such a solid surface, a water drop can fully bounce like a balloon. As a potential application study, the nanotree array surface is found to be self-cleaning.[75]

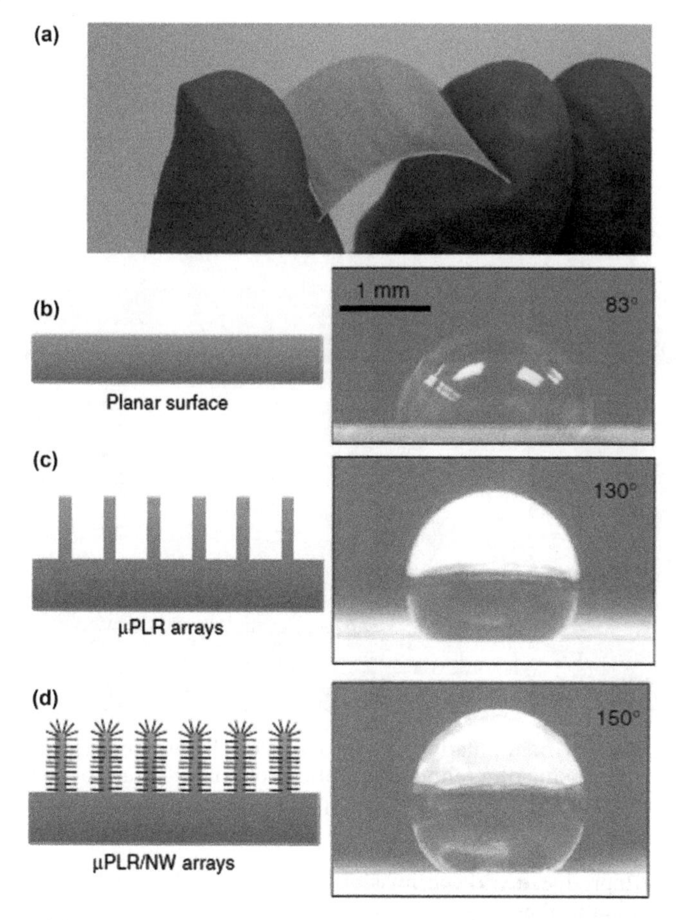

Figure 12.6 (a) Optical image of a PC substrate with hierarchical μPLR/NW arrays on the surface, showing its mechanical flexibility. Contact angle measurements for (b) planar PC substrate, (c) PCμPLR arrays, and (d) hierarchical μPLR/NW structures. The length and pitch of μPLRs are $L_{\mu PLR} = 10$ μm and $P_{\mu PLR} = 6$ μm. In all cases, the surface is coated with a thin layer (~10 nm) of parylene-N by a gas deposition process. Reproduced with permission from ref. 74.

12.2.4 Pool Boiling Enhancement

The large surface area of hierarchical nanostructures also can be used to enhance heat transfer and pool boiling. As the technology in microelectronics design and manufacturing advances, the feature size in many devices is getting smaller and smaller, leading to an explosive increase in manufacturable integrated circuit (IC) chips thereby decreasing the price. However, IC devices should be maintained at low temperature in order to improve reliability and prevent premature failure by means of cooling, since the semiconductor is vulnerable to heat. Due to the rising requirement of

Figure 12.7 Growth manipulation of the ZnO/CuO heterohierarchical nanostructures using different concentrations of $Zn(NO_3)_2$ salt for 10 h. (a) 0.01 M (sample a), (b) 0.025 M (sample b), (c) 0.05 M (sample c), (d) 0.1 M (sample d). The inset in panel d is a side cross-sectional view of an individual ZnO/CuO nanotree.
Reproduced with permission from ref. 75. Copyright 2013, American Chemical Society.

more heat dissipation, such as in CPU cores, pool boiling heat transfer was suggested. It uses the phase change of liquid coolant, where latent heat transfer should be considered, leading to a huge amount of heat dissipated in a specific area compared to other cooling methods without a significant temperature change. Therefore, no significant thermal stress is induced in the devices, increasing device reliability and lifetime. The boiling incipient temperature, heat transfer coefficient (HTC), and critical heat flux (CHF) are important parameters to be considered in pool boiling heat transfer and it was reported that CHF increased with an increase in surface roughness. However, it has been noted that nucleation does not merely correlate to the geometric roughness but also to aging, surface chemistry, surface–liquid combination and wettability, producing limited heat transfer enhancement. In order to increase CHF and HTC, microstructures, such as drilled holes,[97] pin-fin arrays,[98,99] micro-dimples,[100] re-entrant cavities,[101] porous surface structures,[102] finned tubes[102,103] *etc* were suggested and were found to produce high nucleate boiling heat transfer. However, an optimum design of microstructures should be required to satisfy a range of heat flux from the chip. Recently, nanomaterials such as carbon nanotubes,[104] copper nanowires,[105,106] and silicon nanowires[107] were investigated and showed enhanced boiling heat transfer. Lee *et al.*, as shown in Figure 12.8,[108] applied hierarchical ZnO nanowire forests for cooling electronics by means of pool boiling heat transfer. Nanowire forests with branched tree-like hierarchical structures greatly enhanced the surface area for the heat to be dissipated. Furthermore, nanowires facilitated nucleation of bubbles when the coolant started to boil. The testing chips were fabricated by a combination of top-down microfabrication of the heater on one side of the chips and bottom-up hydrothermal synthesis of ZnO nanowires on the other side of the chips. Nanowire forest structures were grown by two methods: lengthwise growth and branched growth. As shown in Figure 12.9, the superheat (subtraction of

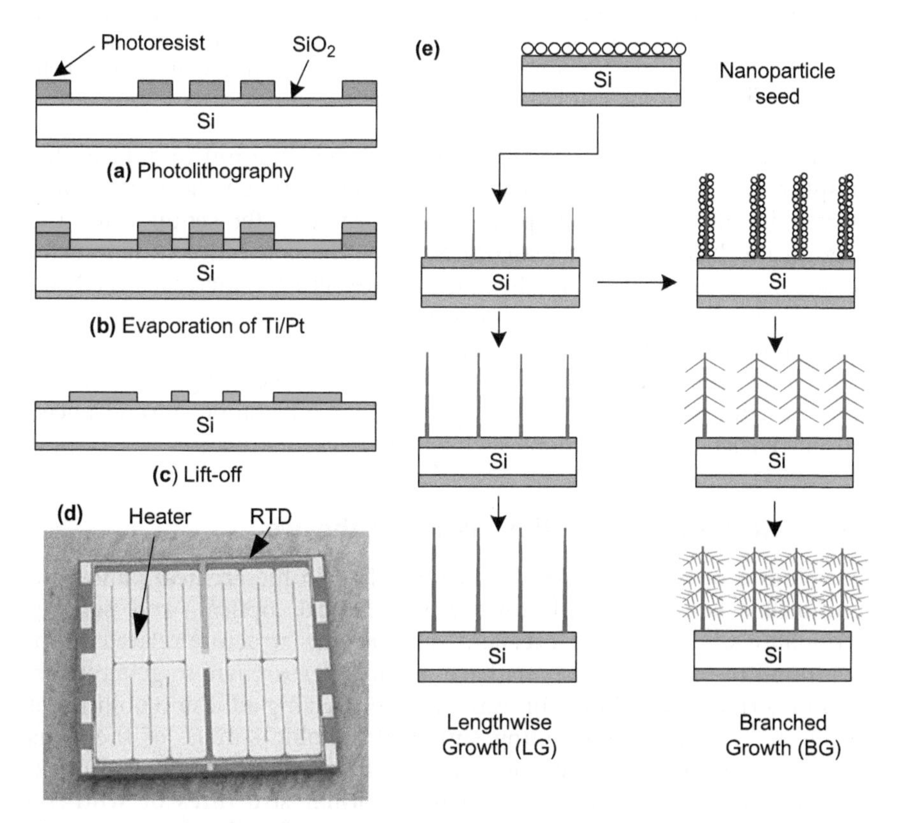

Figure 12.8 Fabrication process of hierarchical nanowire forest pool boiling heat transfer testing chips: (a–c) microfabrication of a platinum heater and resistance temperature detector (RTD) on a silicon wafer, (d) optical image of fabricated heater and detector pattern with a size of 1 cm by 1 cm, (e) synthesis of a ZnO nanowire forest. Both lengthwise (LG) and branched growths (BG) were used in this study.
Reproduced with permission from ref. 108. Copyright 2013, The Japan Society of Applied Physics.

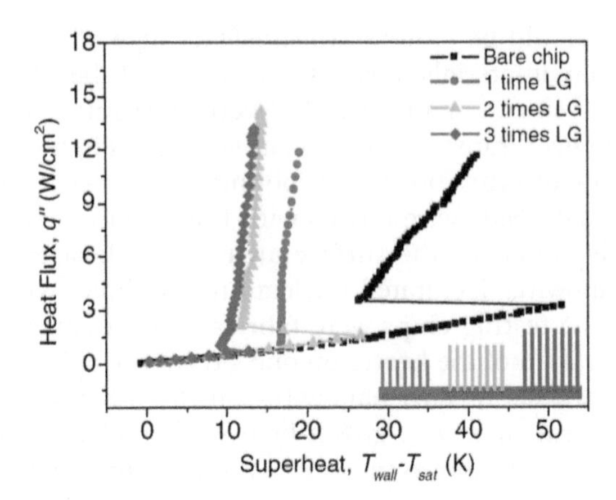

Figure 12.9 The effect of branches on the boiling heat transfer: as the number of branch increases in the hierarchical nanostructures, the CHF increases. Reproduced with permission from ref. 108. Copyright 2013, The Japan Society of Applied Physics.

saturation temperature of coolant from chip temperature) and heat flux (multiplication of voltage supplied to and current through the heater) were evaluated in the custom-built pool boiling heat transfer set-up while the voltage in the heater was increased gradually. The convective heat transfer behavior changed to pool boiling heat transfer, resulting in increased HTC and CHF. Furthermore, we observed a decrease in boiling incipient temperature compared to bare chips. The fabrication method for nanostructures on the IC chips is straightforward and provides an effective way of cooling electronics by boiling heat transfer, which has the potential for integration into advanced electronic package configurations such as 3-D stacked structures.

12.2.5 Gecko-inspired Hierarchical Adhesive

The evolution of biological systems has resulted in hierarchical nano- and microfibrillar structures with diverse mechanical, optical, and sensing functionalities.[109–111] These excellent and unique structure-related properties found in biological systems have inspired researchers to generate artificial materials, mimicking, for example, the amazing adhesion abilities of gecko feet,[112–117] the self-cleaning superhydrophobic surface of the lotus leaf,[118–119] and hairy fluid sensor arrays on the body of fish[120] and crickets.[121] In particular, the hierarchical nanofibrillar structures of synthetic gecko adhesives enable binding to almost any surface by van der Waals interactions.[122] A gecko's superb ability to adhere to surfaces is widely credited to the large attachment area of approximately 220 mm^2 of hierarchical and fibrillar structures on its feet.[123] This combination provides the gecko toe-pad with the means to effectively engage a high percentage of the spatulae, at each step, to any kind of surface topography.[123] The gecko foot has micro/nano-sized hierarchical structures. The fibrils begin from rows of lamellae and each row consists of thousands of primary setal stalks known as setae. Autumn *et al.*[124] estimated that there are 14 400 setae mm^{-2} and each seta is approximately 30–130 μm in length and 5–10 μm in diameter consisting of three levels.[125] The secondary seta is about 20–30 μm in length and 1–2 μm in diameter. At the end of each secondary seta, 100–1000 spatulae with a diameter of 100–200 nm form the points of contact with a surface. The tips of the spatulae are approximately 200–300 nm in width, 5500 nm in length, and 10 nm in thickness.[123] Cumulative van der Waals interactions have been attributed to be the main adhesive mechanism achieved through contact splitting of setae. However, it is the hierarchical topography that allows for an effective compliance to a surface by decreasing the stiffness of each level of seta with the spatula maintaining sufficient mechanical stability.[123] To have the same compliance to a surface without the hierarchical structure, linear β-keratin setae would have to have a length of 160 μm (aspect ratio of 100–160). At this aspect ratio, the mechanical stability of the pillars would be insufficient, resulting in clumping and collapse of the pillars. Clumping of the setae is undesirable as this would reduce the contact points. In addition, these clumped or bunched pillars would have higher stiffness and thus lower compliance.[123]

With the hierarchical nanostructure, Gao *et al.*[126] have shown that the theoretical van der Waals adhesion strength between surfaces can be reached as shown in Figure 12.10. Numerically, Kim *et al.*[127] showed that a hierarchical setae structure provides geckos with the adaptability to have a large effective area of contact with rough surfaces and he further showed that the equivalent stiffness of a three-level hierarchical structure is approximately 40% lower than a one-level linear structure. That stiffness reduction resulted in more than 100% enhancement in adhesion energy.[128]

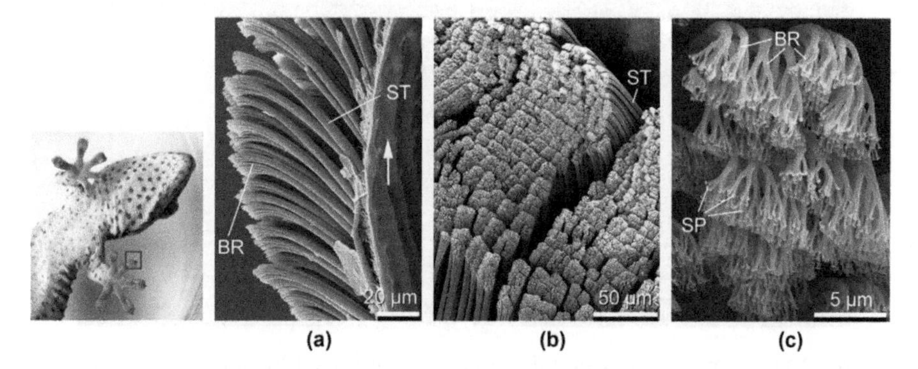

Figure 12.10 The hierarchical adhesive structures of the gecko. A gecko's toe contains hundreds of thousands of setae and each seta contains hundreds of spatulae. (a) and (b): scanning electron micrographs of rows of setae at different magnifications and (c): spatulae, the finest terminal branches of seta. ST: seta; SP: spatula; BR: branch.
Reproduced with permission from ref. 126. Copyright 2013, Elsevier.

Besides the theoretical approach, many fabrication methods of gecko-mimetic structures have been developed to obtain hierarchical structures through various methods[129–132] to replicate as close as possible the topographic features found in geckos and possibly satisfying all seven requisites for dry adhesion proposed by Autumn *et al.*[133] Greiner *et al.*[130] fabricated hierarchical pillars in PDMS with the base pillars which were 50 μm in diameter and 200 μm in length. The top round-ended pillars were 10 μm in diameter with aspect ratios ranging from 0.5 to 2. Contrary to what was expected, the pull-off test showed a pull-off force an order of magnitude lower for these hierarchical structures. This was attributed to the lower packing density of the hierarchical structure when compared to a linear structure. Jeong *et al.*[132] had similar observations to Greiner *et al.*, but Murphy *et al.*[131] reported a two-level hierarchical structure, which has higher adhesion when compared to the linear structures as shown in Figure 12.11 and 12.12. The hierarchical structure might have enabled the reduction of the effective modulus of the material, thus leading to an increase in the contact area experienced by the hemispherical indenter. The materials reported in these investigations were typically soft polymers (elastic modulus of 3–19.8 MPa). Kustandi *et al.*[129] were the only ones who reported the use of a stiff polymer

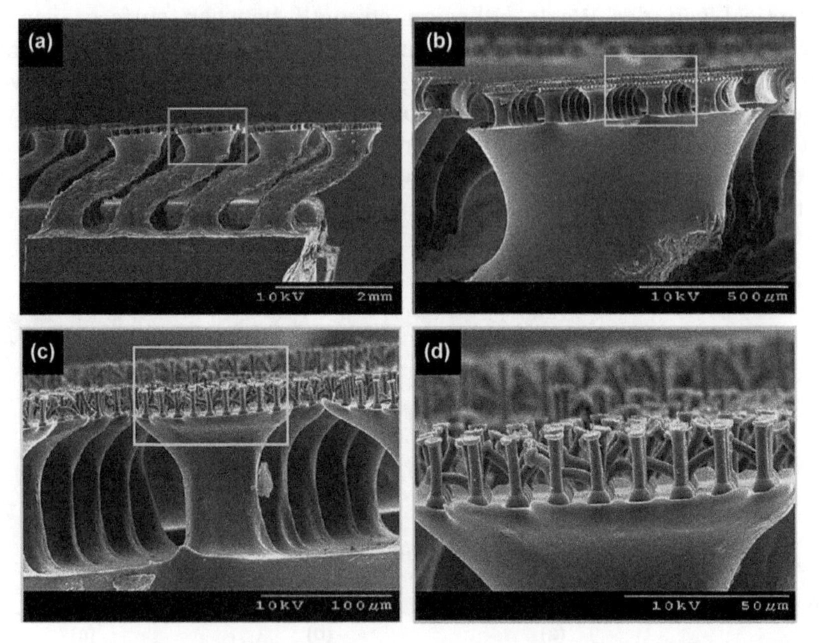

Figure 12.11 Scanning electron micrographs of three-level hierarchical polyurethane fibers: (a) 400-μm-diameter curved base fibers; (b) base fiber tip with midlevel 50-μm-diameter fibers; (c) midlevel fibers in detail; (d) terminal third level fibers at the tip of the midlevel fibers are 3 μm in diameter and 20 μm in height and have 5-μm-diameter flat mushroom tips. Reproduced with permission from ref. 131. Copyright 2013, American Chemical Society.

(PMMA) in their effort to fabricate hierarchical structures by sequentially using two porous alumina templates, but clumping due to the densely packed pillars was observed on the fabricated films and the adhesion force of these structures could not be determined accurately. Ho *et al.*[123] studied several structural variations, however, their results were inconclusive in proving the advantages of a hierarchical structure. They used a multitiered porous anodic alumina template and capillary force assisted nanoimprinting and successfully fabricated a gecko-inspired hierarchical topography of branched nanopillars on a stiff polymer. The hierarchical topography improved the shear adhesion force over a topography of linear structures by 150%. The effective stiffness of the hierarchical branched structure was lower than that of the linear structure and the reduction in effective stiffness favored a more efficient bending of the branched topography and a better compliance to a test surface, hence resulting in a higher area of residual deformation. As the area of residual deformation increased, so did the shear adhesion force. The branched pillar topography also showed a marked increase in hydrophobicity, which is an essential property in the practical applications of these structures for self-cleaning in dry adhesion conditions.[123] As well as the dry adhesion of the gecko-inspired adhesive, Ho *et al.*[123] reported an adhesive that can work in both dry and wet

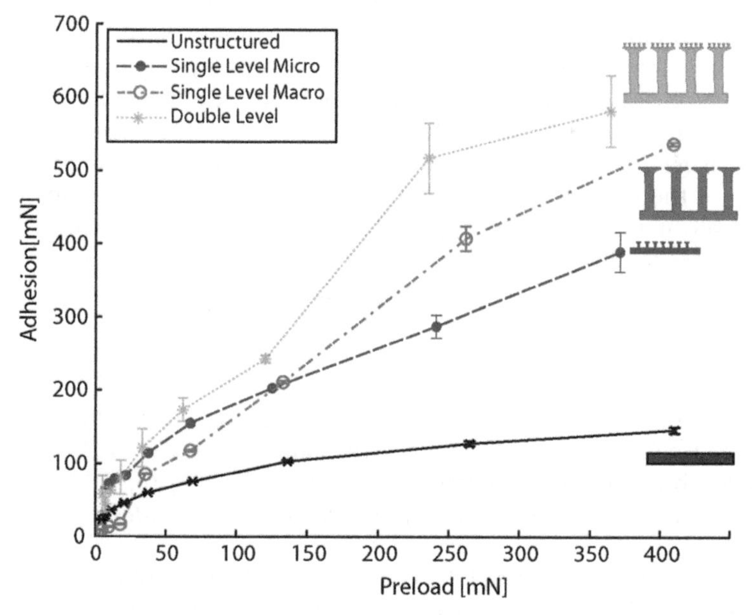

Figure 12.12 Adhesion *vs.* preload data for unstructured, single-level micro, single-level macro, and double-level samples against a 12-mm-diameter glass hemisphere. Error bars represent standard deviations. The double-level fibers generally exhibit the highest adhesion.
Reproduced with permission from ref. 131. Copyright 2013, American Chemical Society.

conditions. They observed the ability of the nanowire connectors to bind strongly even under lubricating conditions, such as a mineral oil, by van der Waals interactions. The superhydrophobic surface of the nanowire connectors enables the wet, self-cleaning of contaminant particles from the surface, similar to the lotus effect. In addition, they examined the effect of nanowire length on the shear adhesion strength, repeated usability, and robustness of the connectors, all critical properties for applications that require reversible binding of components.

12.3 Summary

Hierarchical nanostructured electrical devices have shown dramatic performance enhancement. However, the application of hierarchical nanostructures is not just restricted to electronic devices. The performance enhancement could be attributed to the structural functionality of the hierarchical structuring, which cannot be observed in thin films or bulk material based devices. In the preceding chapters, various hierarchical nanostructures for efficient energy consumption electronics applications were discussed. Among the biggest electronics using nanostructures are displays (Chapter 10) and sensors (Chapter 11). In this chapter, we discussed additional applications of hierarchical nanostructures in (1) highly efficient energy consumption devices including transparent conductors, highly flexible and stretchable electrodes, light emitting diodes, and (2) mechanical hierarchical nanostructures, for example, for superhydrophobic and self-cleaning surfaces, pool boiling enhancement for more efficient heat transfer, and gecko-inspired adhesives. As long as the surface area or surface characteristics are considered as important factors in enhancing the functionality of the devices, applying hierarchical nanostructures is one of the most powerful and promising approaches with a very high degree of design freedom. More diverse applications of hierarchical nanostructures in broader fields will be easy to find in the near future.

References

1. J. H. Lee, P. Lee, H. Lee, S. S. Lee and S. H. Ko, *Nanoscale*, 2012, **4**, 6408.
2. D. S. Hecht, L. Hu and G. Irvin, *Adv. Mater.*, 2011, **23**, 1482.
3. S. Kirchmeyer and K. Reuter, *J. Mater. Chem.*, 2005, **15**, 2077.
4. A. A. Argun, A. Cirpan and J. R. Reynolds, *Adv. Mater.*, 2003, **15**, 1338.
5. Z. C. Wu, Z. Chen, X. Du, J. M. Logan, J. Sippel, M. Nikolou, K. Kamaras, J. R. Reynolds, D. B. Tanner, A. F. Hebard and A. G. Rinzler, *Science*, 2004, **305**, 1273.
6. M. Zhang, S. Fang, A. A. Zahkidov, S. B. Lee, A. E. Aliev, C. D. Williams, K. R. Atkinson and R. H. Baughman, *Science*, 2005, **309**, 1215.
7. D. Zhang, K. Ryu, X. Liu, E. Polikarpov, J. Ly, M. E. Tompson and C. Zhou, *Nano Lett.*, 2006, **6**, 1880.

8. Z. Yu, X. Niu, Z. Liu and Q. Pei, *Adv. Mater.*, 2011, **23**, 3989.
9. G. Eda, G. Fanchini and M. Chhowalla, *Nat. Nanotechnol.*, 2008, **3**, 270.
10. K. S. Kim, Y. Zhao, H. Jang, S. Y. Lee, J. M. Kim, K. W. Kim, J. H. Ahn, P. Kim, J. Choi and B. H. Hong, *Nature*, 2009, **457**, 706.
11. S. Bae, H. Kim, Y. Lee, X. Xu, J. Park, Y. Zheng, H. Balakrishnan, T. Lei, H. R. Kim, Y. I. Song, Y. Kim, K. W. Kim, B. Ozyilmaz, J. Ahn, B. H. Hong and S. Iijima, *Nat. Nanotechnol.*, 2010, **5**, 574.
12. V. C. Tung, M. J. Allen, Y. Yang and R. B. Kaner, *Nat. Nanotechnol.*, 2009, **4**, 25.
13. E. C. Garnett, W. Cai, J. J. Cha, F. Mahmood, S. T. Conner, M. G. Christoforo, Y. Cui, M. D. McGehee and M. L. Brongersma, *Nat. Mater.*, 2012, **11**, 241.
14. D. C. Hyun, M. Park, C. Park, B. Kim, Y. Xia, J. H. Hur, J. M. Kim, J. J. Park and U. Jeong, *Adv. Mater.*, 2011, **23**, 2946.
15. M. G. Kang, M. S. Kim, J. S. Kim and L. J. Guo, *Adv. Mater.*, 2008, **20**, 4408.
16. A. Kumar and C. Zhou, *ACS Nano*, 2010, **4**, 11.
17. J. Y. Lee, S. T. Connor, Y. Cui and P. Peumans, *Nano Lett.*, 2008, **8**, 689.
18. L. Hu, H. S. Kim, J. Y. Lee, P. Peumans and Y. Cui, *ACS Nano*, 2010, **4**, 2955.
19. J. Y. Lee, S. T. Connor, Y. Cui and P. Peumans, *Nano Lett.*, 2010, **10**, 1276.
20. S. De, T. M. Higgins, P. E. Lyons, E. M. Doherty, P. N. Nirmalraj, W. J. Blau, J. J. Boland and J. N. Coleman, *ACS Nano*, 2009, **3**, 1767.
21. A. Madrria, A. Kumar, F. N. Ishikawa and C. Zhou, *Nano Res.*, 2010, **3**, 564.
22. X. Y. Zeng, Q. K. Zhang, R. M. Yu and C. Z. Lu, *Adv. Mater.*, 2010, **22**, 4484.
23. W. Gaynor, G. F. Bukhard, M. D. McGehee and P. Peumans, *Adv. Mater.*, 2011, **23**, 2905.
24. A. R. Rathmell, S. M. Bergin, Y. L. Hua, Z. Y. Li and B. J. Wiley, *Adv. Mater.*, 2010, **22**, 3558.
25. H. Wu, L. Hu, M. W. Rowell, D. Kong, J. J. Cha, J. R. McDonough, J. Zhu, Y. Yang, M. D. McGehee and Y. Cui, *Nano Lett.*, 2010, **10**, 4242.
26. Z. Yu, L. Li, Q. Zhang, W. Lu and Q. Pei, *Adv. Mater.*, 2011, **23**, 4453.
27. J. H. Lee, P. Lee, D. Lee, S. S. Lee and S. H. Ko, *Cryst. Growth Des.*, 2012, **12**, 5598.
28. P. Lee, J. Lee, H. Lee, J. Yeo, S. Hong, K. H. Nam, D. Lee, S. S. Lee and S. H. Ko, *Adv. Mater.*, 2013, **24**, 3326.
29. J. A. Rogers, T. Someya and Y. Huang, *Science*, 2010, **26**, 1603.
30. X. Lu and Y. Xia, *Nat. Nanotechnol.*, 2006, **1**, 163.
31. Y. Sun, W. M. Choi, H. Jiang, Y. Y. Huang and J. A. Rogers, *Nat. Nanotechnol.*, 2006, **1**, 201.
32. R. F. Service, *Science*, 2003, **301**, 909.
33. G. H. Gelinck, H. E. A. Huitema, E. Veenendaal, E. Cantatore, L. Schrijnemakers, J. B. P. H. Putten, T. C. T. Geuns, M. Beenhakkers,

J. B. Giesbers, B.-H. Huisman, E. J. Meijer, E. M. Benito, F. J. Touwslager, A. W. Marsman, B. J. E. Rens and D. M. Leeuw, *Nat. Mater.*, 2004, **3**, 106.
34. J. A. Rogers, Z. Bao, K. Baldwin, A. Dodabalapur, B. Crone, V. R. Raju, V. Kuck, H. Katz, K. Amundson, J. Ewing and P. Drzaic, *Proc. Natl. Acad. Sci. U. S. A.*, 2001, **98**, 4835.
35. H. C. Ko, M. P. Stoykovich, J. Song, V. Malyarchuk, W. M. Choi, C.-J. Yu, J. B. Geddes, J. Xiao, S. Wang, Y. Huang and J. A. Rogers, *Nature*, 2008, **454**, 748.
36. Y. Ahn, E. B. Duoss, M. J. Motala, X. Guo, S.-I. Park, Y. Xiong, J. Yoon, R. G. Nuzzo, J. A. Rogers and J. A. Lewis, *Science*, 2009, **323**, 1590.
37. D.-H. Kim, N. Lu, R. Ma, Y. Kim, R.-H. Kim, S. Wang, J. Wu, S. M. Won, H. Tao, A. Islam, K. J. Yu, T.-i. Kim, R. Chowdhury, M. Ying, L. Xu, M. Li, H.-J. Chung, H. Keum, M. McCormick, P. Liu, Y.-W. Zhang, F. G. Omenetto, Y. Huang, T. Coleman and J. A. Rogers, *Science*, 2011, **333**, 838.
38. T. Someya, Y. Kato, T. Sekitani, S. Iba, Y. Noguchi, Y. Murase, H. Kawaguchi and T. Sakurai, *Proc. Natl. Acad. Sci. U. S. A.*, 2005, **102**, 12321.
39. T. Sekitani, Y. Noguchi, K. Hata, T. Fukushima, T. Aida and T. Someya, *Science*, 2008, **321**, 1468.
40. T. Sekitan, H. Nakajima, H. Maeda, T. Fukushima, T. Aida, K. Hata and T. Someya, *Nat. Mater.*, 2009, **8**, 494.
41. K. Takei, T. Takahashi, J. C. Ho, H. Ko, A. G. Gillies, P. W. Leu, R. S. Fearing and A. Javey, *Nat. Mater.*, 2010, **9**, 821.
42. Z. Fan, H. Razavi, J.-w. Do, A. Moriwaki, O. Ergen, Y.-L. Chueh, P. W. Leu, J. C. Ho, T. Takahashi, L. A. Reichertz, S. Neale, K. Yu, M. Wu, J. W. Ager and A. Javey, *Nat. Mater.*, 2009, **8**, 648.
43. D. J. Lipomi, B. C. K. Tee, M. Vosgueritchian and Z. N. Bao, *Adv. Mater.*, 2011, **23**, 1771.
44. L. B. Hu, J. W. Choi, Y. Yang, S. Jeong, F. La Mantia, L. Cui and Y. Cui, *Proc. Natl. Acad. Sci. U. S. A.*, 2009, **106**, 21459.
45. S. P. Lacour, D. Chan, S. Wagner, T. Li and Z. G. Suo, *Appl. Phys. Lett.*, 2006, **88**, 204103.
46. T. Yamada, Y. Hayamizu, Y. Yamamoto, Y. Yomogida, A. Izadi-Najafabadi, D. N. Futaba and K. Hata, *Nat. Nanotechnol.*, 2011, **6**, 296.
47. X. Wang, H. Hu, Y. Shen, X. Zhou and Z. Zheng, *Adv. Mater.*, 2011, **23**, 3090.
48. S. Kirchmeyer and K. Reuter, *J. Mater. Chem.*, 2005, **15**, 2077.
49. A. A. Argun, A. Cirpan and J. R. Reynolds, *Adv. Mater.*, 2003, **15**, 1338.
50. D. S. Hecht, L. Hu and G. Irvin, *Adv. Mater.*, 2011, **23**, 1482.
51. Z. C. Wu, Z. Chen, X. Du, J. M. Logan, J. Sippel, M. Nikolou, K. Kamaras, J. R. Reynolds, D. B. Tanner, A. F. Hebard and A. G. Rinzler, *Science*, 2004, **305**, 1273.
52. M. Zhang, S. Fang, A. A. Zakhidov, S. B. Lee, A. E. Aliev, C. D. Williams, K. R. Atkinson and R. H. Baughman, *Science*, 2005, **309**, 1215.

53. D. Zhang, K. Ryu, X. Liu, E. Polikarpov, J. Ly, M. E. Tompson and C. Zhou, *Nano Lett.*, 2006, **6**, 1880.
54. Z. Yu, X. Niu, Z. Liu and Q. Pei, *Adv. Mater.*, 2011, **23**, 3989.
55. G. Eda, G. Fanchini and M. Chhowalla, *Nat. Nanotechnol.*, 2008, **3**, 270.
56. K. S. Kim, K. S. Kim, Y. Zhao, H. Jang, S. Y. Lee, J. M. Kim, K. S. Kim, J.-H. Ahn, P. Kim, J.-Y. Choi and B. H. Hong, *Nature*, 2009, **457**, 706.
57. S. Bae, S. Bae, H. Kim, Y. Lee, X. Xu, J. Park, Y. Zheng, J. Balakrishnan, T. Lei, H. R. Kim, Y. I. Song, Y.-J. Kim, K. S. Kim, B. Özyilmaz, J.-H. Ahn, B. H. Hong and S. Iijima, *Nat. Nanotechnol.*, 2010, **5**, 574.
58. V. C. Tung, M. J. Allen, Y. Yang and R. B. Kaner, *Nat. Nanotechnol.*, 2009, **4**, 25.
59. D. C. Hyun, M. Park, C. Park, B. Kim, Y. Xia, J. H. Hur, J. M. Kim, J. J. Park and U. Jeong, *Adv. Mater.*, 2011, **23**, 2946.
60. M. G. Kang, M. S. Kim, J. S. Kim and L. J. Guo, *Adv. Mater.*, 2008, **20**, 4408.
61. A. Kumar and C. Zhou, *ACS Nano*, 2010, **4**, 11.
62. J. Y. Lee, S. T. Connor, Y. Cui and P. Peumans, *Nano Lett.*, 2008, **8**, 689.
63. L. Hu, H. S. Kim, J. Y. Lee, P. Peumans and Y. Cui, *ACS Nano*, 2010, **4**, 2955.
64. J. Y. Lee, S. T. Connor, Y. Cui and P. Peumans, *Nano Lett.*, 2010, **10**, 1276.
65. S. De, T. M. Higgins, P. E. Lyons, E. M. Doherty, P. N. Nirmalraj, W. J. Blau, J. J. Boland and J. N. Coleman, *ACS Nano*, 2009, **3**, 1767.
66. A. Madrria, A. Kumar, F. N. Ishikawa and C. Zhou, *Nano Res.*, 2010, **3**, 564.
67. X. Y. Zeng, Q. K. Zhang, R. M. Yu and C. Z. Lu, *Adv. Mater.*, 2010, **22**, 4484.
68. W. Gaynor, G. F. Bukhard, M. D. McGehee and P. Peumans, *Adv. Mater.*, 2011, **23**, 2905.
69. A. R. Rathmell, S. M. Bergin, Y. Hua, Z. Li and B. J. Wiley, *Adv. Mater.*, 2010, **22**, 3558.
70. H. Wu, H. Wu, L. Hu, M. W. Rowell, D. Kong, J. J. Cha, J. R. McDonough, J. Zhu, Y. Yang, M. D. McGehee and Y. Cui, *Nano Lett.*, 2010, **10**, 4242.
71. Z. Yu, L. Li, Q. Zhang, W. Lu and Q. Pei, *Adv. Mater.*, 2011, **23**, 4453.
72. K. E. Korte, S. E. Skrabalak and Y. Xia, *J. Mater. Chem.*, 2008, **18**, 437.
73. C. Yang, H. Gu, W. Lin, M. M. Yuen, C. P. Wong, M. Xiong and B. Gao, *Adv. Mater.*, 2011, **23**, 3052.
74. H. Ko, Z. Zhang, K. Takei and A. Javey, *Nanotechnology*, 2010, **21**, 295305.
75. Z. Guo, Z. Chen, J. Li, J. Liu and X. Huang, *Langmuir*, 2011, **27**, 6193.
76. L. Feng, S. Li, Y. Li, L. Zhang, J. Zhai, Y. Song, B. Liu, L. Jiang and D. Zhu, *Adv. Mater.*, 2002, **14**, 1857.
77. K. Autumn, Y. A. Liang, S. T. Hsieh, W. Zesch, W.-P. Chan, W. T. Kenny, R. Fearing and R. J. Full, *Nature*, 2000, **405**, 681.
78. W. Barthlott and C. Neinhuis, *Planta*, 1997, **202**, 1.

79. X. Gao and L. Jiang, *Nature*, 2004, **432**, 36.
80. P. Vukusic and J. R. Sambles, *Nature*, 2003, **424**, 852.
81. J. Aizenberg, J. C. Weaver, M. S. Thanawala, V. C. Sundar, D. E. Morse and P. Fratzl, *Science*, 2005, **309**, 275.
82. C. Sanchez, H. Arribart and M. M. G. Guille, *Nat. Mater.*, 200, **4**, 227.
83. M. A. Meyers, P.-Y. Chen, A. Y. Lin and Y. Seki, *Prog. Mater. Sci.*, 2008, **53**, 1.
84. T. Onda, S. Shibuichi, N. Satoh and K. Tsujii, *Langmuir*, 1996, **12**, 2125.
85. W. Chen, A. Y. Fadeev, M. C. Heich, D. Oner, J. Youngblood and T. J. McCarthy, *Langmuir*, 1999, **15**, 3395.
86. Y. Wu, H. Sugimura, Y. Inoue and O. Takai, *Chem. Vap. Deposition*, 2002, **8**, 47.
87. K. Tsujii, T. Yamamoto, T. Onda and S. Shibuchi, *Angew. Chem., Int. Ed. Engl.*, 1997, **36**, 1011.
88. A. Nakajima, K. Abe, K. Hashimoto and T. Watanabe, *Thin Solid Films*, 2000, **376**, 140.
89. J. Bico, C. Marzolin and D. Quere, *Europhys. Lett.*, 1999, **47**, 220.
90. M. Miwa, A. Nakajima, A. Fujishima, K. Hashimoto and T. Watanabe, *Langmuir*, 2000, **16**, 5754.
91. W. Barthlott and C. Neinhuis, *Planta*, 1997, **202**, 1.
92. P. Ball, *Nature*, 1999, **400**, 507.
93. J. Zhai, H. J. Li, Y. S. Li, S. H. Ki and L. Jiang, *Physics*, 2002, **31**, 483.
94. A. Campo and E. Arzt, *Chem. Rev.*, 2008, **108**, 911.
95. H. Li, X. Wang, Y. Song, Y. Liu, Q. Li, L. Liang and D. Zhu, *Angew. Chem., Int. Ed.*, 2001, **40**, 1793.
96. H. Li, X. Wang, Y. Song, Y. Liu, Q. Li, L. Liang and D. Zhu, *Chem. Res. Chin. Univ.*, 2001, **22**, 759.
97. S. Chatpun, M. Watanabe and M. Shoji, *Exp. Therm. Fluid Sci.*, 2004, **29**, 33.
98. K. N. Rainey, S. M. You and S. Lee, *Int. J. Heat Mass Transfer*, 2003, **46**, 23.
99. H. Honda, H. Takamastu and J. J. Wei, *J. Heat Transfer*, 2002, **124**, 383.
100. W. J. Miller, B. Gebhart and N. T. Wright, *Int. Commun. Heat Mass Transfer*, 1990, **17**, 389.
101. R. M. Nowell, Jr., S. H. Bhavnani and R. C. Jaeger, *IEEE Trans. Compon., Packag., Manuf. Technol., Part A*, 1995, **18**, 534.
102. W. Nakayama, T. Daikoku and T. Nakajima, *J. Heat Transfer*, 1982, **104**, 286.
103. L.-H. Chien and R. L. Webb, *J. Heat Transfer*, 1998, **120**, 1042.
104. S. Ujereh, T. Fisher and I. Mudawar, *Int. J. Heat Mass Transfer*, 2007, **50**, 4023.
105. C. Li, Z. Wang, P.-I. Wang, Y. Peles, N. Koratkar and G. P. Peterson, *Small*, 2008, **4**, 1084.
106. Y. Im, Y. Joshi, C. Dietz and S. Lee, *Int. J. Micro-Nano Scale Transp.*, 2010, **1**, 79.

107. R. Chen, M.-C. Lu, V. Srinivasan, Z. Wang, H. H. Cho and A. Majumdar, *Nano Lett.*, 2009, **9**, 548.
108. D. Lee, T. Kim, S. Park, S. S. Lee and S. H. Ko, *Jpn. J. Appl. Phys.*, 2012, **51**, 11PE11.
109. H. Ko, Z. Zhang, Y. Chueh, J. C. Ho, J. Lee, R. S. Fearing and A. Javey, *Adv. Funct. Mater.*, 2009, **19**, 3098.
110. C. Sanchez, H. Arribart and M. M. G. Guille, *Nat. Mater.*, 2005, **4**, 227.
111. M. A. Meyers, P.-Y. Chen, A. Y. Lin and Y. Seki, *Prog. Mater. Sci.*, 2008, **53**, 1.
112. K. Autumn, Y. A. Liang, S. T. Hsieh, W. Zesch, W.-P. Chan, W. T. Kenny, R. Fearing and R. J. Full, *Nature*, 2000, **405**, 681.
113. L. Qu, L. Dai, M. Stone, Z. Xia and Z. L. Wang, *Science*, 2008, **322**, 238.
114. L. Ge, S. Sethi, L. Ci, P. M. Ajayan and A. Dhinojwala, *Proc. Natl. Acad. Sci. U. S. A.*, 2007, **104**, 10792.
115. M. Murphy, B. Aksak and M. Sitti, *J. Adhes. Sci. Technol.*, 2007, **21**, 1281.
116. L. Qu and L. Dai, *Adv. Mater.*, 2007, **19**, 3844.
117. G. Huber, H. Mantz, R. Spolenak, K. Mecke, K. Jacobs, S. N. Gorb and E. Arzt, *Proc. Natl. Acad. Sci. U. S. A.*, 2005, **102**, 16293.
118. W. Barthlott and C. Neinhuis, *Planta*, 1997, **202**, 1.
119. R. Blossey, *Nat. Mater.*, 2003, **2**, 301.
120. J. Chen, Z. Fan, J. Zou, J. Engel and C. Liu, *J. Aerospace Eng.*, 2003, **16**, 85.
121. G. J. M. Krijnen, M. Dijkstra, J. J. van Baar, S. S. Shankar, W. J. Kuipers, R. J. H. de Boer, D. Altpeter, T. S. J. Lammerink and R. Wiegerink, *Nanotechnology*, 2006, **17**, S84.
122. K. Autumn, M. Sitti, A. Peattie, W. Hansen, S. Sponberg, Y. A. Liang, T. Kenny, R. Fearing, J. Israelachvili and R. J. Full, *Proc. Natl. Acad. Sci. U. S. A.*, 2002, **99**, 12252.
123. A. Y. Y. Ho, L. P. Yeo, Y. C. Lam and I. Rodriguez, *ACS Nano*, 2011, **5**, 1897.
124. K. Autumn and A. M. Peattie, *Integr. Comp. Biol.*, 2002, **42**, 1081.
125. R. Rodolfo and V. Ernst, *J. Morphol.*, 1965, **117**, 271.
126. H. Gao, X. Wang, H. Yao, S. Gorb and E. Arzt, *Mech. Mater.*, 2005, **37**, 275.
127. T. W. Kim and B. Bhushan, *J. Adhes. Sci. Technol.*, 2007, **21**, 1.
128. T. W. Kim and B. Bhushan, *Ultramicroscopy*, 2007, **107**, 902.
129. T. S. Kustandi, V. D. Samper, W. S. Ng, A. S. Chong and H. Gao, *J. Micromech. Microeng.*, 2007, **17**, N75.
130. C. Greiner, E. Arzt and A. del Campo, *Adv. Mater.*, 2009, **21**, 479.
131. M. P. Murphy, S. Kim and M. Sitti, *ACS Appl. Mater. Interfaces*, 2009, **1**, 849.
132. H. E. Jeong, J.-K. Lee, H. N. Kim, S. H. Moon and K. Y. Suh, *Proc. Natl. Acad. Sci. U. S. A.*, 2009, **106**, 5639.
133. K. Autumn, *Properties, Principles and Parameters of the Gecko Adhesive System*, Springer-Verlag, Berlin, 2006, pp. 225–255.

CHAPTER 13

Summary

SEUNG HWAN KO

Applied Nano and Thermal Science (ANTS) Lab Department of Mechanical Engineering, Seoul National University, 1 Gwanak-ro, Gwanak-gu, Seoul 151-742, Korea
Email: maxko@snu.ac.kr

Hierarchical nanostructures are usually a combination of multiscale, multidimensional nanostructures such as nanowires, nanoparticles, nanosheets, nanopores and so on. Hierarchical nanostructures are expected to overcome the limitations of simple single-scale nanostructures. However, simple geometric combinations or a mixture of those multidimensional nanomaterials cannot be called true hierarchical nanostructures. To be functional hierarchical nanostructures, smart structuring or organization of each nanomaterial should be carried out to obtain ultimate performance compared with single component nanomaterials. To satisfy those requirements, functional hierarchical nanostructures should not only have a large surface area but also enhanced electrical, chemical, mechanical and optical performance.

This book has introduced recent developments in hierarchical nanostructuring especially for highly efficient energy device applications. Surface and electrical properties are the primary concerns in most energy devices. Maximizing efficiency in energy devices can be achieved by either new material development or functional structuring. Hierarchical functional nanostructuring has rapidly gained interest to achieve an increase in surface area and favourable electrical properties.

The energy devices covered in this book are (1) energy generation devices (solar cells (DSSCs, OPVs), fuel cells, piezoelectric, thermoelectric, water splitting and so on), (2) energy storage devices (secondary battery,

RSC Nanoscience & Nanotechnology No. 35
Hierarchical Nanostructures for Energy Devices
Edited by Seung Hwan Ko and Costas P Grigoropoulos

Published by the Royal Society of Chemistry, www.rsc.org

supercapacitor, hydrogen storage), and (3) energy efficient electronics (display, sensors *etc*). Hierarchical nanostructuring includes highly porous metal–organic frameworks, nanoparticle assembly with defined pore size, and multiple-generation highly branched nanowire trees. These topics are currently among the major issues in society. Energy-related issues have increased since the recent energy crisis. However, widespread use of the next generation of green energy devices is still limited by efficiency and cost. Most energy-related books focus only on new material development. This book covers the fundamentals to state-of-the-art functional hierarchical nanostructuring aspects.

Recent developments of hierarchical nanostructures have initiated considerable research interest in energy devices. In this book, applications of hierarchical nanostructures in the emerging energy conversion and storage areas were highlighted, ranging from solar cells, photocatalysis, photoelectrochemical water splitting to advanced energy storage devices such as Li ion batteries and supercapacitors as well as displays. However, the potential energy applications of hierarchical nanostructures are not only limited to these but also include applications in other research areas such as fuel cells, thermoelectric devices, piezoelectric devices, which are also being explored. Furthermore, using hierarchical structures to harvest various types of ambient power such as thermal, wind, vibration and electromagnetic energy would also be very promising, providing a potentially endless source of energy.[1]

Hierarchical nanostructures are an extension of research into 1D nanostructures to improve the limitation of low dimensional nanostructures. In comparison to the fast progress of 1D nanostructures, the fabrication and applications of 3D hierarchical structures have a relatively short history. Further development in this research field requires improvements in synthetic methods and novel fabrication processes to provide better control of the structural complexity, composition uniformity, surface chemistry and interface electronics, and last but not least, the yield, of branched nanostructures.[1] While the chemical vapour method results in high-quality nanobranches, it relies on expensive equipment and toxic source materials. Therefore, developing simple and environmentally-friendly benign growth methods such as multiple solution growth for the fabrication of 3D semiconductor hierarchical structures are of great interest[1] in a more practical manner.

Hierarchical nanostructure research has a bright future in solving the current limitations in energy devices. The ultimate goal is to push the energy devices towards practical applications, which requires the development of devices with high efficiency, low cost and a long lifespan. Thus, the optimization of the branched nanostructures and the improvement of the energy conversion efficiency will remain the future research focus.[1] On the other hand, using 3D hierarchical nanostructures in energy applications also has drawbacks in that it brings challenges in quantifying the charge transport and recombination loss in terms of solar devices. Understanding the light absorption property,[2–12] interface electronic band structure, and photocarrier dissociation and recombination are key issues to the overall performance.[1] For batteries and supercapacitors, it is also a question of whether the entire branched nanowires maintain their electrical conductivity during repeated cycling. To establish a reliable structure–performance correlation, a wide range of characterizations should be carried out in a self-consistent way.[1] Maybe *in situ* measurements would be ideal to provide more insights. Overall, these are very multidisciplinary topics, for which physics and chemistry experimentalists and theoreticians come together and generate innovative ideas for better smart hierarchical nanostructures.

References

1. C. Cheng and H. J. Fan, *Nano Today*, 2012, 7, 327–343.
2. S. H. Ko, D. H. Lee, H. W. K, K. H. Nam, J. Y. Yeo, S. J. Hong, C. P. Grigoropoulos and H. J. Sung, *Nano Lett.*, 2011, **11**, 666.
3. I. Herman, J. Yeo, S. Hong, D. Lee, K. H. Nam, J. Choi, W. Hong, D. Lee, C. P. Grigoropoulos and S. H. Ko, *Nanotechnology*, 2012, **23**, 194005.
4. Q. Zhang, C. S. Dandeneau, X. Zhou and G. Cao, *Adv. Mater.*, 2009, **21**, 4087.
5. J. Weickert, R. B. Dunbar, H. C. Hesse, W. Widermann and L. Schmidt-Mende, *Adv. Mater.*, 2011, **23**, 1810.
6. A. I. Hochbaum and P. Yang, *Chem. Rev.*, 2010, **110**, 527.
7. C. Cheng and H. J. Fan, *Nano Today*, 2012, 7, 327.
8. M. Law, L. E. Green, J. C. Johnson, R. Saykally and P. D. Yang, *Nat. Mater.*, 2005, **4**, 455.
9. D. I. Suh, S. Y. Lee, T. H. Kim, J. M. Chun, E. K. Suh, O. B. Yang and S. K. Lee, *Chem. Phys. Lett.*, 2007, **442**, 348.
10. J. B. Baxter and E. S. Aydil, *Sol. Cells*, 2006, **90**, 607.
11. C. Y. Jiang, X. W. Sun, G. Q. Lo, D. L. Kwong and J. X. Wang, *Appl. Phys. Lett.*, 2007, **90**, 263501.
12. H. Cheng, W. Chiu, C. Lee, S. Tsai and W. Hsieh, *J. Phys. Chem. C.*, 2008, **112**, 16359.

Subject Index

References to tables and charts are in **bold** type